1543 AND ALL THAT

AUSTRALASIAN STUDIES IN HISTORY AND PHILOSOPHY OF SCIENCE

VOLUME 13

The titles published in this series are listed at the end of this volume.

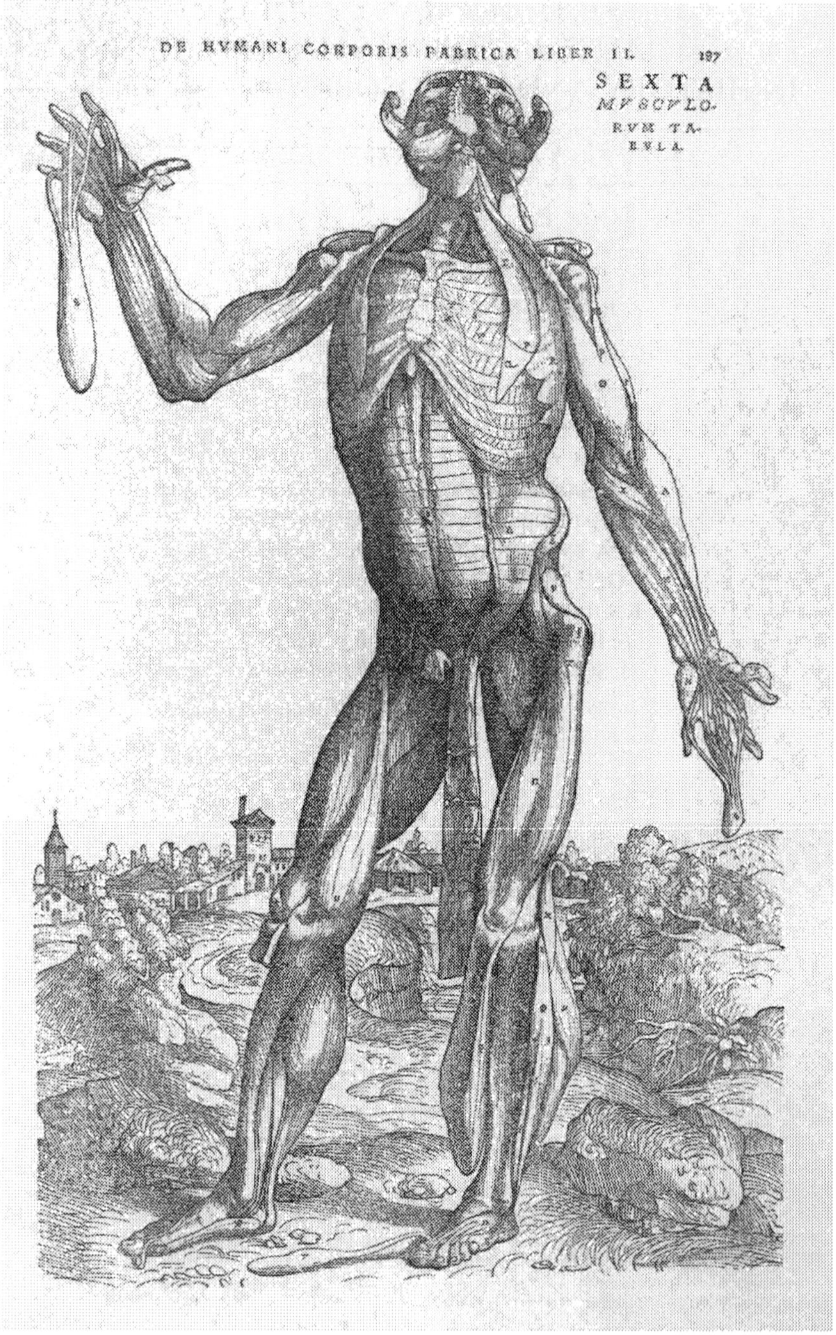

Frontispiece. Andreas Vesalius, *Sixth Plate of the Muscles,* woodcut, designed by Jan Steven van Kalkar, from *De humani corporis fabrica* (Basel, 1543).

1543 AND ALL THAT

Image and Word, Change and Continuity in the Proto-Scientific Revolution

Edited by

GUY FREELAND
ANTHONY CORONES
School of Science and Technology Studies,
The University of New South Wales

KLUWER ACADEMIC PUBLISHERS
DORDRECHT / BOSTON / LONDON

Library of Congress Cataloging-in-Publication data is available.

Published by Kluwer Academic Publishers BV,
PO Box 17, 3300 AA Dordrecht, The Netherlands

Sold and distributed in North, Central and South America
by Kluwer Academic Publishers, PO Box 358,
Accord Station, Hingham, MA 02018-0358, USA

In all other countries, sold and distributed
by Kluwer Academic Publishers, Distribution Center,
PO Box 322, 3300 AH Dordrecht, The Netherlands

ISBN 978-90-481-5302-2

Printed on acid-free paper

TABLE OF CONTENTS

FOREWORD ... ix

LIST OF ILLUSTRATIONS ... xi

ACKNOWLEDGEMENTS ... xv

GUY FREELAND / Introduction: In Praise of Toothing-Stones ... 1

MARTIN KEMP / Vision and Visualisation in the Illustration of Anatomy and Astronomy from Leonardo to Galileo ... 17

JAMES FRANKLIN / Diagrammatic Reasoning and Modelling in the Imagination: The Secret Weapons of the Scientific Revolution ... 53

JOHN SUTTON / Body, Mind, and Order: Local Memory and the Control of Mental Representations in Medieval and Renaissance Sciences of Self ... 117

JAMIE C. KASSLER / On the Stretch: Hobbes, Mechanics and the Shaking Palsy ... 151

GUY FREELAND / The Lamp in the Temple: Copernicus and the Demise of a Medieval Ecclesiastical Cosmology ... 189

ANTHONY CORONES / Copernicus, Printing and the Politics of Knowledge ... 271

NEIL THOMASON / 1543—The Year that Copernicus Didn't Predict the Phases of Venus ... 291

KEITH HUTCHISON / The Natural, the Supernatural, and the Occult in the Scholastic Universe ... 333

KIRSTEN BIRKETT / Early English Reformers and Magical Healing ... 357

BARRY BRUNDELL / Bellarmine to Foscarini on Copernicanism: A Theologian's Response ... 375

NOTES ON CONTRIBUTORS ... 395

INDEX OF NAMES ... 399

FOREWORD

Australia and New Zealand boast an active community of scholars working in the field of history, philosophy and social studies of science. *Australasian Studies in History and Philosophy of Science* aims to provide a distinctive publication outlet for their work. Each volume comprises a group of essays on a connected theme, edited by an Australian or a New Zealander with special expertise in that particular area. In each volume, a majority of the contributors is from Australia or New Zealand. Contributions from elsewhere are by no means ruled out, however, and are indeed actively encouraged wherever appropriate to the balance of the volume in question. Earlier volumes in the series have been welcomed for significantly advancing the discussion of the topics they have dealt with. I believe that the present volume will be greeted equally enthusiastically by readers in many parts of the world.

R. W. Home
General Editor
Australasian Studies in History
and Philosophy of Science

LIST OF ILLUSTRATIONS

Frontispiece. Andreas Vesalius, *Sixth Plate of the Muscles*, woodcut, designed by Jan Steven van Kalkar, from *De humani corporis fabrica* (Basel, 1543). (Photo. Scientific Illustration; repr. by kind permission of the University of New South Wales Library.)

In: *GUY FREELAND, 'Introduction: In Praise of Toothing-Stones'*

Fig. 1. Michael Esson, *Vesalian Interpretation 3* (1992). (Repr. by kind permission of the Artist.)

Fig. 2. *Reliefs, University of Padua.*

In: *MARTIN KEMP, 'Vision and Visualisation in the Illustration of Anatomy and Astronomy from Leonardo to Galileo'*

Fig. 1. Leonardo da Vinci, *Study of the Foetus and the Womb, with Optical and Mechanical Diagrams*. (Windsor, Royal Library, 19102, repr. by kind permission of Her Majesty the Queen.)

Fig. 2. Hans Wächtlin, *Dissection of the Brain, Thorax and Abdomen*, woodcut, 1517, from L. Fries, *Spiegel der Artzny* (Strasbourg, 1518).

Fig. 3. Berengario da Carpi, *Muscle-Man with Rope*, woodcut (by Ugo da Carpi?), from *Commentaria super anatomia Mundini* (Bologna, 1521).

Fig. 4. Charles Estienne, *Dissection of the Abdomen of a Woman*, woodcut, from *La dissection des parties du corps humain* (Paris, 1546).

Fig. 5. Andreas Vesalius, *Tools for Dissection*, woodcut, designed by Jan Steven van Kalkar, from *De humani corporis fabrica* (Basel, 1543).

Fig. 6. Andreas Vesalius, *Skeleton from the Side,* woodcut, designed by Jan Steven van Kalkar, from *De humani corporis fabrica* (Basel, 1543).

Fig. 7. Andreas Vesalius, *Demonstration of a Hinge*, woodcut, from *De humani corporis fabrica* (Basel, 1543).

Fig. 8. Andreas Vesalius, *Muscles of the Upper and Forearm and Tendons of the Wrist*, woodcut, designed by Jan Steven van Kalkar, from *De humani corporis fabrica* (Basel, 1543).

Fig. 9. Bartolommeo Eustachio, *Superficial Dissection of the Muscles from the Front*, engraving with annotations in ink, from *Tabulae anatomicae* (Rome, 1722). (Repr. by kind permission of the Library, University of St Andrews.)

Fig. 10. Petrus Apianus, *Chorography Compared to Pictures of the Eye and Ear*, from *Astronomicum caesareum* (Ingolstadt, 1540).

Fig. 11. Nicholas Copernicus, *Diagram of the Orbits of the Earth and Planets*, woodcut with underlining in pen, and paste marks, from *De revolutionibus orbium coelestium* (Nuremberg, 1543). (Repr. by kind permission of the Library, University of St Andrews.)

Fig. 12. Nicholas Copernicus, *Diagram of the Motions of the Pole around a Mean Position*, woodcut, from *De revolutionibus orbium coelestium* (Nuremberg, 1543).

Fig. 13. Petrus Apianus, *Torquetum*, woodcut, from *Cosmographicus liber* (Landshut, 1524).

Fig. 14. Tycho Brahe, *Mural Quadrant* or *Tychonicus,* engraving from *Astronomiae instauratae mechanica* (Wandesburg, 1598).

Fig. 15. Tycho Brahe, Lesser Quadrant, engraving from *Astronomiae instauratae mechanica* (Wandesburg, 1598).

Fig. 16. Tycho Brahe, *Elevation and Plan of the Palace of Uraniborg on the Island of Hven*, engraving from *Astronomiae instauratae mechanica* (Wandesburg, 1598).

Fig. 17. Johannes Kepler, *Temple of the Astronomers*, from *Tabulae Rudolphinae* (Ulm, 1627).

Fig. 18. Johannes Kepler, *Demonstration of the Orbits of the Planets*, from *Mysterium cosmographicum* (Tübingen, 1596).

Fig. 19. Johannes Kepler, *Demonstration of an Orbit by Analogy to a Boat in a Stream*, from *Astronomia nova* (Heidelberg, 1609).

In: *JAMES FRANKLIN, 'Diagrammatic Reasoning and Modelling in the Imagination: The Secret Weapons of the Scientific Revolution'*

Fig. 1. *Villard's Wheel of Fortune*, from T. Bowie (ed.), *The Sketchbook of Villard de Honnecourt* (Bloomington, Ind., 1959). (Repr. by kind permission of Indiana University Press.)

Fig. 2. *Apuleius' Square of Opposition.*

Fig. 3. *Classification of pictures.*

Fig. 4. *Pacioli's Classification of Ratios*. F.B. Geijsbeek, *Ancient Double-Entry Bookkeeping* (Denver, Colorado, 1914, repr. Osaka, 1975). (Repr. by kind permission of Nihon Shoseki Ltd.)

Fig. 5. *Stevin's Wreath of Spheres*, from S. Stevin, *De beghinselen der weeghconst* (Leiden, 1586).

Fig. 6. *Euclid I.1.*

Fig. 7. *Euclid XI.31*, from *The 'Heiberg' Manuscript.*

Fig. 8. *Duccio's 'Perspective'.*

Fig. 9. *Table of developments.*

In: *JAMIE C. KASSLER, 'On the Stretch: Hobbes, Mechanics and the Shaking Palsy'*

Fig. 1. *Bass viol*, from M. Mersenne, *Harmonie universelle* (Paris, 1636/7).

In: *GUY FREELAND, 'The Lamp in the Temple: Copernicus and the Demise of a Medieval Ecclesiastical Cosmology'*

Fig. 1. *Ground plan of San Marco, Venice*. Adapted from plan in *Opera di San Marco*, in O. Demus, *The Mosaic Decoration of San Marco, Venice* (Washington, DC, ©1988, Dumbarton Oaks). (Repr. by kind permission of Dumbarton Oaks.)

Fig. 2. *Christ in Majesty*. Apse of St Michael and All Angels, Copford, Essex. (Repr. by kind permission of the Rector and Churchwardens of St Michael and All Angels Church, Copford).

Fig. 3. *Christos Helios*, based on the mosaic in the Chapel of the Fisherman, grottoes of St Peter's, Rome. (Drawing by J. Weiner.)

Fig. 4. Cimabue and others, *Christ Pantocrator*, based on the mosaic in the apse of Pisa Cathedral. (Drawing by J. Weiner.)

Fig. 5. *Half figure Pantocrator*, based on the apse mosaic of Cefalù Cathedral, Sicily. (Drawing by J. Weiner.)

Fig. 6. *Theotokos and Child*, based on the apse mosaic of Hosios Lucas, Phocis. (Drawing by J. Weiner.)

Fig. 7. *Cosmogram*, based on a detail of the apse mosaic of the Transfiguration, Sant' Apollinare in Classe. (Drawing by J. Weiner.)
Fig. 8. *Cosmogram*, based on the mosaic on the barrel-vault of Hagia Sophia, Thessaloniki. (Drawing by J. Weiner.)
Fig. 9. *Cosmogram*, based on the mosaic above the apse in San Vitale, Ravenna. (Drawing by J. Weiner.)
Fig. 10. *Christ as the Sun Surrounded by the Seven Acts of Mercy*. All Saints', North Street, York.
Fig. 11. *Christ Emmanuel*, west arm of the north wall, San Marco, Venice. (Photo E. Ritter, in O. Demus, *The Mosaic Decoration of San Marco, Venice* (Washington, DC, ©1988, Dumbarton Oaks)). (Repr. by kind permission of Dumbarton Oaks.)
Fig. 12. *The Ascension Dome and the Lamp*, San Marco, Venice. (Photo. E. Ritter, in O. Demus, *The Mosaic Decoration of San Marco, Venice* (Washington DC, ©1988, Dumbarton Oaks)). (Repr. by kind permission of Dumbarton Oaks.)
Fig. 13. *The Ascended Christ in Majesty*, detail of the Ascension Dome, San Marco, Venice. (Photo. E. Ritter, in O. Demus, *The Mosaic Decoration of San Marco, Venice* (Washington, DC, ©1988, Dumbarton Oaks)). (Repr. by kind permission of Dumbarton Oaks.)

In: *NEIL THOMASON, '1543—The Year that Copernicus Didn't Predict the Phases of Venus'*
Fig. 1. *Comparison of the Ptolemaic (left) and Copernican (right) systems with respect to the appearance of Venus.*

ACKNOWLEDGEMENTS

The Editors wish to thank, in particular, Jo Wodak for her generous assistance with proof-reading; Jim Endersby, for his assistance with technical matters; Soula Georgiadis, for secretarial assistance; our referees, for undertaking an essential, but anonymous task; Rod Home for his fatherly care of the project and for making sure that our volume was eventually completed; and finally, our contributors for their heroic patience and unfailing co-operation.

GUY FREELAND

INTRODUCTION: IN PRAISE OF TOOTHING-STONES

For the history of science, 1543 is—by virtue of general consent and plain historiographical logic alike—the veritable *annus mirabilis* of the sixteenth century. It is not simply the fact that Copernicus' *De revolutionibus orbium coelestium* (*On the Revolutions of the Heavenly Spheres*) and Vesalius' *De humani corporis fabrica* (*On the Fabric of the Human Body*) were published in 1543 that renders the year remarkable, but that it marks the epiphany (rather than the nativity, much less the conception) of what might usefully be termed the 'Proto-Scientific Revolution'; that period, essentially High Renaissance in character, which makes straight the way for the Scientific Revolution.

In addition to its two great volumes, 1543 also saw the publication of Tartaglia's important edition of Archimedes, in Moerbeke's translation. Another edition of Archimedes in fact appeared in 1543, which contained Greek texts of Archimedes together with a Latin translation by Jacopo da Cremona. These Archimedean texts were destined to have a significant impact on seventeenth century mathematics and mechanics. Also the year saw the publication of Tartaglia's influential edition of Euclid, Maurolico's *Cosmographia* and possibly Canano's *Musculorum humani corporis picturata dissectio* (an alternative date is 1541), famous for its early copperplate engravings. A work of a different kind of the same date, which nevertheless was significant in relation to the development of scientific method, was Petrus Ramus' *Aristotelicae animadversiones*, which mounted a vitriolic attack on Aristotle and the curricula of the universities—a sign that the times were a-changing. Other important works were published around our year. For example, 1542 had seen the publication of Fernel's Galenist, *De naturali parte medicinae* and Fuchs' famous herbal, *De historia stirpium* (a German edition followed in 1543). 1543 also saw the establishment of the first school of clinical medicine, that established by Montanus at the University of Padua. *Annus mirabilis*, indeed.

The present volume had its genesis in the earliest stage of planning for the 25th Annual Conference of the Australasian Association for the History,

G. Freeland and A. Corones (eds.), 1543 and All That, 1 – 15

Philosophy and Social Studies of Science (A^2HPS^3) held in July, 1992; an event which also marked the Silver Anniversary of the foundation of A^2HPS^3 itself in 1967. It was decided that there would be much to be said for including in the programme a symposium, titled '1543 And All That', which would herald the four hundred and fiftieth anniversary of the publication of Copernicus' *De revolutionibus* and Vesalius' *De fabrica* and which might in due course provide the basis for a volume of papers.

From the start, we were predisposed to think in terms of something more wide-ranging than just another 'Copernican Revolution' symposium. One important factor which influenced early discussions was the fact that the University of New South Wales Library was in possession of a copy of the first edition of Vesalius' *De fabrica*. This led to the suggestion that the host of the Conference, the School of Science and Technology Studies at the University, might seek to invite a distinguished overseas expert on the illustrations of Renaissance scientific works to give the opening paper of the proposed Symposium. Happily our first choice, Professor Martin Kemp, then of the University of St Andrews, accepted the invitation and generous financial support was made available by the British Council and the Faculty of Arts and Social Sciences of the University of New South Wales for his visit to Australia. The possibilities of holding some sort of an exhibition in conjunction with the Conference, focussing on the Vesalius volume—once the property of the historian of science, Charles Singer—occurred to several people more or less simultaneously and led to a happy cooperation between the School of Science and Technology Studies, the Association, two artists teaching at the University's College of Fine Arts, Michael Esson and Alun Leach-Jones (who expressed great interest in putting together an exhibition of their drawings inspired by the Vesalius plates), the U Committee of the University of New South Wales and the University of New South Wales Libraries. The result was the exhibition 'Vesalian Interpretations', which combined exhibition of the book itself with drawings by the two artists, and, incidentally, christened the Library's new exhibition room.

The volume which, belatedly, now makes its appearance is, however, far from being the proceedings of the '92 A^2HPS^3 Symposium. Despite the fact that six of our authors—Martin Kemp, Jamie C. Kassler, Neil Thomason, Keith

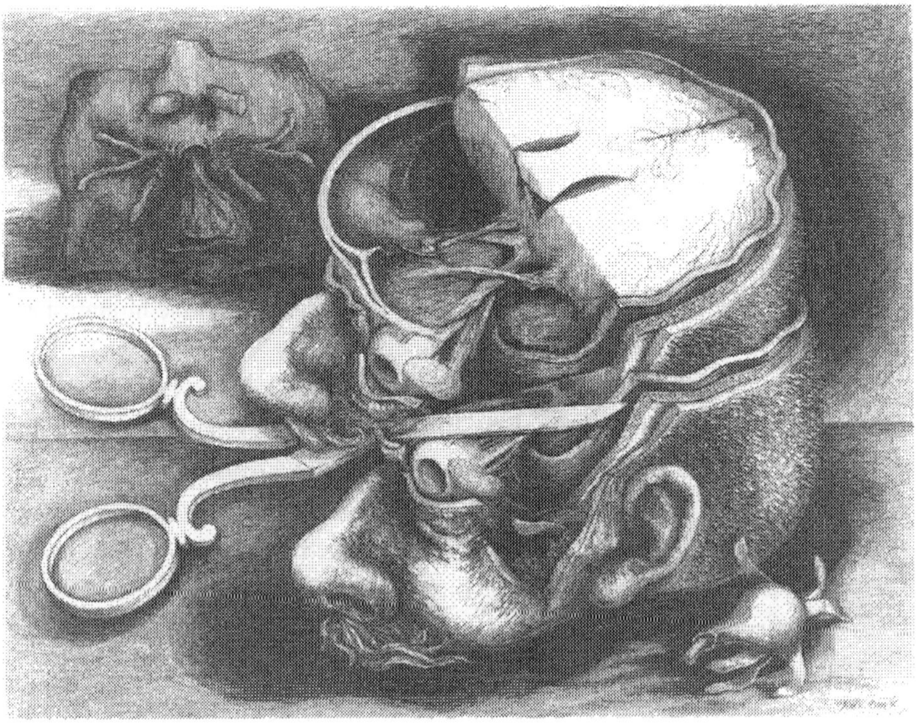

Figure 1. Michael Esson, *Vesalian Interpretation* 3 (1992). Graphite on paper, 105 × 135 cm.

Hutchison, Kirsten Birkett and Guy Freeland—contributed to the original symposium, only three of the present papers actually derive from it, those of Martin Kemp, Jamie C. Kassler (significantly revised) and Kirsten Birkett. The remaining seven articles have been collected up over subsequent years.

Given their origins, it was inevitable that the Symposium and the resultant volume should not only be wide-ranging—placing 1543 in its scientific context of the Proto-Scientific Revolution and its cultural context of the Renaissance—but should focus on the image as much as on the word. It is only recently that historians of science in general have been paying more than a perfunctory attention to the illustrative material of scientific works, or indeed to scientific books *qua* books. One seminal treatise which did much to stimulate a greater interest in the image was Elizabeth Eisenstein's, *The Printing Press as an Agent of Change*, 2 vols (Cambridge, 1979). In demonstrating how important the fifteenth-century revolution in printing was for the development of sixteenth and seventeenth century science, she, *ipso facto*, demonstrated just how important plates, diagrams and tables of data are to science. Indeed, Eisenstein's book alone should be sufficient to establish the truth that the Proto-Scientific Revolution did not simply owe a major debt to the Renaissance but that it was an intrinsic strand of the High Renaissance as such. But printing and publishing are only part of the story; the new science owed a great debt also to the development of linear perspective, Renaissance naturalism, the humanist re-evaluation of Classical Antiquity and much else.

In speaking of the 'Proto-Scientific Revolution' we have, of course, committed ourselves to a periodisation. But the division of time into periods has come under strong historiographical attack. Periods, it is said, are the invention (perhaps even the fantasy) of historians and their use only distorts and over-simplifies the complexities of the historical record and leads historians to overlook continuities and gradual evolutionary change. In the case of the Renaissance, some historians have gone to the lengths of rejecting the concept of 'renaissance' altogether. Others, while accepting the usefulness of the concept, have seen a continuity between the Renaissance of the twelfth century and the Italian Renaissance. Even historians who do distinguish two Renaissances (or two largely discontinuous phases of a European Renaissance) disagree as to what date should be taken as marking the commencement of the Italian Renaissance: should it be 1300, 1400, 1450, or some other date? The fourteenth century, for example, is variously labelled by different historians as, or as belonging to: the High Middle Ages, the Late or 'Waning' Middle Ages, the Early Renaissance, or the Proto-Renaissance. In like manner, while some historians characterise, as we do, the sixteenth century as 'High Renaissance', others (including some historians of science who see the Scientific Revolution as an indivisible movement commencing in 1543) fuse it with the seventeenth century and label both alike as 'Early Modern'. It is all very confusing, and historians, we have to conclude, have some reason to distrust periodisation. But, of course, the fact of the matter is that, dislike the use of periods as they may, most historians, particularly those working on the pedagogic coalface or writing for a general public, have found periods just too convenient to be thrown out. And surely in this they are right.

Historians have an obligation (as do scientists) to bring some order into a chaos of particularities. The case of the historian is in reality not so very different from that of the geologist or of the archaeologist. True, geological and archaeological strata can literally be seen in the earth in a way that historical periods cannot be seen in the records historians gather, but, nevertheless, as long as we stand far enough back, we can surely see in the historical data a certain stratification. In neither geology nor archaeology, be it noted, does the use of periods conflict with the recognition of continuity and evolutionary change. Indeed, both sciences work within a general evolutionary framework.

A concept which might help to resolve some of the seeming conflicts over historical periodisation is one to be found in Machiavelli's, *The Prince* (published posthumously in 1532); itself, of course, a High Renaissance work. The concept is that of a 'toothing-stone'. Toothing-stones are projecting stones left at the end of a freestanding wall or wall of a building when a phase of construction comes to an end, so that it can be extended at some later date. In *Il Principe*, Machiavelli uses the analogy of toothing-stones to explain why memory of innovations tends to disappear; one innovation invariably leaves a toothing-stone for the next: '... in the antiquity and persistence of his [the 'natural' or hereditary prince's] rule memories of innovations and the reasons for them disappear; because one change always leaves a toothing-stone for the next' (trans. G. Bull, *The Prince* (Harmondsworth, 1961) p. 34). Machiavelli's concept can, we suggest, be a useful one in analysing linkages without violating such integrity as historical periods might possess; and consequently might do something to help historians feel happier about the periods they dislike but cannot do without. The builder's toothing-stones were usually left because work had to stop because of some misfortune which struck—funds were expended, war broke out, the patron died, the tower or nave collapsed, or whatever. When building did recommence, the toothing-stones might well be utilised for constructing an extension in a very different style than the original.

One must not push the analogy too far, but awareness of toothing-stones might help us resolve what is perhaps the biggest difficulty with periodisation. Scholars who reject the notion of the Italian Renaissance altogether, or who extend the Renaissance back one, two or even three centuries before 1400 usually do so because the distinctive features of the Italian Renaissance—linear perspective and linear time, printing, the development of capitalism, humanism, individualism, naturalism, and so forth—all have roots going back into the Middle Ages. The rise of verism in art and the development of rule-of-thumb perspective techniques trace back to the end of the thirteenth century, as, almost certainly, does the invention of the weight-driven mechanical clock; the shift to the equinoctial hours which replaced the variable seasonal hours occurred in the more developed parts of Europe around the mid-fourteenth century; Thomist philosophy and theology, which were, *mutatis mutandis*, to triumph at the Council of Trent, date back to the mid-thirteenth century, with roots stretching back to Anselm, Abelard and others; the fifteenth century, and subsequent, debates over scientific method are continuous with those of the late Middle Ages. Humanism can be seen in a Petrarch and Proto-Protestantism in a Wycliffe. Commercial practices were developing from the High Middle Ages, as was the industrialisation based on water-power spear-headed by the Cistercians. The most important feature of Vesalius' *De fabrica*, the illustrations, have nothing to do with movable type, but were produced by block printing; and books printed with wooden blocks had been produced during the fourteenth century. Moreover, a commercial book trade, employing production-line techniques involving teams of secular scribes, had also been developing apace during the course of the fourteenth century. Even such innovations as polyphonic music in the liturgy trace back to the late Middle Ages. All of these factors can be regarded as toothing-stones, which project into the fifteenth and sixteenth centuries and are there developed or extended much further to give distinctive colour to a new age. (Of course, one of the points where the analogy with building breaks down is that cultural toothing-stones can only be detected with hindsight; another is that the culture and mentality of the previous age in general, and not just the toothing-stones it generates, tend to persist into the following age, as with the strong residual medievalism of the Renaissance.) If one concentrates on these creative elements one naturally tends to see continuity and development more than change. However, if one looks to the larger canvas it is change which tends to strike one rather than continuity. The culture and mentality of Northern Italy at the time Copernicus was there (or at least the culture and mentality of the circles within which he would have moved) were indeed very different from the culture and mentality which prevailed, say, three hundred years earlier in France.

It is here that the concept of toothing-stones can really help us. To push our analogy a little further, we can see that there will be those who will be so struck by the differences in architectural styles between two building programmes that they ignore the toothing-stones which link the two, while there will be those who are so impressed by the toothing-stones that they fail to stand back from the building, and so miss the fact that the two programmes are stylistically distinct. Clashes between 'Big Picture' and 'Little Picture' historiographies, which erupt from time to time—in fact, there has been a recent eruption in the history of science (mentioned by Anthony Corones in his paper)—are but little more than Tweedledum and Tweedledee battles. Little Picture historiography is of highly limited socio-cultural value unless it links into the Big Picture, which invariably contributes much to the definition of the cosmos of any historically conscious people. But, at the same time, the real stuff of history undeniably resides in the individual episodes of Little Picture historiography. Big Picture and Little Picture historiography should not be seen as antagonistic rivals but as twin aspects of the unitary enterprise we call history.

If one takes into account that one period always leaves toothing-stones for the next, then it is possible to give full weight to linking continuities while at the same time acknowledging the distinctive stylistic features and genuine innovations of different periods. Of course, it is possible to focus on the one aspect or the other, but if we take into account toothing-stones then we can appreciate how it is possible to use two different labels for the same span of years. If we focus on the toothing-stones which link the medieval and Renaissance periods then we might well be led to use the label for the fourteenth century 'Proto-Renaissance'; if we focus on the decline of medieval culture then we might be inclined to use the label 'Waning Middle Ages'. The fourteenth century is indeed both. It is, of course, the virtual collapse of the medieval world during the disastrous fourteenth century which constitutes the discontinuity between the medieval and Renaissance periods. This collapse was partly brought about by the disruptive effects (within the medieval culture) of those very innovations which were to provide the toothing-stones for the

Renaissance, but probably even more by the series of calamities—notably the Black Death, almost incessant warfare, papal schisms and the near collapse of the feudal system—which wreaked havoc during the course of the century.

We adopted the label 'Proto-Scientific Revolution', of course, on analogy with such well-established expressions as 'Proto-Renaissance'. It would have been possible simply to have used the expression 'Renaissance Science', or the expression 'Scientific Renaissance', but our intention was not just to characterise sixteenth century contributions to science in terms of their distinctively Renaissance character and style, though that also was an aim, but to concentrate on those features of the science of the period which fed directly into the seventeenth-century Scientific Revolution. (In utilising the term 'Revolution' we are merely following convention; we leave the reader to supply their own connotation.) Although the Proto-Scientific Revolution coincides with the Renaissance, it has a different epicentre than that of the Renaissance as a whole. The epicentre of the Renaissance as a general artistic and cultural movement is arguably situated within the decades around the mid-fifteenth century, but the epicentre of the 'Proto-Scientific Revolution' (although it could be held to penetrate back into the fifteenth century) can be pinpointed to a specific year a century later, 1543. Although, what is arguably the greatest contribution of the *quattrocento*, the theory of linear perspective, can be regarded as a scientific theory as to the nature of visual space, and inspired the projective geometry of Desargues, it is deeply embedded in the practice of Renaissance artists. Indeed, during the fifteenth century the hub of scientific activity is to be found in the studios and workshops of the *perspectivi* (which is the reason why it has been all but invisible to the eye of many an historian of science). True there are developments elsewhere, such as the seminal cosmological insights of Nicholas of Cusa; but Cusanus' cosmology is grounded in a philosophical and theological discourse. It is during the High Renaissance that science begins to emerge as a vibrant field of activity more in its own right in universities such as Padua (where Vesalius, and later Galileo, taught, and where Copernicus, and later William Harvey, were medical students for a while) and in the courts.

Giving due weight to the existence of toothing-stones entails that one cannot focus on a specific period without taking note of the toothing-stones that it received from an earlier era and that it would be destined to leave for a future period. Contributors to this volume have not, therefore, been constrained to roam only within a defined span of years—whether this be understood as the narrow limits of the sixteenth century or a more generous span of from, for example, 1435, the year Alberti commenced work on *Della pittura*, or 1452, the birth of Leonardo da Vinci, to 1642, the death of Galileo and the birth of Newton. (The delayed diffusion of the Renaissance into Northern Europe could certainly be cited as grounds for extending the Proto-Scientific Revolution into the earlier decades of the seventeenth century.) On the contrary, contributors have been invited to trace connections both back into the medieval period, and earlier, and forward into the Early Modern period. So while some contributors do have their sights set firmly on or around 1543, others look forward and/or back, ranging over a wider span of history as the toothing-stones direct them.

If 'change and continuity' gives us one polarity of the Proto-Scientific Revolution, 'image and word' gives us another. The Renaissance saw the reconstitution of both the forms and roles of image and word, and clearly this reconstitution impacted on the development of science in a multitude of complex ways. The topic is one of massive dimensions and we certainly would claim to do no more than air but a few of the relevant issues in the present collection of papers.

Undoubtedly, one of the most important toothing-stones bequeathed to the Italian Renaissance was the turn to verism, with the development of rule-of-thumb techniques of linear perspective, by painters of the Waning Middle Ages such as Giotto, Duccio and Cimabue. With the development of linear perspective proper in the first half of the fifteenth century a precision tool became available which could be directed to serve the ends of natural science. Just how, and to what extent, the tool was taken up within different branches of High Renaissance science is a matter that needs to be determined through careful investigation of the actual evidence (and here Martin Kemp has made a significant contribution). There can be little doubt, however, that the progressive decline of the influence of medieval canons of iconography, combined with the development of focussed perspective, during the Proto-Renaissance and *quattrocento* promoted not only artistic verism *per se* but naturalism in general. The systematic observation of natural phenomena, plants, animals and lifeless forms by Leonardo da Vinci and other artists was not something added on to the new artistic practices but was part and parcel of those practices and a consequence of the adoption of a new (whatever the Ancient anticipations in fact were) theory of the geometrical nature of visual space. Armed not only with new techniques and new geometry but with new concepts of linear space and linear time, the artist could now seize the experienced moment with an immobile eye occupying a totally (that is, uniquely) privileged location in linear space (that is, as opposed to Aristotelian spherical space) and a totally privileged position along the equably flowing stream of equinoctial linear time.

If there were changes in the forms of pictorial images, and the uses to which they were put, during the course of the Renaissance, so there were changes in the structuring of language and the deployment of words. In traditional Christian understanding, the word of the scriptures was complemented by the language of an essentially (but not exclusively) ecclesiastical iconography. In a highly visual, and also (despite the fact of a not insignificant level of literacy) oral, culture the written word read out, particularly within a church setting, interacted with the iconography and traditional pictorial images. All that was to change. Well before the Protestant iconoclasts got to work with crowbars (or muskets) and whitewash pail the traditional iconography was being overtaken by the new art of the *perspectivi*. But the Reformation, with its cry of *sola scriptura* and its emphasis on the plain grammatical sense of the text, permanently changed the relative roles of word and image. As can be seen in the case of Bellarmine (see Barry Brundell's paper), the Counter-Reformation, despite a triumphalism centred on the defiant image of the elevated transubstantiated host in its glittering sunburst monstrance, reflected in attenuated form a number of the attitudes of the Reformers, not least as regards the emphasis on the plain grammatical sense of scripture. In iconography, the Counter-Reformation secured the triumph of the Baroque, which administered the *coup de grâce* to an already dying traditional iconography which, in its essentials, the West had once shared with the East (see Guy Freeland's paper).

It is widely accepted that one of the factors that made possible the Reformation in the sixteenth century, or at least made possible the Reformation in the form which it took, was the invention of movable type printing in the previous century. Without the wide availability of Bibles in the vernacular, the Protestant call to individuals to read and interpret the Bible for themselves in the light of the Holy Spirit, without the mediation of ecclesiastical authority, would have been but empty rhetoric. Private reading was by no means unknown in the Middle Ages; with some individuals, even silent reading. Nevertheless, reading usually meant reading aloud, and frequently reading aloud in a group. With the invention of movable type printing, however, reading became less of a public, communal activity (although we must not forget such Prostestant practices as daily family Bible readings) to become more of a private, individual activity (even if that might have entailed for the poor a visit to the chained-Bible in the parish church). But if movable type printing had a massive impact on religion (it affected Catholicism as well as Protestantism) and brought about changes in the modes of communication of knowledge, and a new ethic of self-help, it also had a major impact on the development of science (see Anthony Corones' article). Printing—both movable type and improved block printing—constituted a major toothing-stone which the *quattrocento* provided for the benefit of the High Renaissance and Early Modern periods.

We invite our readers to keep one eye constantly open for the toothing-stones which link the late Middle Ages with the Renaissance or the Renaissance with the Early Modern period and signal continuity and non-saltatory evolutionary development. Equally, an eye should be kept open to spot the lack of a toothing-stone, which might signal the advent of mutative revolutionary change. In endeavouring to order the papers to best advantage, the Editors have, however, found the polarity of image and word more helpful than that of change and continuity. There is, accordingly, a general drift through the volume from an emphasis primarily on visualisation and the pictorial or diagrammatic image, to an emphasis primarily on the word and the text.

In our first paper, Martin Kemp sets the scene for the whole volume by comparing and contrasting the nature, function, and relationship to visualisation of the pictorial image in the two sciences which are pre-eminently those of 1543, astronomy and anatomy, taking as his period the span of time from Leonardo da Vinci to Galileo. Kemp reaches the conclusion that the relationships between visualisation and pictorial representation are different in the two sciences he considers. James Franklin continues and extends the analysis of the visual image by looking specifically at the nature and function of diagrams and diagrammatic reasoning, casting his eye over a wider time span. The mental training occasioned by the diagrammatic reasoning of the medieval and Renaissance periods, Franklin argues, prepared the ground for the Scientific Revolution. Franklin's article also serves an important role in the economy of the volume by drawing attention to a third significant ingredient of the rich mix whose coction was to fashion 1543 as the *annus mirabilis* of Rennaisance science, mathematics.

John Sutton's paper comments directly on the papers of Martin Kemp and James Franklin, both of which make reference to the mental representation of visual images. Visual and verbal representations have usually been assumed to be distinct, but Sutton finds in medieval, Renaissance and Early Modern ideas concerning psychological control indirect evidence for a form of mental representation which is neither linguistic nor visual. These speculations, he argues, link with ideas in contemporary cognitive science concerning the possibility of what is called 'superpositional storage'. Sutton thereby pleasingly makes a move in the direction of unifying our seeming dichotomy of word and image. Jamie C. Kassler's paper, which follows on neatly from the three earlier papers, examines one specific image, that of the stretched musical string, from the lyre of Apollo to Thomas Hobbes' bass viol. This image constitutes a long enduring toothing-stone of science, one which, as Kassler shows, has its origins in Antiquity, notably in Stoic science, and extends through Vesalius and the Proto-Scientific Revolution to Thomas Hobbes in the seventeenth century. The focus on Hobbes, the pitiful plight of whose own internal strings are sympathetically described by the author, might be considered to 'stretch' the Proto-Scientific Revolution beyond breaking point, but Hobbes' mechanics and physiology were, as Kassler shows, deeply embedded in the medical tradition of Vesalius and Renaissance Padua, which spawned William Harvey and so many other English physicians of the seventeenth century. And, of course, Hobbes himself, by virtue of his longevity (he was already in his twelfth year at the turn of the century), links the two periods of scientific history.

Guy Freeland's contribution is the first of three papers which focus on Copernicus. Freeland returns to those much worried metaphors Copernicus

Figure 2. *Reliefs, University of Padua*. Repeating motifs on the colonnade of the sixteenth century courtyard of Il Bo (the Ox, so named because it had earlier been an Inn of that name). The juxtaposition of the lute, with its stretched strings (see Jamie C. Kassler's paper) and the armillary sphere is revealing of High Renaissance culture. The rosettes might well represent the sun (see Guy Freeland's article). On the ceiling below can be seen some of the coats of arms of alumni which adorn the colonnade, including (though not in this illustration) the arms of William Harvey.

uses in relation to his heliocentric universe, those of the cosmos as a temple, of the sun hanging as a lamp in the midst of the temple, and of the sun enthroned as a monarch at the centre of the world. In arguing that a possible source, in addition to others, for these striking metaphors is to be found in an ancient Christian ecclesiastical cosmology, Freeland at the same time illustrates certain aspects of the earlier Christian iconographic language which was largely swept away by the new art of the Renaissance.

Our sixth contribution, that of Anthony Corones, marks a transition from papers which are principally concerned with the image to those which are largely focussed on the word. Corones places *De revolutionibus* firmly into the context of the print revolution. He tells of the deep distrust which Copernicus had of publication and of his attempt, in his prefatory letter addressed to Pope Paul III, to delimit the intended readership of the book and of the way in which it should be read. The failure of his apologetic strategy was a consequence, Corones argues, of the insufficient understanding which Copernicus had of the complex consequences movable type printing had for the transmission of knowledge. Our next paper, by Neil Thomason, is of a rather different kind, as it considers a specific passage of the text itself only as a starting-point for an investigation into the treatment it has received over the years from historians and philosophers of science; the correct and the incorrect readings, and the failure to read the text itself at all. Thomason's topic is that of two parallel traditions concerning the phases of Venus. The facts are that neither Copernicus himself nor, seemingly, Copernicus' followers (with the sole apparent exception of Galileo's student Castelli) predicted, from Copernican premises, the phases of Venus which Galileo was to observe with the telescope, and which were to provide him with ammunition for his Copernican apologetics. Thomason shows that alongside a tradition which gets the story right there is another which gets it wrong. He draws some sobering conclusions concerning the value of the history of science for philosophers of science.

Our last three papers, which form a chronological sequence, are concerned with the changing climate of theological thought from the late Middle Ages through to the early seventeenth century, and some of the effects which these changes had on the development of science. Keith Hutchison deals with the 'before', scholasticism. As the author observes, historical change can only be understood by comparing something old with something new. Hutchison's topic is that of the shifting understanding and boundaries of the categories of the natural and the supernatural (an issue already touched upon *en passant* by Freeland), particularly in relation to matter theory and occult qualities. The change occurs with the rejection by mechanical philosophers in the seventeenth century (Gassendi being a partial exception) of the scholastic 'secondary causes', the inherent powers implanted by God at the creation into the natural order. Scholasticism, a development essentially of the late Middle Ages, became a toothing-stone for the Renaissance. Although Hutchison does not mention

Vesalius, this is well demonstrated by the 'living anatomy' of *De fabrica*, in which the musclemen, in particular, are depicted as retaining the inherent vital powers of the 'fabric' of their bodies even in the most advanced stages of what Kemp calls 'their myological striptease'. But as Hutchison shows, Protestant theology in the sixteenth century rejected the scholastics' secondary causes. And, of course, thereby the Reformers constructed a toothing-stone for mechanical philosophers of the following century.

Kirsten Birkett, in her article, looks in detail at one strand of Protestant theology, that of the English Reformers Tyndale and Cranmer. She shows that the Reformers, led by the doctrine of *sola scriptura*, went much further than rejecting scholastic matter theory; they denied the whole sacramental theology of the Catholic Church. The reformers maintained that the matter of the sacraments—the bread and wine of the Mass, the baptismal waters, the holy oils—underwent no change as a result of their consecration. 'The Protestant God himself', she maintains, 'did not appear in natural objects'. Belief in the efficacy of the matter of the sacraments—and *a fortiori* the sacramentals and the relics of saints etc—was, thus, virtually equated with belief in the efficacy of substances subjected to the practices of magic. All belief in special powers of matter was dismissed as superstition. Miracles could only be performed by God directly (or with the assistance of the angels at His bidding) or by the devil. Protestant doctrine, Birkett argues, had much to do with the growing rejection of magic during the sixteenth and seventeenth centuries. Barry Brundell's paper complements Kirsten Birkett's paper by taking a look at Counter-Reformation thought as it impacted upon the fate of Copernicanism in the person of Cardinal Bellarmine. He shows that Bellarmine's condemnation of Copernicanism in 1615 was conditioned by the contingencies of his time and place. Although above all wishing to assert the authority of the Church over the interpretation of scripture, against the attacks of the Protestants, Bellarmine nevertheless shares with his opponents, Brundell shows, a strong tendency to equate the literal level of Biblical texts with the plain grammatical sense of the words.

School of Science and Technology Studies, University of New South Wales

MARTIN KEMP

VISION AND VISUALISATION IN THE ILLUSTRATION OF ANATOMY AND ASTRONOMY FROM LEONARDO TO GALILEO

> 'The words or the language, as they are written or spoken, do not seem to play any role in my mechanism of thought. The psychical entities which serve as elements in thought are certain signs and more or less clear images which can be 'voluntarily' reproduced or combined ... The above mentioned elements are, in my case, of visual and some of muscular type. Conventional words or other signs have to be sought for laboriously only in a secondary stage'. (Einstein)[1]

The conjunction of the rise of the printed book as a prime means of transmitting information and the Renaissance reformulation of the means of visual representation was clearly an integral part of what we call the scientific revolution. On one level, it seems perfectly obvious that to be able to represent (say) a plant in a convincingly naturalistic manner in a printed botanical treatise would serve to provide straightforward instruction and to transmit checkable information to students of the natural world. Indeed, the polemic in favour of illustration by Leonhart Fuchs, introducing his great book on botanical science in 1542, provides early support for this view. He confronts those who 'will cite the most insipid authority of Galen that no one who wants to describe plants should try to make pictures of them'.[2] Fuchs asks rhetorically, 'who in his right mind would condemn pictures which can communicate information much more clearly that the words of even the most eloquent men?' In a similar manner, Leonardo, that most fervent advocate of visual communication, had already demanded, 'O Writer, with what words will you describe with such perfection the entire configuration which the drawing here does?'[3] And, comparably if somewhat more unexpectedly, Michael Mästlin's referee's report on Kepler's *Mysterium cosmographicum* for Tübingen, suggests that 'Kepler might provide a diagram and numerical tabulations [of the order and sizes of the spheres

G. Freeland and A. Corones (eds.), 1543 and All That, 17 – 51

according to Copernicus], because the subject is absolutely incomprehensible without a diagram'.[4]

It has been claimed that the new techniques of systematic naturalism in the visual arts—above all the artists' new science of perspective—are inseparable from the 'search for truth' in Renaissance science. A nice formulation of this view is provided by Alistair Crombie:

> The conception of the *virtuoso*, the rational artist aiming at reasoned and examined control alike of his own thoughts and intentions and actions and of his surroundings, seems to me to be the essence of European morality, meaning both habits and ethics, out of which the European scientific movement was generated and engineered. In this context the rational artist and the rational experimental scientist appear as exemplary products of the same intellectual culture.[5]

For present purposes, it matters not whether this intellectual culture is seen as triumphantly progressive or (as Foucauldians would have us believe) imperialistically oppressive. The complementarity of the cultural symptoms remains essentially the same. Indeed, to go even further in forging the conceptual alliance between art and science, perspectival representation has been seen in the Panofskian tradition as the 'symbolic form' of the Renaissance—as the conceptual model through which vision was radically redirected, the world was made to look different to the observer, and the transmission of knowledge was reformed.[6] Specific incidents have been adduced—as discussed later in this study—to show that the interpretation of new visual phenomena, such as those revealed by the telescope, were most effectively conducted by observers who were literate in the painters' methods of three-dimensional design, particularly in the sciences of cast shadows and perspective.[7]

I find a sharp contrast between these big claims about visual representation and the levels of understanding we have achieved about the roles actually played by depiction at each stage in the processes which lie behind the making of an illustrated scientific text.[8] These processes potentially involve, in a complex and not necessarily sequential manner, variant combinations of observation, visualisation, graphic modelling, publication, communication and reception. Furthermore, the framework within which a particular combination of processes is realised will differ substantially over time and even within the same period. Our habit of assuming certain kinds of role for representation in our various modern sciences may provide us with very misleading criteria when we approach the texts and images of past eras. I remain sufficient of an empiricist to believe that the characterisation of the role of representation in science cannot be adequately achieved without a close study of how illustrations actually functioned in their particular historical environments. The present paper is designed to take *the* two sciences of 1543, anatomy and astronomy, sciences which apparently rely upon very different modes of visualisation and representation, and to look at how illustrative material functioned in relation to the agendas of the scientists. By choosing such different sciences, we will also be able to broach if not to answer the question as to the extent to which the visual representations as realised on the page provide access to the conceptual models in the scientists' minds—the kinds of non-verbal models of which Einstein spoke.

Since I have asserted airily that anatomy and astronomy are very different in their visual characteristics, I think it is only fair that I give at this stage some general idea of what I mean—though this idea will necessarily depend upon some crunching generalisations. Anatomy is *par excellence* a descriptive science, at least in its modern sense, and its primary subjects of interest can be viewed to good effect with the naked eye, even if other techniques of examination (including microscopy) have in the post-Renaissance period amplified the scope of observation. Linked to the physical process of dissection, anatomical illustration lends itself to sequential, step-by-step exposition in which the visual presentation acts as a surrogate for the eye-witness experience or as a visual summation of many eye-witness experiences. In the hands of Vesalius and many of his successors, anatomical illustration lent itself to what I will call the 'rhetoric of reality'; that is to say the use of recognisable visual signals of uncompromising naturalism to convince the viewer that the forms are portrayed from life. These visual signals were frequently accompanied by texts or captions which emphasised the concrete situations and procedures by which the representations were generated, and by visual references to the act of dissection itself, through such devices as the display of tools.

In Astronomy, by contrast, the plain description of the appearance of the heavens to an unaided eye at a single moment would serve little purpose, and even a series of sequential pictures would generate forecastable patterns and little else. The appearance of the heavens only becomes eloquent to the enquirer after structure when coupled with systematic measurements in which the eye serves as just one component in an instrumental system of controlled recording over a period of time. The translation of these measurements into coherent visual form involves the representation of things that cannot literally be seen, such as the orbs that enclose the paths of the planets, the points that mark the centres around which they turn, or the circles that map out the invisible spheres (crystalline or notional) which determine the motions of the celestial mechanism. The rhetoric in this case is very different. It is the 'rhetoric of irrefutable precision', conveyed by tables of figures and flat geometrical diagrams. Yet it is this very translation of the visual phenomena into mathematical schemata remote from immediate sensory and physical experience that contributed to the vulnerability of the representations, since a particular geometrical diagram of the cosmos maybe just one of a number of analogue models which can be contrived to fit the appearances. It was this long-recognised dilemma that gave Osiander his licence in the foreword to *De revolutionibus* to argue that the heliocentric theory was a fruitful new hypothesis rather than a representation of the physical actuality of the universe—a licence which could draw some partial support from Copernicus' argument that relative motions produce a 'reversible agreement', though Copernicus casts his arguments in predominantly realist terms.[9] However, as we will see, the new breed of astronomers found alternative ways to build the 'rhetoric of the real' into their presentations.

In looking at the sciences of 1543, it seems wise to begin with anatomy, since it apparently presents the simpler case, and anatomical illustration has been more widely discussed in the existing literature than the role of illustrations in astronomy. This is not to say, however, that extensive discussion necessarily results in adequate understanding. Even recent histories of anatomical illustration show a notable reluctance to discard the traditionally triumphalist view in which the central purpose of the historical narrative is to outline the inevitable progress in depictions of the body according to the procession of perfectible naturalism.[10] To my mind, this remains a valid narrative within its own limited frames of reference, but it casts aside all those factors which might explain the nature of the imagery in its broader social, intellectual and aesthetic aspects. Even on its own terms, the narrative of perfected representation causes problems when set within the history of observational science, since the logical consequence of any insistence upon observing the real thing is that illustration is at best a limited substitute for the primary experience and at worst a dangerous evasion of the obligation to undertake first-hand observation. It should not come as too much of a surprise to find Vesalius, the second authentic hero of the standard story, asserting that,

> I believe it is not only difficult but entirely futile and impossible to hope to obtain an understanding of the parts of the body or the use of simples from pictures or formulae alone, but no one will deny that they assist greatly in strengthening the memory in such matters.[11]

Vesalius' reference to memory is unlikely to have been casual, given the prominent emphasis upon the need to cultivate the art of memory in an era in which the continued cost and limited availability of books and manuscripts meant that much information had necessarily to be carried around in the mind.

The first authentic hero of the conventional story is, of course, Leonardo da Vinci, who would not have been inclined to accept Vesalius' qualification on the limits of the understanding of anatomy which could be gleaned from illustrations. Indeed, he emphasised that his drawings were superior to the witnessing of a single dissection, given the considerable practical problems of dissecting and the need to combine results from many dissections. However, just taking one of Leonardo's drawings—one of his most famous (fig. 1)—we will be readily able to see how much more complicated are the visual and intellectual factors than his own claims for representation might lead us to assume. The study of a foetus in the womb, with related diagrams and notes, demonstrates all his skills as a draftsman in conveying the three-dimensional presence of objects and his extraordinary inventiveness in devising methods of demonstration—most notably in the upper diagrams of the interdigitations of the placenta and uterus wall.[12] Yet underlying his personal rhetoric of reality—both in the drawings and in the discussions of dissections in the sets of related notes—are a series of complex dialogues with various kinds of tradition and meaning. Most obviously, as consistent with Galenism, he has incorporated features from animals, as in the cotyledonous placenta derived from his study of ungulates. One of his notes speculates on the traditional question of the relationship of the souls of the mother and foetus, so that 'something desired by the mother is often found imprinted on the limbs of the infant'—a concept based on the notion of the soul as the 'form' (or form-generating agency) of the body. The whole set-up of the image, particularly as revealed in the small sketches of the enclosing coats of the womb, assumes its full effect in the context of his theory of the microcosm, in which the constituent parts of nature express the profound analogies within the whole. In this case the parallel is between the womb and an opening bud or seed-case. At centre right is an entirely diagrammatic figure exploring the behaviour of a spherical body with a heavy weight at its periphery on an inclined plane, which may have been occasioned by his thinking about the orientation of the foetus with its heavy head in the womb. In the bottom right corner is an optical diagram and note which explains 'why a picture seen with one eye will not demonstrate such relief as the relief seen with both eyes'—which indicates that even for Leonardo the illusion of three dimensions on a two-dimensional surface possessed inherent limitations compared with the viewing of the real thing. However, the assertive language of objectivity spoken by the drawings is not such as to encourage the spectator to be openly aware of the limits and pitfalls of naturalistic representation.

The earliest published illustration that lays overt claims to be a true picture of an actual dissection makes a startlingly direct assertion of presenting the unvarnished truth. This is the print by Hans Wächtlin (or Wechtlin) of a dissection by Dr. von Brackenau of a hanged man in Strasburg in 1517 (fig. 2), first published by Lorenz Fries a year later.[13] As befits an illustration by a printmaker in the succession of Dürer, who himself depicted plants and animals with uncompromising attention to their individual peculiarities and accidental damage, the criminal is portrayed with tortured face and savagely twisted right arm. The fact that the man was a vile criminal is underlined in the caption as a strategy to sanction the gory display—and to set it in a nexus of German imagery which would include Hans Baldung Grien's macabre iconography of death.[14] It is, I think, no coincidence that Wächtlin's rawly direct style should (like the successor images) have been used to illustrate a book in the German vernacular rather than learned Latin. The relatively unobtrusive labelling, which encroaches on the main image as little as possible, is designed to enhance the sense that we are looking at a true picture. Successive derivations of this much copied image show its translation into more schematised formats, as in Lorenz Fries' 1519 treatise, or in its adaptation as a blood-letting figure in 1540.[15]

Even Wächtlin's apparently direct image, however, raises problems about how the anatomical content entered the representation. To take just one feature,

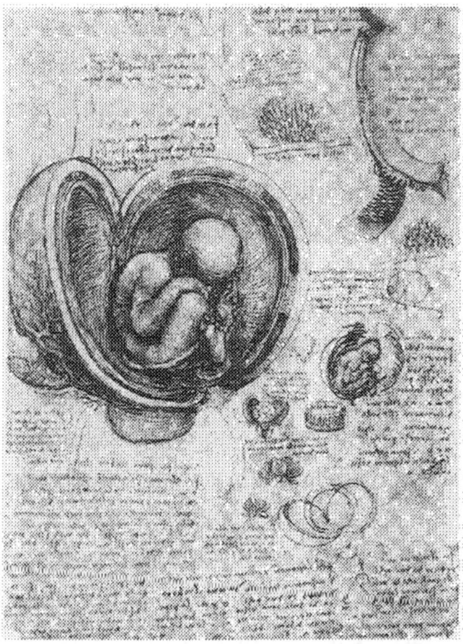

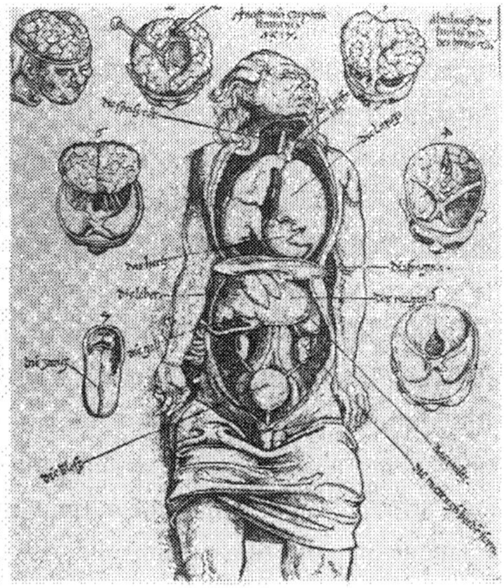

Figure 1. (*Left*) Leonardo da Vinci, *Study of the Foetus and the Womb, with Optical and Mechanical Diagrams*, Windsor, Royal Library, 19102.

Figure 2. (*Right*) Hans Wächtlin, *Dissection of the Brain, Thorax and Abdomen*, woodcut, 1517, from L. Fries, *Spiegel der Artzny* (Strasbourg, 1518).

the lobed liver corresponds to stock accounts and representations (Magnus Hundt, 1501).[16] We must assume some kind of mechanism by which the schemata of traditional anatomy were available to the draftsman and provided a visual foundation for his representation of features. It should be remembered that in an inevitably messy dissection 'seeing' would certainly not have been readily translated into 'knowing'. The apparent naturalism does not mean that the image is necessarily to be more trusted than the earlier woodcut, but it does mean that it is making implicit and explicit *claims* to be trusted. The same point can be made by looking at one of Leonardo's drawings of the muscles of the abdomen, which, even on a small and summary scale, conveys something of the conviction of his draughtsmanship.[17] For all its air of objective directness, the diagonally criss-cross muscles depend closely upon Pietro d'Abano's *Conciliator* of 1496 and upon his desire to emphasise graphically that 'every muscle uses its force along the line of its length'.[18]

Viewed in the light of such complications, the traditional reservations about illustrations in anatomical texts appear more understandable. Thus Berengario da Carpi, whose *Commentaria* of 1521 and *Isagogae breves* of 1522 are the first anatomical books in which illustrations make a really substantial impact on the tone of the whole production, warned the reader his figure of the vertebrae, for example, 'does not exhibit their true likeness... [and] their actual form is better seen in dried vertebrae in cemeteries'.[19] His much admired muscle-men serve strictly limited anatomical functions with respect to his text, and whenever he mentions his illustrations he does so in terms that restrict their role. However, as someone who was a prominent figure in the Medicean Rome of Pope Leo X, as the recipient from Raphael of a painting of *St. John the Baptist*, and as a pupil of Aldus Manutius, Berengario was well placed to understand the value of stylish illustrations in making his book effective in its social, intellectual and commercial environment. His poised *écorché* holding a noosed rope is the participant in an implied *historia* in the setting of Berengario's demonstration of the *theatrum* of the body (fig. 3). If Wächtlin's rhetoric of reality was of a rustic nature, Berengario's tends towards the nobly Roman.

Not surprisingly, in the humanist orbits of medical science in the Renaissance, it was the nobly tragic which became the dominant mode of illustration. The School of Fontainbleau stylishness of the illustrations in Charles Estienne's *De dissectione* (originally published in 1545 in Latin and translated into French a year later) has often been mocked for overwhelming their anatomical content, but the fancy presentation is far from gratuitous or merely decorative.[20] The anatomised men and women (fig. 4), performing the assigned roles as dying warriors or violated Lucretias, testify to the drama of human beings who have been placed in the world by God to contemplate the heavens, to 'investigate the divine works of nature' and to give due purpose to the creation through their deeds. If we read the introduction to the first book of *De dissectione*, 'containing the argument of the whole work', in which he debates the purpose of man with Anaxagoras, with due references to Chrysippus and Zeno, we gain a sense of the Stoic foundation of his enterprise, in which

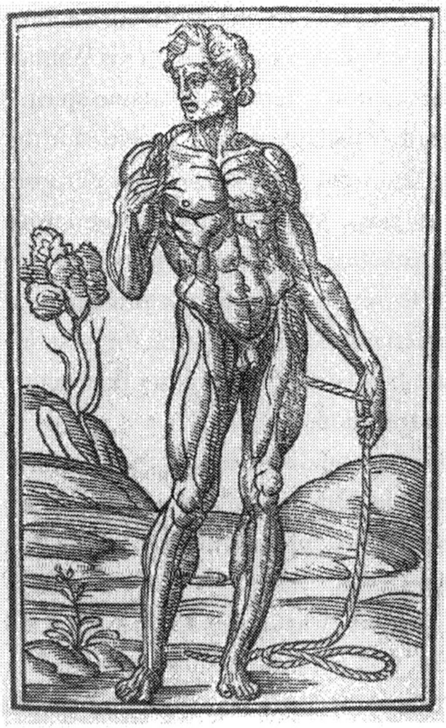

Figure 3. (*Left*) Berengario da Carpi, *Muscle-Man with Rope*, woodcut (by Uga da Carpi?), from *Commentaria super anatomia Mundini* (Bologna, 1521).

Figure 4. (*Right*) Charles Estienne, *Dissection of the Abdomen of a Woman*, woodcut, from *La dissection des parties du corps humain* (Paris, 1546).

man as observer and as the 'measure of all things' gives value to God's creation through perception of His divine plan.

Not the least of Vesalius' achievements was to embody all the existing varieties of the rhetoric of reality into a wonderfully functioning and complex whole. The title page of the *Fabrica* obviously sets the anatomist in the context of a great *historia*, in which the 'house of the soul, as Plato has it' is explored in a *all'antica* temple or *theatrum* of anatomy.[21] But the underlying message of Vesalius descending from the professorial throne to conduct the dissection with his own hands, aligns him with the German directness of Wächtlin and von Brackenau. His insistent emphasis upon first-hand dissection, a practice in which Vesalius must have possessed remarkable skills, is visually underlined by the cluttered still life of instruments (fig. 5), many of which were common or garden tools used by other trades. Hans Baldung Grien's illustration for Walther Ryff in 1541 had already included comparable tools in much the same spirit.[22] The illustration of a tethered pig on a board, 'which we usually provide for the administration of vivisections', appears at first sight to serve a similar purpose, but the text provides a rather different gloss, since it is concerned with Vesalius' conscious adoption of Galen's practice of vivisections of pigs for physiological investigations.[23] This serves to remind us that the principles of anatomical investigation enunciated by Galen provided inspiration for Vesalius to study form in rigorous detail through first-hand dissections, rather than acting (as so often believed) as the dead hand of tradition.

The famous muscle-men sustain this air of actual dissection, as they perform their myological striptease (see frontispiece), and the tone of the accompanying notes talks the spectator through the various procedures in much the same way as Vesalius must have done in the dissecting room.[24] Thus on the seventh plate he informs us that the rope from which the cadaver was suspended 'was diverted back to the occiput because of the muscles that are conspicuous in the neck'. However, the overall presentation is remote from the German manner, and clearly adopts and extends the more heroic mode of Berengario's Italian woodcuts. The frieze of gesturing figures in their continuous landscape act out a grand drama, gesturing like Old Testament prophets or collapsing in martyr-like death. Such a heroic presentation is fully justified as an appropriate (i.e. decorous) way to present 'the ingenuity and workmanship of the supreme artisan' (*sumi opificis solertiam artificumque*).[25] With the illustrations of the skeleton (fig. 6), which Vesalius acknowledges 'contribute more to display than to instruction', the pseudo-history becomes literal, as we are informed that while 'genius lives on, all the rest will perish'—a motto taken from Virgil's *Elegiae in Maecenatem* that makes particular sense in the context of Vesalius' self-conscious bid for enduring fame in his hugely ambitious project.[26]

No book was ever planned more meticulously to effect an enduring reform of both the subject and its mode of presentation. The letter to Oporinus, published in the opening matter, is insistent both about the necessary visual quality—'nowhere neglect the significance of the pictures' (*nusquam picturae ratione ... neglecta*)—and about following his intricate system of text, indices of figures, labelling, commentary and cross references. The variety, insight and maturity with which different kinds of visual material are exploited is

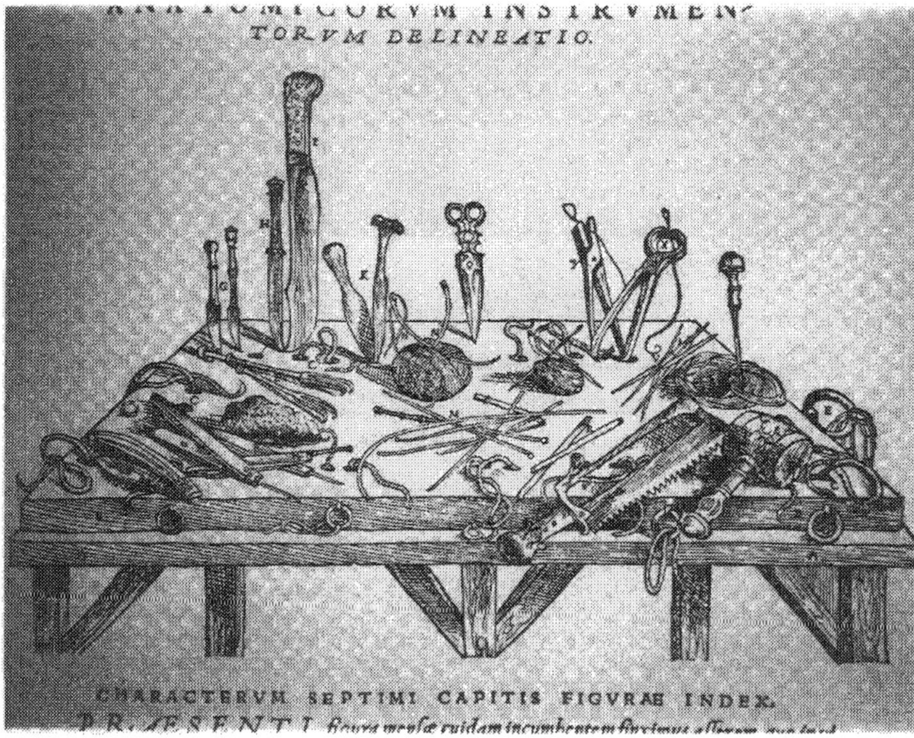

Figure 5. Andreas Vesalius, *Tools for Dissection*, woodcut, designed by Jan Steven van Kalkar, from *De humani corporis fabrica* (Basel, 1543).

astonishing. In addition to the large, pictorial representations of the main components in the fabric of the body, a series of small inset illustrations (fig. 7) graphically demonstrate structural principles. Some of the diagrams, such as the hinge, appear more than once, with a full annotation reserved for the first appearance. Sometimes the demonstration is diagrammatic in the most schematic sense, using what he called delineations in a 'purfunctory', 'rough' or 'rudimentary' (*rudis*) manner.[27] In his *Venesection Letter* of 1539 he introduced his illustration of the veins with the words, 'in this accurate though rather rudimentary figure' (*in haec vera, quamvis rudiori figura*), which shows clearly that he recognised the way in which the 'truth' of a particular illustration is dependent upon a correct reading of its conventions in relation to its

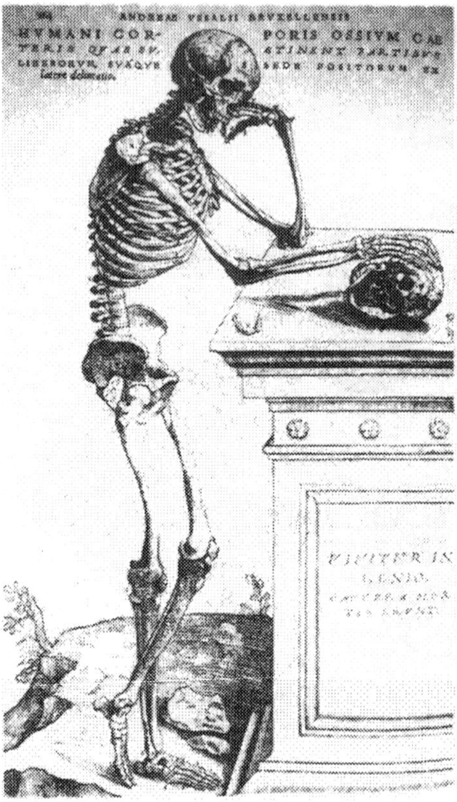

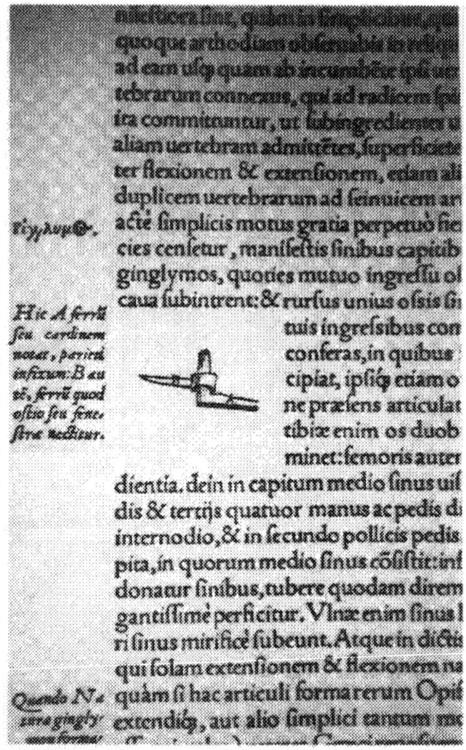

Figure 6. (*Left*) Andreas Vesalius, *Skeleton from the Side*, woodcut, designed by Jan Steven van Kalkar, from *De humani corporis fabrica* (Basel, 1543).

Figure 7. (*Right*) Andreas Vesalius, *Demonstration of a Hinge*, woodcut, from *De humani corporis fabrica* (Basel, 1543).

designated function.[28] A particularly nice example of his discussion of conventions is when he tells us that his section of the eye shows the forms 'in the manner in which we habitually depict the heavens and four elements on a flat surface' (*atque hoc quoque modo caelos & quatuor elementa in plano depingere solemus*).[29] The following page shows how the three dimensional components may be built up like a piece of precious jewellery by a 'supreme artisan'. He was also alert to the problem of exactly what woodcut lines represent in the more pictorial illustrations. Parallel lines could, after all, stand as shading or serve to indicate linear structures. Generally the lines serve to outline major contours and to shade within these contours, but on one occasion—specifically to demonstrate the muscle fibres (fig. 8)—Vesalius stresses that the lines signal the linear appearance of the forms rather than serving as elements in the artistic modelling of relief.[30]

In this one great book, Vesalius essentially tested all the illustrative types which were to be available to anatomists during the sixteenth century. The only variations left were the choice of medium, such as the copperplate engravings used by Valverde, and the devising of different systems of reference and labelling. One of the most interesting of these variants was devised by Bartolommeo Eustachio for his anatomical tables, which were not finally published until 1714.[31] The style of the illustrations is consciously synthetic—that is to say presenting the forms in a simple and clear manner which abstracts them from the flesh-and-blood reality of dissection—and the overall presentation is self-consciously poised (fig. 9). Eustachio has done away with even Vesalius' reticent letters, leaving the figure totally unmarked. Reference to individual parts is achieved through a Ptolemaic system of coordinates in the marginal scales, which necessitates the use of two straight edges, just as demonstrated in Apianus' *Cosmographicus*.[32] This ultra-cool and cerebral system of mapping the topography of the body did not prove popular, and one of the owners of the copy I consulted obviously ran out of patience with it, adding labels in a conventional manner.

The actual 'look' (or visual quality) of anatomical illustrations in the sixteenth century is, as these necessarily few examples have shown, far from being simply determined by the need to portray the forms accurately. However, the core of the endeavour does reside in a belief in the value of the veridical portrayal of tangible objects in space and in due proportion—so that there could potentially be a direct process of visual matching between the actual forms and their depictions. I emphasised at the outset that such a system of veridical portrayal could not stand at the heart of the astronomer's strategy of research and exposition. However, many of the figurative and metaphorical images evoked by Renaissance writers on astronomy rely upon exactly the same kinds of criteria of formal structure as characterise the treatises by anatomists. The most famous of these is the much-cited bodily simile in Copernicus' dedication of *De revolutionibus* to Pope Paul III in 1543. He characterised those astronomers who had relied upon the proliferation of such devices as epicycles, eccentrics and equants as failing to:

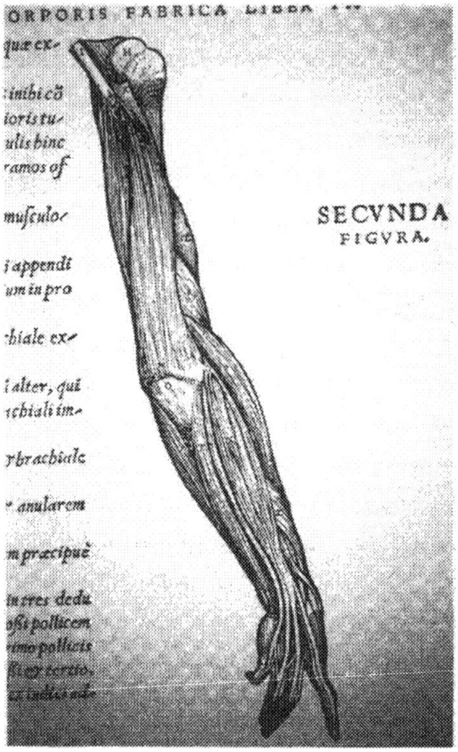

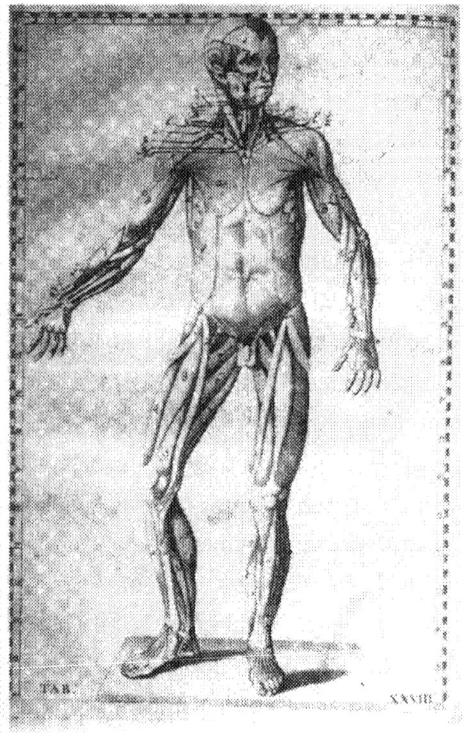

Figure 8. (*Left*) Andreas Vesalius, *Muscles of the Upper and Forearm and Tendons of the Wrist*, woodcut, designed by Jan Steven van Kalkar, from *De humani corporis fabrica* (Basel, 1543).

Figure 9. (*Right*) Bartolommeo Eustachio, *Superficial Dissection of the Muscles from the Front*, engraving with annotations in ink, from *Tabulae anatomicae* (Rome, 1722).

elicit or deduce from the eccentrics the principal consideration, that is the structure of the universe and the true symmetry of its parts. On the contrary, their experience was just like some one taking from various places hands, feet, a head, and other pieces. Very well depicted, they may be, but not for the representation of a single person. Since these fragments would not belong to one another at all, a monster rather than a man would be assembled. Hence in the process of demonstration or 'method' as it is called, those who have employed eccentrics are found either to have omitted something essential or to have added something extraneous and wholly irrelevant.[33]

To some extent, the resort to a bodily analogy was a standard strategy in Medieval and Renaissance thought, just as Apianus in 1524 followed Ptolemy in comparing geography to the portrayal of a complete head, while chorography was equivalent to the portrayal of an eye or ear in isolation (fig. 10). But, considered in the context of its humanist audience in the Rome of the Farnese Pope (who was a notable patron of Michelangelo), this passage is full of weighty allusions.[34] The idea of a body perfectly proportioned according to the principle of *symmetria* was by this time deeply embedded in Renaissance aesthetics, and was an integral part of the key doctrine of decorum, according to which every part should be appropriate to the form and significance of the whole. This doctrine could be gleaned directly from ancient poetics and rhetoric, or from such Renaissance authors as Leon Battista Alberti, whose *De re aedificatoria* had transmogrified the Roman ideas of Vitruvius into Renaissance form. As someone who had participated in humanist poetics as a translator of Theophylactus Simocatta from the Greek (including a letter which centres upon Parrhasius' portrait of Helen of Troy) and as someonc who was reputed to have been sufficiently competent in painting to produce a self-portrait (a version of which was once owned by Tycho Brahe), Copernicus is unlikely to have been using visual analogies in an innocent manner.[35]

Indeed, in the next paragraph, he reinforces the meaning by emphasising that the 'movements of the world machine' were 'created for our sake by the best and most orderly artisan of all' (*ab optimo et regulariss. omnium opifice*), and he later refers to *divina haec Opt. Max. fabrica*.[36] The terms *opifex* and *fabrica* are already familiar to us from the Vesalian lexicon of significant words. Rheticus, that faithful promoter of Copernican ideas, glosses the bodily analogy by direct reference to Galen's *De usu partium*, to the effect that 'Nature does nothing in vain'.[37] He then asks rhetorically, 'should we not attribute to God, the creator of nature, that skill which we observe in the common maker of clocks? For they carefully avoid inserting in the mechanism any superfluous wheel'. The idea of perfection such that nothing can be added or taken away without detriment to the *symmetria* of the whole conforms to the standard Renaissance concepts of visual beauty and structural necessity, expressed in their canonical forms by Alberti.[38] It was from such a standpoint that Copernicus decried those who 'either ... omitted something essential or ... admitted something extraneous and wholly irrelevant'.[39]

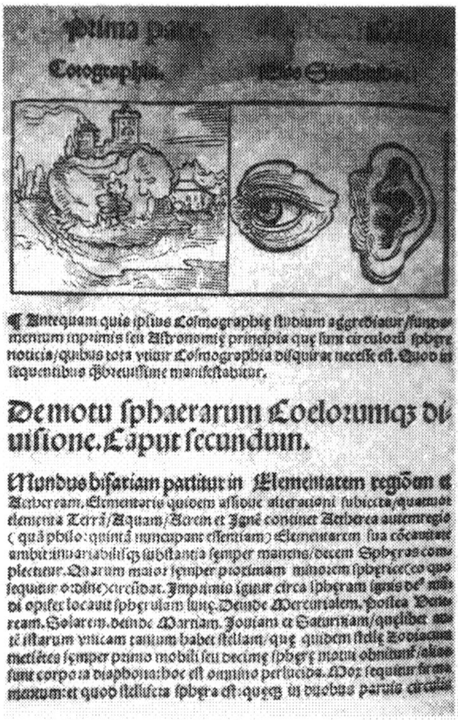

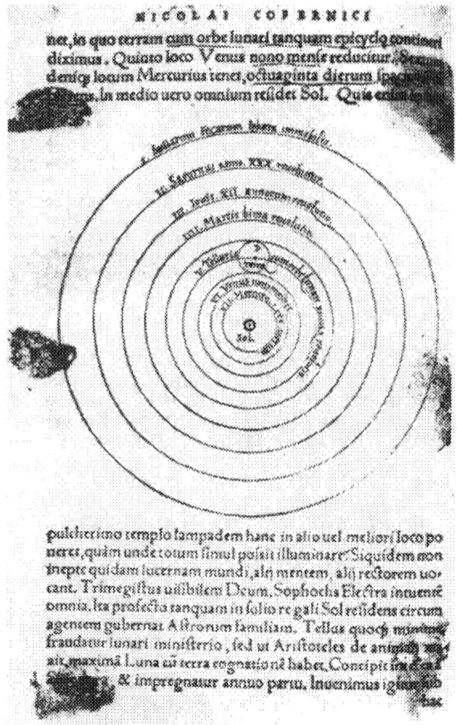

Figure 10. (*Right*) Petrus Apianus, *Chorography Compared to Pictures of the Eye and Ear,* from *Astronomicum caesareum* (Ingolstadt, 1540).

Figure 11. (*Left*) Nicholas Copernicus, *Diagram of the Orbits of the Earth and Planets,* woodcut with underlining in pen, and paste marks, from *De revolutionibus orbium coelestium* (Nuremberg, 1543).

On the page of the manuscript of the first book of *De revolutionibus*, facing his key visual statement of his new system of orbits (fig. 11), he further extends the visual analogies into specifically architectural and social contexts:

> At rest ... in the middle of everything is the sun. For in this most beautiful temple, who would place this lamp in another or better position than that from which it can light up the whole thing at the same time? For the sun is not inappropriately called by some people the lantern of the universe, its mind by others, and its ruler still by others. The thrice Greatest [Hermes] labels it a visible god, and Sophocles' Electra, the all-seeing. Thus indeed, as though

seated on a royal throne, the sun governs the family of the planets revolving around it.[40]

If we put together the principles that are emerging—*symmetria*, decorum, perfect economy and necessity of design, hieratic social order with respect to supreme authority, and man as the observer through which the whole system becomes apparent—we are in precisely the kind of world enunciated in Alberti's writings, and signalled in a less sustained way in Estienne's preface. It is essentially a neo-Stoic vision, laced with Neoplatonic idealism and Pythagorean metaphysics. I am not concerned here to debate whether the Copernican system actually lived up to its ideal of perfect economy and *symmetria*—which seems doubtful—but rather to characterise the nature of the vision that underlay his aspirations.[41]

Not the least important of these ideas was the conception of the central role of man as the observer, and, indeed, in the form commonly formulated in the Renaissance, as the reason why the whole set-up had been created by God. The paradox at the heart of the Copernican system was, of course, the fact that this central observer had been removed from the physical centre of the system. The inhabitants of the earth were now in a position where all planetary motions were relative to the motion of the body on which they were standing. To make this point, Copernicus quotes from Virgil's *Aeneid*: 'forth from this harbour we sail, and the land and the cities slip backwards'.[42] This notion of the appearance of shapes and motions as irredeemable relative to the position of the observer was essential to the Renaissance revolution in the depiction of the visible world. A theorist like Leonardo could stress that the point at which parallel lines appear to converge (the 'vanishing point') moves with any motion of the observer's point of view, and that two horses running away from us along parallel tracks appear to be converging.[43]

However, this relativity was not taken to mean that visual experience must collapse in subjective confusion. Rather, the science of perspective leads to a rational understanding of the principles of systematic depiction such that the true shape, position and motion of an object can be determined unambiguously from proper analysis of the image. Thus, the centrality of the observer, is if anything strengthened by his or her role within a system of relative perceptions. There are clear signs in Copernicus himself and in some of his more realistically inclined successors, that the position of the astronomer on a mobile body was seen as presenting an opportunity to record the motions of the whole system around the static sun in such a way that the astronomer could capture the physical reality rather than merely formulating mathematical hypotheses which were analogous to the appearances. It was much in this sense that Kepler asked and answered his question, 'in what manner were the earth's dimensions adapted to the size of the solar globe?'.[44] He answered that it was 'in terms of vision. For the earth would be the home to the contemplative creature, and it was for him that the entire universe had been created'.

Although such concepts as the proper visual principles of the body or temple of the universe and the role of man as the 'mean and measure of all things' are deeply shared by the two sciences of 1543, the relationship of the overall vision and the illustration of the phenomena was necessarily quite different in each discipline. Any Copernican could not but be aware of the obvious problem that what we actually 'see' is the sun rising, moving across the skies, and setting. We may understand the point of relativity, but, in terms of how our perception actually works, our eyes and body do not bear obvious witness to the motion of the earth. This was the dilemma which Kepler endeavoured to overcome in his paper written as a student, which postulated an observer on the moon, and in his posthumously published *Dream*, in which it is the earth which appears to move from a station point on the planet Levania (i.e., the moon).[45] To make the point visually, however, required a different strategy. The obvious one was to represent the system diagrammatically as if characterised by an Olympian viewer who could stand outside the system. This was of course the stock method adopted for the geocentric system in earlier publications, and Copernicus' diagram contained no new mode of visual presentation. In fact, his manuscript could hardly be more unadventurous in its visual presentation, containing inset or marginal diagrams of an entirely linear and traditional kind. In this format he does no more than to show the basic geometrical components of the motion of the system in a sequential and accumulative manner.

When he did attempt to characterise one of the more complex, compound motions—that of the pole around the mean position (I)—the 'twisted line' is not easy to read in terms of the resulting motion, and in the printed edition is mistakenly transposed into two separate ovals (fig. 12).[46] Copernicus had admitted earlier in the same chapter, 'that these matters are not easily explained adequately with words. Hence they will not be understood when heard, I am afraid, unless they are also seen by the eyes. Therefore let us draw on a sphere the ecliptic ABCD...'.[47] But the diagrammatic resources available to him were not visually eloquent to anyone who had not already cultivated an ability to visualise in the mind in non-verbal form (as described by Einstein) the complex consequences of the relative motions of bodies moving in orbits and epicycles with eccentrics. I think it is fairly clear that an astronomer of Copernicus' and Kepler's levels of visionary insight must possess abilities of spatio-temporal visualisation of an astounding order—at least astounding to me—if they are to envisage in a coherent manner what would happen if 'any part thereof were to be moved from its place' in such a way that it would not produce 'confusion in all the other parts and of the Universe as a whole'.[48] But this spatio-temporal visualisation is not reflected in visual form in Copernicus' illustrations.

This is not to say that the key diagram of the orbits lacked a certain kind of potency. It is significant that at least two astronomers, including Kepler, tabulated the distances of the planets on this same page in their copies of Copernicus, since it is easier to envisage the numerical values in juxtaposition to a visual kcy.[49] And a copy of the first edition in St. Andrews University shows clear signs of paste marks of a sheet once stuck over the heliocentric system (fig. 11), presumably to substitute the printed universe by a less offensive arrangement.[50]

When we consider more generally the kinds of visualisation demanded by either Ptolemaic or Copernican astronomy, we may assume that professional practitioners were acquainted with a wide range of visual sciences and their associated instrumental devices. The nexus of required learning is neatly encapsulated in the one-page catalogue issued by Regiomontanus in Nuremberg in 1474, containing books already available and titles he was intending to publish.[51] The range extends from pure geometry, such as Euclid's *Elements*, Archimedes on the sphere and cylinder, Apollonius' *Conics* and a treatise on the five regular bodies, through the *scientiae mediae* of music, astronomy (Peurbach, Ptolemy, Proclus *et al.*) and optics (Witelo and Ptolemy), to the practical sciences of engineering and instruments. The availability of actual instruments for the practice of astronomy is signalled in large print at the base of the prospectus.

It was through the use of astronomical instruments that the essential mediation between the observed phenomena and their geometrical analysis could be accomplished, and it was through astronomical models that representation could best be achieved for the purposes of instruction.

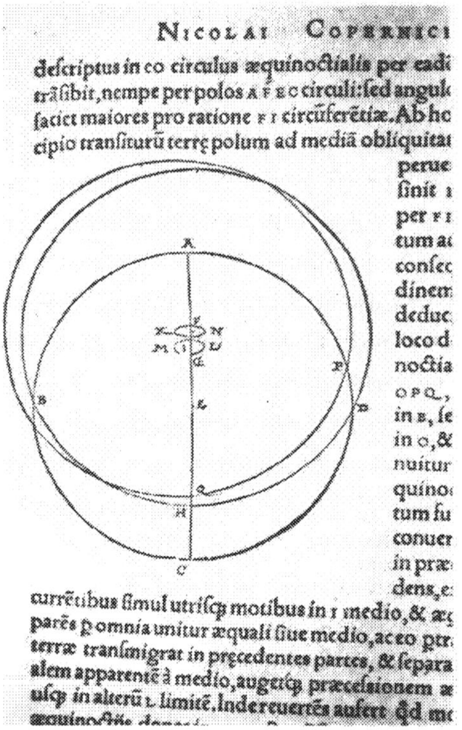

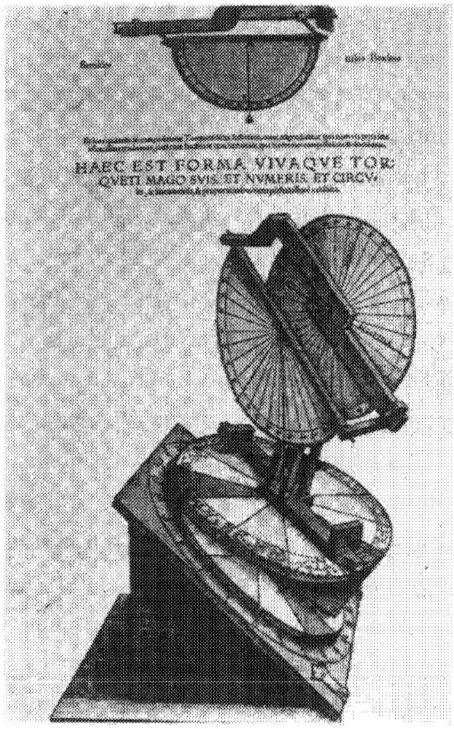

Figure 12. (*Left*) Nicholas Copernicus, *Diagram of the Motions of the Pole around a Mean Position*, woodcut, from *De revolutionibus orbium coelestium* (Nuremberg, 1543).

Figure 13. (*Right*) Petrus Apianus, *Torquetum*, woodcut, from *Cosmographicus liber* (Landshut, 1524).

Instruments such as armillary spheres, orbaria and torqueta, as illustrated by Apianus (fig. 13), could provide aids to spatial understanding in a way that was impossible with Copernicus' illustrations, although the schematic orbits could still only deal with the rudiments of the system rather than the full complexity of apparent motions.[52] Most instruments were not of course direct attempts to model the celestial machine in fully spatial terms, but a number may be seen as serving as kinds of analogue models. The most common was the astrolabe, which results from a conical projection from the centre of the North celestial pole in such a way that the *rete* acts as a star map, which is laid over coordinates, lines of equal latitude and (often) hour lines. Astrolabes were designed for practical observation and mathematical calculation, and were not well suited to serve as aids to the visualisation of the actual spatial configurations, any more than Copernicus' diagrams had been—however visually compelling and 'concrete' a finely-made astrolabe may at first sight seem as a model of the physical set up. Such physical models also in their turn stood in a symbiotic relationship to depictions, since they were themselves subject to explanatory illustration in a variety of diagrammatic and perspectival techniques, most spectacularly, as will be seen, in the publications of Tycho Brahe.

Copernicus was of course well familiar with the varieties of highly appealing models available in the Renaissance, and a magnificent set of instruments, comprising an astrolabe, torquetum and celestial globe were presented to the Jagiellonian University in Cracow by Martin Bylica in 1494, the final year in which Copernicus was a student.[53] The astronomical globe is particularly nice as an aid to visualisation since it can be adjusted so that the rotation of the globe models the observed rotation of the stars around the celestial pole at the particular location in which it is used, while the astrolabe can be used to work out related problems of an astronomical or astrological nature. These are luxury instruments, which could only be made by a supreme *opifex* of the worldly variety—in this case probably Hans Dorn. The quality underlines the aesthetic of economy and perfection which Copernicus shared with humanist theorists of the visual arts, such as Alberti and Leonardo, for whom the 'fittingness' of form and function was a keystone to the understanding and representation of nature. The expensive perfection of such devices made them especially suited to flourish in the courtly culture which did so much to ensure the triumph of humanism across Europe. The technical success of instruments in modelling the motions of the heavens obviously played a major role in what I have called the 'rhetoric of irrefutable precision', but they also gave astronomers a kind of opportunity of participating in the 'rhetoric of the real' that was not open to them through veridical depiction.

In fact, for one of the major contributors to the reform of astronomy, Tycho Brahe, instruments became the keystone in the construction of the real edifice of the heavens, and the chief means of personalising astronomy in terms of the heroic observer. No one had ever placed such weight upon the explanation and illustration of his instruments. Apianus' *Instrumentum primi mobilis* in 1534 and *Astrononicum caesareum* in 1540 provide only very partial precedents for the way that instruments are described in Tycho's *Astronomiae instauratae mechanica* in 1598 and in his other publications.[54] By demonstrating the mechanisms by which his observations were achieved, Tycho was certifying his practice in terms of the concrete reality of his personal procedures—'so that certainty of the form and use of the instruments might be apparent', as he said in his treatise on the new star of 1572.[55] He explained that 'the construction and use of the instrument is understood by careful study of the accompanying figure quicker than through more elaborate verbal explanation'.[56] The level of his personalisation of instruments was equally strong. He cherished a parallax device once owned by Copernicus, and which was, 'it was said, made by him with his own hand'. Although it was wooden and not convenient to use, Tycho recorded that 'I was so delighted because it reminded me of the great master', and he was moved to compose a heroic poem in its honour.[57]

The personalising of his own equipment is vividly apparent. His great *Mural Quadrant* or *Quadrans Tychonicus* (fig. 14) served as a great emblem of his endeavour. He explains that the pictorial adornments, including his own portrait by Thobias Gemperlin and landscape by Johannes of Antwerp, were 'only added for the sake of ornament, and in order that the space in the middle should not be empty and useless', but this should not lead us to underrate their significance to Tycho's agenda, any more than ornament would have been regarded as redundant in rhetoric.[58] The whole set-up casts Tycho in the role of a new Ptolemy, or perhaps even more ambitiously as a personification of astronomy itself. The emphasis upon the instruments for observation stresses the reality of his procedures while the mobile brass globe in the niche signals the process of envisioning which lead to his own peculiar conflation of the Ptolemaic and Copernican systems. He explained that a large celestial globe he had constructed allowed him to 'determine mechanically, with very little trouble and without calculations, all the details concerning the doctrine of the sphere'. It was the invention of new instruments, upon which Tycho set such store, that permitted some Renaissance thinkers to claim most decisively that the ancients had been both emulated and surpassed.

Throughout his account of his instruments, Tycho explains their manufacture and use in highly individualised terms. Not infrequently, he outlines the iconography of their ornamentation in some detail. Thus he explains that his *Quadrans minor* (fig. 15), which was mercury gilded 'so that it stays beautiful and clean', was adorned so that it might 'offer some instruction'—in this case through an allegory which contrasts a life of higher contemplation (his own life by implication) with the vanity of worldly things.[59] The first of the

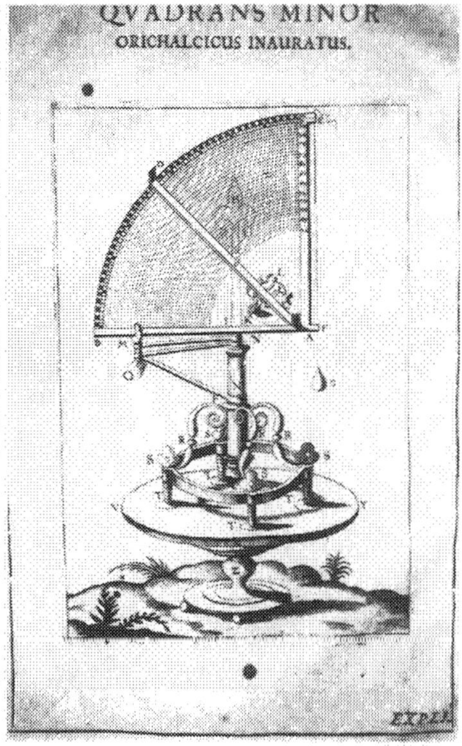

Figure 14. (*Left*) Tycho Brahe, *Mural Quadrant* or *Tychonicus*, engraving from *Astronomiae instauratae mechanica* (Wandesburg, 1598).

Figure 15. (*Right*) Tycho Brahe, *Lesser Quadrant*, engraving from *Astronomiae instauratae mechanica* (Wandesburg, 1598).

accompanying inscriptions—*Vivimus ingenio coetera mortis erunt*—is virtually identical to the Virgilian tag in Vesalius' illustration of the skeleton, while the other says that 'in Christ we live, all the rest perishes'. Another device, his *Armillae equatoriae* was adorned with paired portraits of Ptolemy and Albategnius (al-Battani), and Copernicus and Tycho, as a way of underlining his place in the heroic succession.[60] However, as a corrective to the ideal picture conveyed by this contemplative and productive life of observation, he warns that the vagaries of patronage are such that the astronomer should always ensure that the instruments can be dismantled for transport to another site. As he says, 'the astronomer, as well as a student of other branches of knowledge, has to be a citizen of the world'.[61]

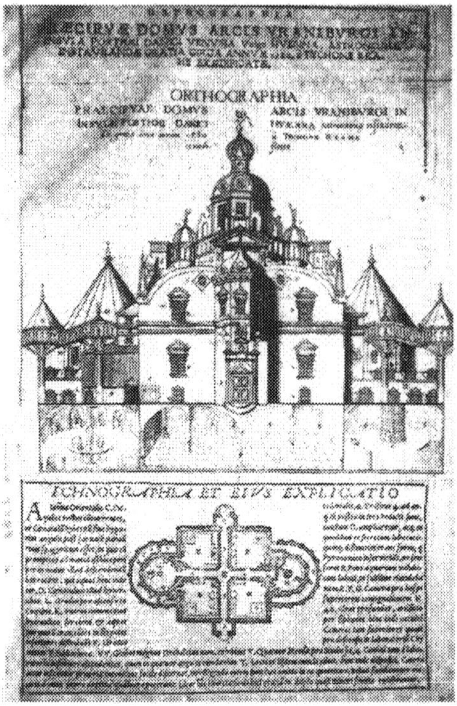

Figure 16. (*Left*) Tycho Brahe, *Elevation and Plan of the Palace of Uraniborg on the Island of Hven*, engraving from *Astronomiae instauratae mechanica* (Wandesburg, 1598).

Figure 17. (*Right*) Johannes Kepler, *Temple of the Astronomers*, from *Tabulae Rudolphinae* (Ulm, 1627).

The most complete expression of Tycho's world in visual terms was of course his remodelling of the island of Hven. His castle of Uraniborg (fig. 16), with its surrounding plantations and ponds was contrived as a microcosm of the universal harmonies.[62] The central building, as he explained, was 'strictly symmetrically arranged, as required with architecture if the work is to be executed in a proper manner according to the rules of art'.[63] The key rule, here as in Copernicus' vision of the body of the universe, was *symmetria*—a rule which Tycho saw as embodied in Dürer's books on human proportion—though the architectural vocabulary in which the *symmetria* was expressed by 'my architect, Johannes Stenwickel [or Steenwinckel] of Emden', is actually remote from the strict requirements of Renaissance theorists and practitioners. A more architecturally literate realisation of an *all'antica* temple for the Danish astronomer's muse was provided for Kepler's publication of Tycho's *Rudolphine Tables* (fig. 17).[64] Kepler himself seems to have been notably literate in the visual arts, an expert in stereometric estimation—the highly useful merchant skill in the visual judging of volumes—and a decently accomplished draftsman in his own right, as the sketch for the frontispiece to the *Tables* suggests.[65] In the printed version, incorporating changes apparently demanded by Tycho's heirs, the slow perfecting of astronomical science is represented by an architectural progress from rustic supports at the rear, through crude piles of stone blocks and archaic Doric columns, to the more polished Tuscan pillar of Copernicus and climaxing in the beautiful Corinthian column, beside which Tycho points to a Copernican diagram of the heavenly system as engraved on the ceiling of the temple.

Kepler's own manner of astronomical visualisation represents the climax and consolidation of the various strands of visual modelling we have seen progressively developing in the writings of Copernicus and illustrations of Tycho. His most famous visual shaping of the planetary system was, of course, the characterisation in his *Mysterium cosmographicum* of the ratios of the orbs in the Copernican system as corresponding to the arrangement of a set of Platonic solids nesting one inside the other (fig. 18).[66] We know from his own account in *De stella nova* that the idea came to him visually in 1595, when he was drawing 'quasi-triangles, in the same circle, in such a way that the end of one was the beginning of the next' for the instruction of his students.[67] The full scale visualisation is presented in a folding plate as a perspectival picture of considerable sophistication, in which the system is characterised in terms of an elaborate piece of mannerist metalwork, of just the kind that his noble patrons enjoyed.[68] In fact, the dedication of this plate to Duke Frederick is closely related to his unavailing attempts to fabricate the system in three dimensions. He promoted his *inventum* to his patron by explaining that

> the whole work and the demonstration thereof can be fittingly and gracefully represented in a drinking cup of an ell in diameter which would be a true and genuine likeness of the world and model of the creation in so far as human research may fathom.[69]

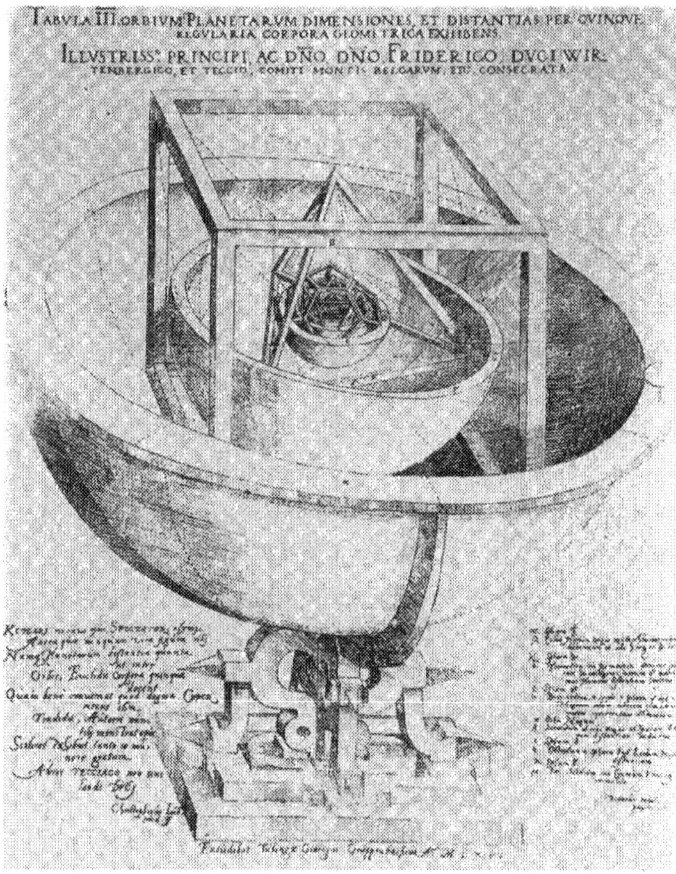

Figure 18. Johannes Kepler, *Demonstration of the Orbits of the Planets,* from *Mysterium cosmographicum* (Tübingen, 1596).

The hollow armatures of the Platonic solids were each to be filled with appropriate beverages, which could be drawn off through taps at the rim. This bizarre scheme was dropped in favour of a plan for a model operated by clockwork, and he hoped to find a master *opifex* who could construct one with such precision that it would have an error of only one degree in a hundred years.[70] Even if this ambitious object was never to be realised, his dedication of the plate to his noble protector did have one fortunate consequence. Mästlin reported that theologians were deterred from voicing open criticism of Kepler's Copernicanism by the identification of the scheme with Duke Frederick.[71]

For Kepler, the conceit of remaking the universe in a working, physical model was no mere intellectual and technical game. At the heart of his

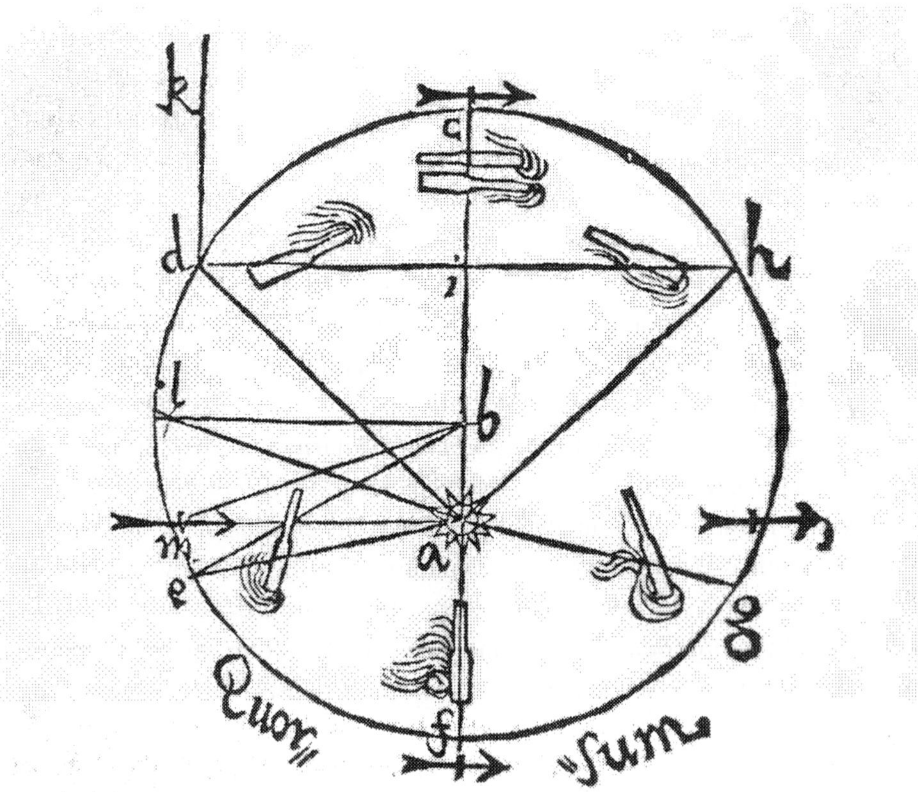

Figure 19 Johannes Kepler, *Demonstration of an Orbit by Analogy to a Boat in a Stream*, from *Astronomia nova* (Heidelberg, 1609).

enterprise—and of his discovery of the elliptical orbits—lay a desire to harmonise the Platonising geometry which he valued above all other forms of mathematical truth with an understanding of the physical mechanics of the motions of the planets. Metaphysics alone would not suffice: 'the celestial machine is not so much a divine organism but rather a clockwork'.[72] It was in this spirit that he transformed one of the stock metaphors of astronomy into a functioning analogy in mechanics. This metaphor, used by Ficino amongst others, envisaged the heavenly bodies steered as by a pilot.[73] The title page to Sebastian Münster's *Organum uranicum* of 1536 picks up this metaphor in visual form.[74] Kepler, in one of the diagrams of a planet orbiting the sun in his *Astronomia nova* depicts a pilot in the 'magnetic' stream emblematically, and in another (fig. 19) adds schematic oars with rippling waves to the diagram which explains the physical geometry of the orbit. Here the process of visualisation is joined to a sense of physical action which is very like the muscular empathy described by Einstein—even if the consequence of the physical analogy in this case hardly shows Kepler at his most efficacious.

In all this discussion of visualisation in astronomy, however strong the visual model adopted for particular purposes, we have seen nothing to compare with the direct practice of veridical representation in anatomy. Such representation only became effective in astronomical science when the celestial bodies could be observed as bodies, that is to say as objects with discernible, individual features. It is this condition that explains why the invention of the telescope occasioned a new branch of visual astronomy, namely that concerned with the actual anatomy of the individual planets and the sun. The two key episodes—the dispute over the apparent irregularities in the surface of the moon, and the nature of the spots observed on (or not on) the sun—have been discussed elsewhere, but it might help towards the conclusion of this paper to remind ourselves of the rather different nature of seeing and knowing which the new sights down the telescope occasioned.[75] The first controversy involved how to interpret the pronounced lights and darks on the moon, particularly at the interface between the shaded and illuminated portions. Galileo, well versed in the science of perspective and the artist's systematic understanding of cast shadows, was able to argue that the most rational way to interpret the changing patterns of light and dark was in terms of shadows cast by huge topographical features, including mountains. The other incident concerns the patches which were seen to progress across the image of the supposedly immaculate sun. Galileo argued from the perspectival foreshortening of the spots as they neared the edge of the sun that they were integral parts of the surface and not shadows of intervening bodies. His method of argument, as he explained, was '*in virtù di perspettiva*'. Galileo's advanced understanding of the principles of artistic representation, which informed his method of analysis and exposition in these two cases, is undoubtedly important more generally to his theory and practice of observation, but the accompanying techniques of veridical representation could still only be brought to bear upon a very narrow range of problems in astronomy as a whole. Galileo's innovations in other of his sciences, such as dynamics and statics, were conducted with quite different forms of visualisation, experimentation and proof, and he did not sustain the pictorial mode in his own later work in astronomy.[76]

Looking back over this necessarily selective survey, what conclusions can be drawn about the role of illustrations and its relationship to the process of visualisation?

For astronomers in the Renaissance, the fundamental processes of visualisation do not seem to have been essentially different from those of Ptolemy or his Islamic successors. The visual qualities of the illustrations bore only schematic relationships to the visualisation demanded of the astronomer. Scientific instruments come closer to the hypothetical mental models, but only with respect to the gross characteristics of the arrangement of the basic armature of the celestial machine. Where more specifically Renaissance modelling can be discerned is in the humanist metaphors and analogies used to characterise form and function, relying upon beauty, economy and decorum (intellectual, visual and social). There was also a re-characterisation of the heroic observer, in which the objects were defined relative to the observing subject—a move which was crucial if the Copernican and Keplerian systems were to become acceptable. One field in which the new practice of perspectival representation became crucial was the depiction of instruments. The publications of Apianus and Tycho Brahe gave astronomers a chance to participate in the kind of broadcasting of secrets and marvels that had become typical of the prestigious books of mechanical devices. The other major aspect of astronomy that was radically affected by new pictorial means was the depiction of celestial bodies as viewed in the telescope. The new features, such as the topography of the moon, involved the new vocabulary of perspective and light and shade, but they remained somewhat peripheral to the major changes in astronomical science.

For anatomists, the visual power of naturalistic representation was a powerful and central tool in the rhetoric of the real, and could be used as an expression of the impulse to reconstruct the fabric of the body on the basis of direct, hand-on experience. The representations served as a powerful form of visual pointing, both to their own features and, potentially, to those of the actual object. However, we should remain alert to the fact that this visual pointing could draw apparently convincing attention to what was not there, and that the process of matching expectation to experience was (if anything) rendered more complex and challenging rather than less so. We should also remain continually aware of the way in which the representation of the human body, in the eyes of its major investigators, was designed to serve to demonstrate the wonderful artifice of the maker of the bodily 'temple' for the soul.

In sum, I do not see any obvious prospect of a grand, unifying theory based on new forms of representation as corresponding directly to (or precipitating) some great overarching reform of the means of visualisation. The relationship between illustration and visualisation seems quite different in the various sciences, though we can frequently observe intricate conjunctions in the structure of metaphor, analogy and 'aesthetics' that is used to locate a specific field of study within its broader intellectual, theological and social nexus. I have to say, as far as I am concerned, the lack of conformity to a grand theory makes matters more interesting to me as a historian of visual representation rather than less so.[77]

Department of the History of Art, University of Oxford

NOTES

[1] Einstein's letter to Jacques Hadamard, quoted by J. Hadamard, *The Psychology of Invention in the Mathematical Field* (Princeton, 1954) pp. 142–3.

[2] L. Fuchs, *De historia stirpium* (Paris, 1542) pp. x–xi.

[3] Leonardo da Vinci, Windsor 19071r; K. Keele and C. Pedretti, *Leonardo da Vinci. Corpus of Anatomical Studies in the Collection of Her Majesty the Queen at Windsor Castle*, 3 vols. (London and New York, 1979) no. 162r.

[4] J. Kepler, *Gesammelte Werke*, ed. W. von Dyck, M. Caspar *et al.*, 20 vols. 1938-88, vol. 13, p. 85; trans. E Rosen, 'Kepler and the Lutheran attitude towards Copernicanism in the context of the struggle beweeen science and religion', *Kepler. Four Hundred Years*, ed. A. Beer and P. Beer, *Vistas in Astronomy*, XVIII (Oxford and New York, 1975) p. 325.

[5] A. Crombie, 'Science and the arts in the Renaissance: the search for truth and certainty, old and new', in J.W. Shirley and D. Hoeniger (eds), *Science and the Arts in the Renaissance* (Cranbury, New Jersey, 1985) pp. 15–6. Compare J. Ackerman, 'The involvement of artists in Renaissance science', *Science and the Arts in the Renaissance*, pp. 94–129 and 'Early Renaissance 'naturalism' and scientific illustration', in A. Ellenius (ed.), *Natural Sciences and the Arts* (Uppsala, 1985) pp. 1-17; and R. Root-Bernstein, 'Visual thinking: The art of imagining reality', *Transactions of the American Philosophical Society* 75 (1985) 50–67.

[6] E. Panofsky, *Die Perspektive als 'symbolishe Form'*, Vorträge der Bibliothek Warburg, 1924–5 (Berlin, 1927); *Perspective as Symbolic Form*, trans. C. Wood (New York, 1991). For reworkings of the Panofskian standpoint, see especially S.Y. Edgerton Jnr., 'The Renaissance artist as quantifier', in M. Hagen (ed.), *The Perception of Pictures* (New York, 1980) I, pp. 179–212; 'The Renaissance development of scientific illustration', *Science and the Arts in the Renaissance*, pp. 168-97; and *The Heritage of Giotto's Geometry. Art and Science on the Eve of the Scientific Revolution* (Ithaca and London, 1991).

[7] S.Y. Edgerton Jnr., 'Galileo, Florentine 'disegno', and the strange spottednesse of the moon', *Art Journal* 44 (1985) 225-48; and M. Kemp, *The Science of Art. Optical Themes in Western Art from Brunelleschi to Seurat*, revised ed. (London and New Haven, 1992) pp. 94–6.

[8] For recent contributions that make some inroads into these matters, see particularly R. Westman, 'Nature, art and psyche: Jung, Pauli, and the Kepler-Fludd polemic', in B. Vickers (ed.), *Occult and Scientific Mentalities in the Renaissance* (Cambridge, 1984) pp. 177–229; M. Mahoney, 'Diagrams and dynamics: Mathematical perspectives on Edgerton's thesis', *Science and the Arts in the Renaissance*, pp. 198–220; W. Ashworth Jnr., 'Light of reason, light of nature: Catholic and Protestant metaphors of scientific knowledge', *Science in Context* 3 (1989) 89–107; E. Tufte, *The Visual Display of Quantitative Information* (Cheshire, Connecticut, 1983), and *Envisioning Information* (Cheshire, Connecticut, 1990); M. Lynch and S. Woolgar (eds), *Representation in Scientific Practice* (Cambridge, Mass., 1990); and M. Winkler and A. Van Helden, 'Representing the heavens: Galileo and visual astronomy', *Isis* 83 (1992) 195–217. An illustrated panorama is provided by B.J. Ford, *Images of Science* (London, 1992).

[9] For Osiander's foreword, see *De revolutionibus orbium coelestium* (Nuremberg, 1543) preface, iv–vi; and the manuscript version (Cracow, Jagiellonian University, Library, MS BJ 10,000), trans. and commentary by E. Rosen in *Nicholas Copernicus. Complete Works*, 3 vols. (London, Warsaw and Cracow, 1972) II, p. xvi. Compare *De revolutionibus*, II, intro., p. 27; trans. Rosen p. 51. See O. Gingerich, 'From Copernicus to Kepler: Heliocentrism as model and as reality', *Proceedings of the American Philosophical Society* 117 (1973) 513–22.

[10] K. Roberts and J. Tomlinson, *The Fabric of the Body: European Traditions of Anatomical Illustration* (Oxford, 1992).

[11] Andreas Vesalius, *Tabulae sex* (Venice, 1436), letter of dedication to Narcissus Parthenopeus; quoted J. Saunders and C. O'Malley, *The Illustrations from the Works of Andreas Vesalius of Brussels* (New York, 1950) p. 233.

[12] Keele and Pedretti, *Corpus*, no. 198r. More generally for artists and anatomy in the Renaissance, see B. Schultz, *Art and Anatomy in Renaissance Italy*, in *Studies in Fine Arts: Art Theory*, No. 12, ed. D. Cuspit (Ann Arbor, 1985); and M. Cornell, *Artists and the Study of Anatomy in Sixteenth-Century Italy* (PhD. Thesis, Warburg Institute, University of London, 1992).

[13] L. Fries, *Spiegel der Artzny* (Strasbourg, 1518).

[14] See R. Koch, *Hans Baldung Grien: Eve, the Serpent and Death* (Ottawa, 1974); L.H. Boudreau, *Hans Baldung Grien and Albrecht Dürer: A Problem in Northern Mannerism* (Ph.D. thesis, University of North Carolina, 1978); and *Hans Baldung Grien: Prints and Drawings* (exhibition catalogue, National Gallery of Art, Washington, 1981).

[15] Roberts and Tomlinson, *The Fabric*, plate 9, for the illustration from L. Fries, *Spiegel der Artzny* (Strasbourg, 1519); and H. von Gersdorf, *Feldtbuch der Wundartzney* (Strasbourg, 1517) p. 262.

[16] M. Hundt, *Antropologium* (Leipzig, 1501), illustrated by Roberts and Tomlinson, *The Fabric*, plate 7.

[17] Windsor 12636; Pedretti and Keele, *Corpus*, no. 111r.

[18] Pietro d'Abano, *Conciliator* (Venice, 1496), CXCIX, in which Pietro argues that the depiction of the diagonal muscles is incorrect—an argument Leonardo appears not to have grasped.

[19] Berengario da Carpi, *Commentaria ... super anatomia Mundini* (Bologna, 1521); and *Isagogae breves ...* (Bologna, 1522); trans. L. Lind, *A Short Introduction to Anatomy (Isagogae breves)* (Chicago, 1939) p. 160. See R. French, 'Berengario da Carpi and the use of commentary in

anatomical teaching', in A. Wear, R. French and I. Lonie (eds), *The Medical Renaissance of the Sixteenth Century* (Cambridge, 1985) pp. 42–74; and M. Kemp, "The mark of truth': Looking and learning in some anatomical illustrations from the Renaissance and eighteenth century', in W. Bynum and R. Porter (eds), *Medicine and the Five Senses* (London, 1993) pp. 85–121.

[20] C. Estienne (Carolus Stephanus), *De dissectione partium corporis humani* (Paris, 1545); French trans. as *La Dissection des parties du corps humain* (Paris, 1546). For the artistic sources, see C.F. Kellett, 'Perino del Vaga et les illustrations pour l'Anatomie d'Estienne', *Aesculapius* 37 (1964) 74–9; and M. Kornell, 'Rosso Fiorentino and the anatomical text', *Burlington Magazine* 81 (1989) 842–7.

[21] *Tabulae sex*, letter of dedication, trans. Saunders and O'Malley, *The Illustrations*, p. 234.

[22] See also J. Dryander, *Anatomia* (Marburg, 1537).

[23] A. Vesalius, *De humani corporis fabrica* (Basel, 1543) VII, xix, p. 661.

[24] M. Kemp, 'A Drawing for the *Fabrica*; and some thoughts upon the Vesalian muscle-men' *Medical History* 14 (1970) 277–88; and 'The mark of truth'.

[25] *Tabulae sex*, letter of dedication; trans. Saunders and O'Malley, *The Illustrations*, p. 234.

[26] Appendix Vergiliana, *Elegiae in Maecenatem*, I, 38: '*vivitur ingenio, cetera mortis erunt*' (reference kindly provided by Professor H. Hine).

[27] E.g., *Fabrica*, I, xx, p. 93, and III, i, p. 358.

[28] Saunders and O'Malley, *The Illustrations*, pp. 230–1.

[29] *Fabrica*, VII, xiv, p. 643.

[30] For a discussion of such linear conventions, see Kemp, 'The mark of truth', pp. 100–1.

[31] B. Eustachio, *Tabulae anatomicae* (Rome, 1722); see Roberts and Tomlinson, *The Fabric*, pp. 188–93.

[32] P. Apianus, *Cosmographicus liber* (Landshut, 1524).

[33] *De revolutionibus*, preface, iii; trans. Rosen, vol. II, p. 4. For Copernicus generally, see *Copernicus Yesterday and Today*, ed. A. Beer and K.A. Strand, in *Vistas in Astronomy*, XVII (Oxford and New York, 1975); and J-P. Verdet, 'L'Astronomia dalle origini a Copernico', in W. Shea (ed.), *Storia delle scienze. Le scienze fisiche e astronomiche* (Milan, 1991) pp. 38–109.

[34] For an instructive interpretation of Copernicus in the tradition of Renaissance rhetoric, see R. Westman, 'Proof, poetics, and patronage', in D. Lindberg and R. Westman (eds), *Reappraisals of the Scientific Revolution* (Cambridge, 1990) pp. 167-207. See also P.L. Rose, 'Universal harmony in Regiomontanus and Copernicus' *Avant, avec, après Copernic: la representation de l'univers et ses conséquences épistémologiques* (Paris, 1975) pp. 153–8.

[35] For the translations from the Greek, see *Complete Works*, III, p. 31. For further discussion of Copernicus's humanism, see K. Hutchison, 'Copernicus, Apollo and Herakles', in S. Gaukroger (ed.), *The Uses of Antiquity: The Scientific Revolution and the Classical Tradition* (Dordrecht, 1991) pp. 1–23; and 'Harmony and authority: The political symbolism of Copernicus's personal seal', in R.G. Mazzolini (ed.), *Non-Verbal Communication in Science prior to 1900* (Florence, 1993). For the standard likeness of Copernicus, see *Vistas in Astronomy*, XVII, figs.5–7. The evidence regarding the possible self-portrait(s) is assessed by Westman, 'Proof, poetics, and patronage', pp. 184–6.

[36] *De revolutionibus*, preface iiiv, and I, x, p. 10; trans. Rosen, pp. 4 and 22.

[37] G.J. Rheticus, *Narratio prima* (Danzig, 1540, and Basel, 1541); quoting Galen, *De usu partium*, X, 14; trans. E. Rosen, *Three Copernican Treatises* (New York, 1971) p. 137.

[38] L.B. Alberti, *De re aedifactoria* (Florence, 1486; also Paris ed. of 1512 and Strasbourg, 1541); trans. J. Rykwert, N. Leach and R. Tavernor, *On the Art of Building in Ten Books* (Cambridge, Mass. and London, 1988), especially the prologue and the introductions to books I and VI.

[39] *De revolutionibus*, preface, iiiv; trans. Rosen, p. 4. Compare *De revolutionibus*, p. 9r; trans Rosen, p. 22, where it is asserted that nature 'avoids producing anything superfluous or useless'.

[40] *De revolutionibus,* I, 10, pp. 9v–10r; trans. Rosen, p. 22. See H.P. Nebelsick, *Circles of God. Theology and Science from the Greeks to Copernicus* (Edinburgh, 1985) pp. 200–73; G. Hatfield, 'Metaphysics and the new science', *Reappraisals of the Scientific Revolution,* pp. 93–166; and, in a post-moderm vein, F. Hallyn, *The Poetic Structure of the World: Copernicus and Kepler* (New York, 1990). For a more circumscribed interpretation of this passage, see S. Drake, 'Copernicanism in Bruno, Kepler and Galileo', *Vistas in Astronomy,* XVII, pp. 177–90, esp. p. 184.

[41] For a detailed assessment of Copernicus's astronomy, see N. Swerdlow and O. Neugebauer, *Mathematical Astronomy in Copernicus's 'De Revolutionibus'*, 2 vols. (New York, 1984).

[42] *De revolutionibus*, I, 8, p. 6r; trans. Rosen, p. 16; Virgil, *Aeneid*, III, 72.

[43] Leonardo da Vinci, MS A 36r and MS K^3 120v, in M. Kemp (ed.), *Leonardo on Painting*, trans. M. Kemp and M. Walker (London and New Haven, 1989) p. 55.

[44] J. Kepler, *Epitome astronomiae Copernicae* (Linz, 1618); in *Gesammelte Werke*, ed. W. von Dyck, M. Caspar *et al*, 20 vols. 1938–88, VII, p. 277.

[45] J. Kepler, *Somnium*, published posthumously by L. Kepler (Frankfurt, 1634); trans. E. Rosen, *Kepler's Somnium* (Madison and London, 1962). In his *Astrononomia nova* (Heidelberg, 1609), Kepler envisages an observer on Mars; *Gesammelte Werke*, III, p. 22.

[46] *De revolutionibus*, III, 3, fol.66v, p. 124. The MS illustration is on fol.74. See O Neugebauer, 'On the planetary theory of Copernicus', in A. Beer (ed.), *Vistas in Astronomy* (Oxford and New York, 1968) p. 96.

[47] *De revolutionibus*, III, 3, p. 66r (trans. Rosen, p. 124).

[48] *De revolutionibus*, III, 3, p. 66r, (trans. Rosen, p. 123).

[49] R. Westman, 'Three responses to the Copernican theory: Johannes Praetorius, Tycho Brahe and Michael Maestlin', in R. Westman (ed.), *The Copernican Achievement* (Berkeley, Los Angeles and London, 1975) pp. 318–9, for annotations by Kepler and the Scottish philosopher, Duncan Lidell.

[50] The annotated Copernicus in St. Andrews was, as an insciption indicates, the property of the 'German Nation' in the University of Padua in 1626.

[51] Illustrated by O. Gingerich in *Vistas in Astronomy*, XVII, fig. 70.

[52] For reviews of such instruments, see E. Zimmer, *Deutsche und Niederländische astronomische Instrumente des 11.–18. Jahrhunderts* (Munich, 1967); A. Turner, *Early Scientific Instruments* (London, 1987); J. Bennett, *The Divided Circle: A History of Instruments for Astronomy, Navigation and Surveying* (Oxford, 1987); and G. Turner (ed.), *Storia delle scienze. Gli strumenti* (Turin, 1991).

[53] Illustrated and discussed by F. Maddison in *Ca 1492: Art in the Age of Exploration* (exhibition catalogue, National Gallery, Washington, 1991) nos. 120–2. For a suggestive discussion of the nature and use of such instruments, see J.V. Field, 'What is scientific about a scientific instrument?', *Nuncius*, III.2, 1988, pp. 3–26. For the courtly context, see T. DaCosta Kaufmann, 'The *Kunstkammer*, politics and science', in *The Master of Nature. Aspects of Art, Science and*

Humanism in the Renaissance (Princeton, 1993) esp. pp. 188–93. See also M. Kemp, "Intellectual ornaments': Style, function and society in some instruments of art', in J. Pittock and A. Wear (eds), *Interpretation and Cultural History* (London, 1991) pp. 135–52.

[54] P. Apianus, *Instrumentum primi mobilis* (Nuremberg 1534); and *Astronomicum caesareum* (Ingolstadt, 1540); see O. Gingerich, 'Apianus's *Astronomicum caesareum* and its Leipzig facsmile', *Journal for the History of Astronomy* 2 (1971) 168-77. T. Brahe, *Astronomiae instauratae mechanica* (Wandesburg, 1598); see *Tycho Brahe's Description of his Instruments and Scientific Work as Given in Astronomiae instauratae mechanica, Wandesbeck 1598, Nuremberg 1602*, trans. H. Raeder, E. Strömgren and B. Strömgren (Copenhagen, 1946). The illustrations of instruments are also found in Brahe's *Progymnasmata*, ed. J. Kepler (Prague, 1602). For Tycho's career and achievements, see V. Thoren, *The Lord of Uraniborg. A Biography of Tycho Brahe* (Cambridge, 1990).

[55] T. Brahe, *De nova et nullius aevi memoria prius visa stella* (Copenhagen, 1573); trans. *His Astronomical Conjecture of the New and Much Admired Star which Appeared in the Year 1572* (London, 1623).

[56] *Mechanica*, p. 67.

[57] *Mechanica*, pp. 44-5. The poem is in *Opera omnia*, VI, p. 266ff.

[58] *Mechanica*, p. 30.

[59] *Mechanica*, p. 13.

[60] *Mechanica*, p. 57.

[61] *Mechanica*, p. 27.

[62] See Thoren, *Lord of Uraniborg*, pp. 106–13 for the architecture, though somewhat overestimating the classicism and Palladianism of the enterprise.

[63] *Mechanica*, p. 131.

[64] J. Kepler, *Tabulae rudolphinae* (Ulm, 1627).

[65] The sketch in the Archiv der Kepler-Kommission, Munich, is illustrated in *Gesammelte Werke*, X, p. 279, and *Kepler: Four Hundred Years*, ed. A. Beer and P. Beer, *Vistas in Astronomy*, XVIII (Oxford and New York, 1975) fig.3.8. For Kepler's stereometry, see his *Nova stereometria doliorum vinariorum*, 1615. For Kepler generally, see M. Caspar, *Kepler* (London and New York, 1959); J.V. Field, *Kepler's Geometrical Cosmology* (Chicago, 1988); *Kepler: Four Hundred Years*; and W. Shea, 'La rivoluzione scientifica', *Le scienze fisiche e astronomiche*, pp. 168–233.

[66] J. Kepler, *Mysterium cosmographicum* (Tubingen, 1596); *Gesammelte Werke*, I, pp. 3–80. Kepler's treatise is also unusual in that he represents the actual paths of the planets in addition to their orbs.

[67] Field, p. 47. See also pp. 45–51 for an outline of the way Kepler reached his formulation.

[68] For apsects of the 'aesthetics' of the Platonic solids in perspectival depiction, see M. Kemp, 'Geometrical bodies as exemplary forms in Renaissance space', in I. Lavin (ed.), *World Art. Themes of Unity in Diversity*, 3 vols. (University Park, Pennsylvania; and London, 1989) I, pp. 237–42.

[69] *Gesammelte Werke*, XIII, pp. 50–3. See F. Prager, 'Kepler als Erfinder', in F. Krafft *et al.* (eds), *Internationales Kepler-Symposium Weil der Stadt 1971* (Hildesheim, 1973) pp. 385–405.

[70] *Gesammelte Werke*, XIII, p. 218ff.

[71] *Gesammelte Werke*, XIII, p. 151.

[72] Letter to Hewart von Hohenburg, 10 February 1605; *Gesammelte Werke*, XV, p. 146.

[73] Marsilio Ficino, *De sole*, 1493, cap.XIII, p. 255.

[74] S. Münster, *Organum uranicum* (Basel, 1536) title page.

[75] *Sidereus nuncius* (Venice, 1610); trans. and intro. A. Van Helden, *Sidereus nuncius or the Sidereal Messenger* (Chicago and London, 1989).

[76] Winkler and Van Helden, 'Representing the heavens'.

[77] A recension of this paper, here printed in its original form, has been published as follows: 'Temples of the body and temples of the cosmos: Vision and visualization in the Vesalian and Copernican revolutions', in B.S. Baigrie (ed.), *Picturing Knowledge: Historical and Philosophical Problems Concerning the Use of Art in Science* (Toronto, 1996).

JAMES FRANKLIN

DIAGRAMMATIC REASONING AND MODELLING IN THE IMAGINATION: THE SECRET WEAPONS OF THE SCIENTIFIC REVOLUTION

Tartaglia's Italian Euclid of 1543 is geometry in the narrow sense. But the big two books of 1543, Copernicus' *De revolutionibus* and Vesalius' *De humani corporis fabrica* are also geometry, if a slightly wider sense of the term is allowed. Though Copernicus writes on physics, he does not speak of forces, energies, masses or the like: there are only the appearances of the heavens from certain points of view. Though Vesalius is biology, there is little physiology, or mechanical analogy, or discussion of causes: the emphasis is on how parts of the body look from suitable points of view. But the three books share more than just pictures, and it is this extra element that is the focus of this article. Euclid's *Elements* is not a picture book of shapes. The point of Euclid is to *reason* about the diagrams, and expose the necessary interrelations of the spatial parts. So it is with Copernicus and Vesalius. The text of Copernicus is an exercise in reasoning about which geometrical scheme will best fit the sequences of spatial points recorded in the astronomical tables. Vesalius uses the best of the discoveries of artists to make easy for the reader inference about how the systems of the body look in isolation, and in relation to one another. The difference between a Vesalian diagram and a photograph is exactly that the former allows one to work out structural facts which are almost invisible in the photograph. The plates also allow muscles in the partially dissected cadaver to be drawn with the natural tensions of the living body—obviously impossible with a photograph.

An exclusively geometrical focus persisted in science in the work of Kepler, in both his ellipse theory and his Platonic solids theory of the planets[1] (as well as in his optical theory). Galileo's first success, the *Sidereus nuncius*, is, amongst other things, about inferring the shape of the moon's surface from optical considerations of light and shadow.[2] Galileo's famous saying that the universe is written in the language of mathematics, which when quoted in

G. Freeland and A. Corones (eds.), 1543 and All That, 53 – 115

isolation makes us think, for example, '$s = \frac{1}{2} gt^2$', continues in the original, 'its characters are triangles, circles, and other geometrical figures, without which it is humanly impossible to understand a single word of it'.[3] Descartes still believes that the way forward for science is to ignore all properties of matter except the purely geometrical, but he is already starting to be old-fashioned. By then the next generation had begun to invest science with the many things beyond pure geometry that we now regard as essential to it, and which it was the achievement of the Scientific Revolution to put together—experiment, for example, and forces, and algebraic formulas.

Since *a* Scientific Revolution was plainly under way by 1610, and since at that time it had achieved almost nothing except in geometry, the thesis, 'The Scientific Revolution was kick-started by geometry' is well-supported. This article is intended to explain this fact, by tracing how medieval and Renaissance reasoning with diagrams, both physical and mental, trained Europeans to think adequately to do science.

Part 1 deals with physical diagrams, part 2 with mental ones. The two parts are not as distinct as may appear at first glance, since diagrams are pictures that are especially simplified to cause a mental construct stripped of irrelevancies, while an imagined diagram is imagined as *like* a physical picture.

1. DIAGRAMMATIC REASONING

1.1 Image and Diagram

First, a little precision is desirable on the entities like 'diagrams' and 'images'. This is especially necessary because of the very wide and imprecise usage of the word 'image'. 'Image' is commonly used as a dead metaphor, meaning something like 'implicit theory', so that a phrase like 'the medieval image of the world' has no specially visual or spatial content.[4] Or take titles which do claim to refer to pictures, like, 'Images of the other in incunabular woodcuts: Renaissance constructions of ethnicity and gender'. The post-modernist *frisson* makes the author's commitments clear enough, but in the process neglects certain necessary distinctions among 'images'. These are the distinctions that are common ground in artificial vision research and satellite image processing, where 'images' are classified according to the amount of cognitive processing incorporated into them. One distinguishes at least the following three stages:

1. The raw image, such as a photograph, or the moving dots on a TV screen. Or perhaps the Shroud of Turin, to take a medieval example.

2. A line drawing, like a Dürer rabbit. The edges and regions are explicitly identified, but there is as yet no labelling of the objects in the scene.

3. A map or plan. Here there are symbols attached, identifying some line as a road, or some dot as a church. The symbols are on the map in places that preserve the spatial relations between the entities symbolised—all the spatial relations if the map is 'to scale', or only some of them in examples like the usual plans of the London Underground, where only spatial order is preserved.

Beyond these three there are purely textual descriptions of space, as with a list of objects in a scene, or a description of how to get to the East. The representation of information can be compressed and informative, but also subject to alarming errors of scale, of the kind made famous by Columbus and the Children's Crusade.

'Diagram' is a wider term than 'map' or 'plan', in that what a diagram represents need not be spatial. A diagram is a picture, in which one is intended to perform inference about the thing pictured, by mentally following around the parts of the diagram. The famous saying, 'A picture is worth a thousand words' is true especially of diagrams, those pictures streamlined for inference by removal of irrelevancies (such as, usually, shading and real colour). (A photograph may not be worth any words at all, as a beginning student of biology realises when confronted with slides of the view down a microscope.) 'Worth a thousand words' for what purpose? As a store of information, from which inference can proceed.[5]

Consider the common medieval diagram of the Wheel of Fortune[6] (fig. 1 shows Villard de Honnecourt's especially abstract representation of it[7]). One is supposed to start at any of the positions (most naturally the top), then follow the circle clockwise to infer the order of events, and eventually return to the starting point. The (spatial) circle of the diagram thus represents time in the cycle of events. The Wheel is in fact a particularly bad diagram of fortune, because of the incorrect inferences it encourages. For example, it suggests that reversals of fortune happen inevitably, and even at inevitable intervals, as the

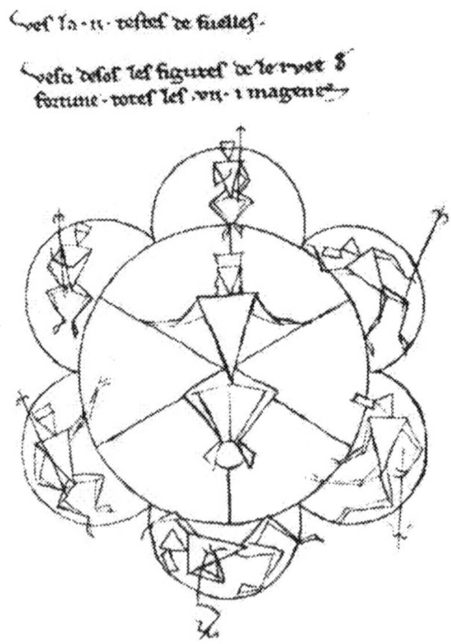

Figure 1. *Villard's Wheel of Fortune. The Sketchbook of Villard de Honnecourt.*

Wheel turns inexorably. Seeing fortune in terms of the Wheel must, then, inhibit any thinking of it in terms of randomness, where the time until a reversal is completely unknown, as it is in real life. From the present point of view, this simply casts light on what a diagram is: it represents spatially the structure, or what is believed to be the structure, of something, in a way that facilitates the reaching of conclusions about it. A diagram thus contrasts with a single drawn figure (of a saint, for example), in which there is no intention that one should follow from part to part and infer anything. It also contrasts with such things as Islamic geometrical decoration, where there are complex arrangements of geometrical parts, but parts that have no meaning.

A diagram of higher quality, where the inference is good, is the Square of Opposition in logic (fig. 2). Here, lines in space represent logical relations between propositions. The Square is apparently due to Apuleius of Madaura, in the second century A.D. The original text contains not a diagram, but a description of how to draw one.[8] The logical relation between any proposition and any other can be read off immediately from the diagram.

Before going further, it will be useful to have a rough classification of the kinds of pictorial representation, which will help organise the topics to follow. Let us represent the classification itself with the usual diagram, a tree (fig. 3).

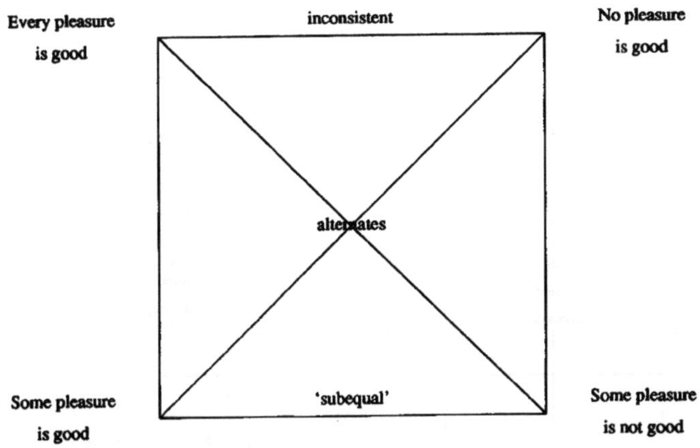

Figure 2. *Apuleius' Square of Opposition.*

The dotted line indicates the connection that gives perspective its peculiar fascination: by drawing something according to geometrical rules (and filling in with colours) one has something recognisably like a photograph. (On the controversy as to whether images literally resemble their objects, the present article accepts the affirmative side.[9]) On the connections between geometrical diagrams and perspective, more will be said later.

Medieval texts are often rich in diagrams, even if that is not always clear from modern editions which criminally leave them out.[10] Our survey begins from the bottom left of the tree below.

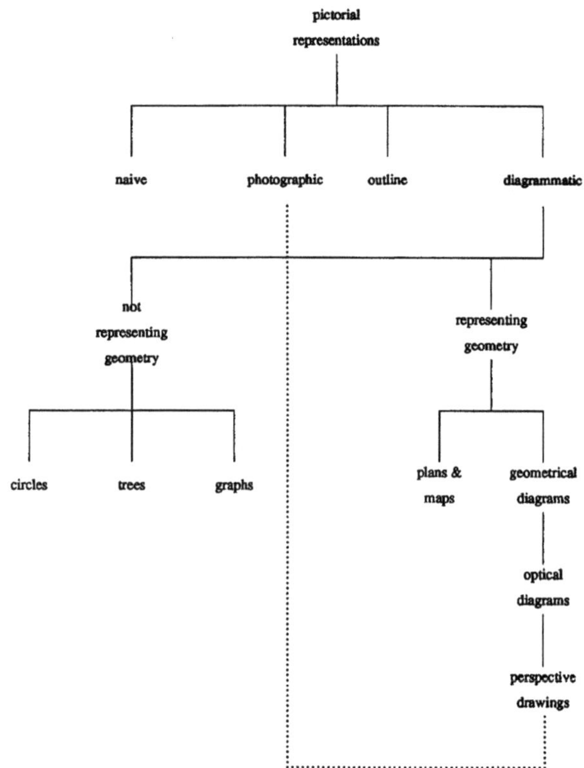

Figure 3. *Classification of pictures.*

1.2 Circle Diagrams

The Wheel of fortune is far from the only circle diagram common in medieval texts. Isidore of Seville's *Etymologies*, the standard medieval encyclopedia, is sometimes called the *Liber rotarum* on account of the number of *rotae*, or wheel diagrams. Among the most popular were ones linking the four elements and the four humours, and those linking the zodiac, months and seasons.[11]

There are a few medieval diagrams in which the wheels are physically separate pieces joined to the page with string, so that the rotation can be done in reality and not just in the imagination. A fourteenth century book on divination includes two geared wheels; the smaller one is rotated on a peg, and the larger one comes to rest at a number which is interpreted according to the instructions in the book. Ramon Lull's *Art* (a method for demonstratively confuting the heathen and recovering Jerusalem) used rotating discs (as well as various trees).[12] The various complicated wheels illustrating parallels between the seven virtues, the seven vices, the seven Beatitudes and the seven gifts of the Holy Spirit are perhaps better imagined than pictured.[13] The English phrase 'by rote' probably derives from the ubiquity of wheel diagrams in medieval education (or if not, it is from the French 'route', which refers to another diagrammatic way of organising facts).[14]

1.3 The Rank Growth Of Trees

One of the most widely visible of medieval diagrams, in the full sense of something geometrical on which one performs inference by following the diagram around, was the Jesse tree. A Jesse tree is the family tree of Christ, going back to Jesse, represented in stained glass or some other medium. Typically, it is rather abbreviated, but some examples have up to fifty personages in the leaves.[15] The reference is to *Isaiah* 11:1, 'A shoot springs from the stock of Jesse, a scion thrusts from his roots'. Whether an actual family tree is envisaged in the original is not entirely clear—the Bible is not rich in visual imagery, except for the apocalyptic books. The ancient sources mention family trees displayed by Roman families in their homes,[16] but no examples are known. Medieval texts, on the other hand, are full of them. There are family trees of kings;[17] also of gods.[18]

The legal world was also familiar with family trees. A kind of generic family tree was common in legal texts to illustrate family relationships, and the degrees of consanguinity to be inferred from them; an example available to almost everyone literate is in Book 9 of Isidore of Seville's *Etymologies*.[19] The same information on family relationships can be depicted by quartering a coat of arms. Sir Anthony Wagner's *Historic Heraldry of Britain* explains the complexity of the inferences involved in interpreting arms:

> A husband impales his wife's Arms with his own, unless she be her father's heir or coheir in blood (that is to say, if she has no brothers), when he displays them on an escutcheon of pretence, superimposed in the centre of his own shield. In the latter case only, the children acquire a right to quarter the Arms of their mother's family (and any quarterings previously acquired in the same way by that family) with their father's. In this way a shield with many quarterings indicates a succession of heiress marriages, so that the bearer represents in blood all the families whose Arms he quarters.[20]

These Gothic extravagances belong especially to the very late medieval period examined in Huizinga's *Waning of the Middle Ages*. Huizinga acutely observes how the excessive growth of trees is part of 'symbolism in its decline', and tends to substitute for serious causal thought:

> The world unfolds like a vast whole of symbols, like a cathedral of ideas. It is the most richly rhythmical conception of the world, a polyphonous expression of eternal harmony ... All notions of one thing proceeding from another took the naive form of procreation or ramification. The image of a tree or pedigree sufficed to represent any relations of origin or cause. An *arbor de origine juris et legum*, for example, classified all law in the form of a tree with numerous branches ... From the causal point of view, symbolism appears as a sort of short-circuit of thought ... all mental association based on any casual similitude whatever will immediately set up the idea of an essential and mystic connexion.[21]

Still, there is structure other than the causal, which can be worthwhile but difficult to investigate. Some trees, at least, summarise genuine information. One is the logical diagram known as Porphyry's Tree, which classifies the kinds of being. In accordance with Stigler's Law of Eponymy,[22] this is not due to Porphyry, but to one of his Latin translators, at the latest Boethius.[23]

A tree diagram (fig. 4) which, unlike Porphyry's Tree, is for the classification of something more or less useful, is the division of the kinds of proportion in Pacioli's *Summa de arithmetica*, of 1494.[24] Pacioli notes that the diagram continues indefinitely off the bottom of the page. The modern mathematician is unlikely to find any but the top few levels of any assistance in understanding ratios, but the basic plan is reasonable enough.

Some historical perspective may be gained by noting that diagrams of essentially the same type have since proved widely useful in the sciences. 'Tree of life' diagrams are important in Darwin's early notebooks, for example.[25] In this century, there are theorems in logic like 'the propositional calculus is a distributive lattice', 'diagram-chasing' in category theory,[26] and 'semantic net' or 'frame' representations in artificial intelligence.[27] 'Venn' diagrams (due to Euler[28]) rely on the fact that the transitive relation of set inclusion is represented geometrically by the obviously transitive relation of containment of areas.

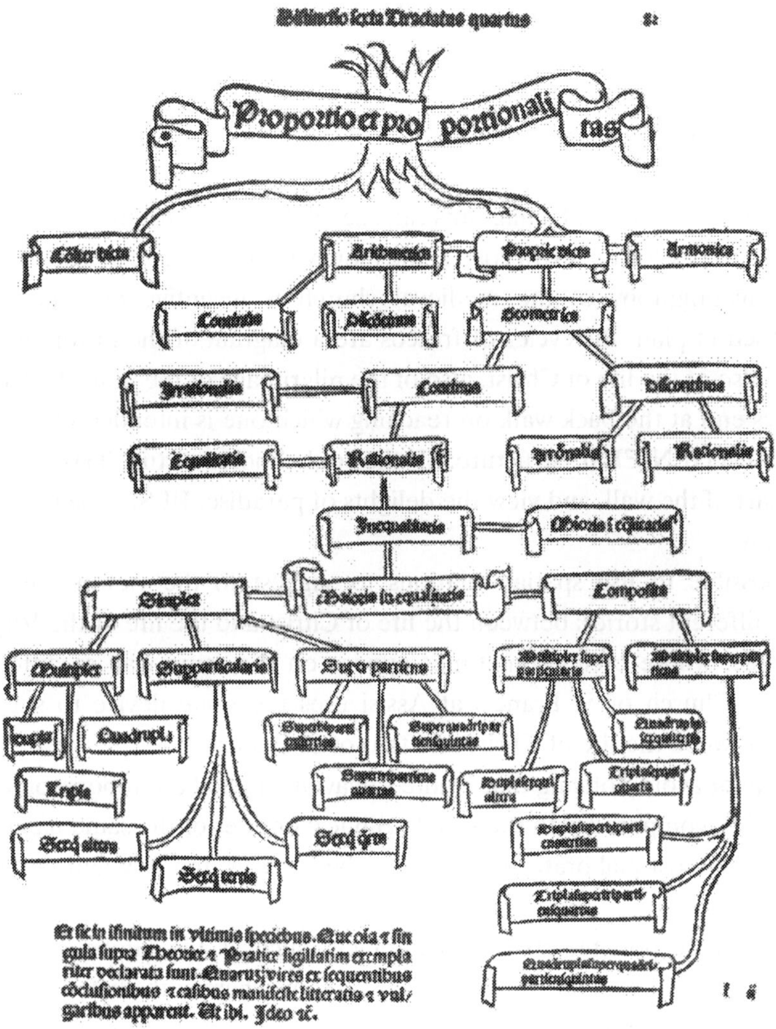

Figure 4. *Pacioli's Classification of Ratios.*

Networks of nodes and connections, which are in general more complicated than trees in that they contain cycles, can be very useful in analysing telecommunications links, games and so on. The first mathematically sophisticated use of such a network was in Vandermonde's study of the knight's tour in chess, in 1771.[29] He used the nodes to represent positions of the knight, and connected with a line those nodes which can be reached from one another by a legal knight's move. It is clear, then, that the medievals were essentially correct in seeing tree diagrams as suitable for the representation of abstract structure.

1.4 Diagrams Of Everything

Space is three-dimensional. Diagrams can use all three dimensions, if someone is prepared to pay for the extra cost. If the money is there, they can also be very big. The iconography of the medieval church is inventive in detail, but standardised in plan: the cycles of frescos are a diagram of the history of the universe, also of the life of Christ, also of the pilgrimage of the individual soul. The cycles end at the back wall, on reaching which one is intended to impose a logical IF-THEN-ELSE structure: IF the soul chooses virtue, THEN go to the top part of the wall, and view the delights of paradise; ELSE, see the lurid detail below.

It is possible to use spatial relations to suggest to the viewer parallels between different stories: between the life of Christ and the life of the Virgin, for example, whose cycles sometimes appear on the same wall at different levels. The Church of St Francis at Assisi uses the same device to suggest parallels between the life of Christ and that of St Francis.[30]

One cannot avoid noticing that Dante's heaven and hell are topographically very complex, compared with the simple Biblical entities of the same name, or even with the most elaborate Biblical described spaces, the Ark and the New Jerusalem.[31]

More abstruse possible structural parallels, such as the Globe Theatre's alleged representation of the universe,[32] remain somewhere in the limbo between the speculative and the proved. It can at least be said that the medieval and Renaissance mind would have regarded a complex building that did not represent the zodiac, or the virtues, or the macrocosm, or all of these at once, as a crying waste of representational possibilities.

When funds did not permit grandiose construction, there was always the human body available for moralising as a diagram or 'microcosm' of the universe.[33]

1.5 Graphs

Graphs, as in graphs of yearly profits, daily temperatures and the like, are one of the few mathematical inventions since ancient times that are needed to read the newspapers. The important thing in reasoning with a graph is that at least one of the dimensions should represent a quantity that is not spatial. Most commonly, the horizontal axis is reserved for time, and the vertical axis for some quantity that varies with respect to time, such as profits, temperature, or the distance travelled by a falling body. Graphs appear to be an invention of Oresme, about 1350, earlier diagrams that look something like graphs being only maps of purely spatial quantities. His horizontal axes can represent time or space, and his vertical axes any quantity that varies in intensity, such as velocity, intensity of colour, or joy. But his graphs have no scales on the axes: he is solely concerned with the differences in the shapes of different graphs.[34] A printed edition of 1486 is liberal with the diagrams.[35]

Consider a graph with time on the horizontal axis and pitch (of sound) on the vertical axis. Discretize both axes; that is, cut up time and pitch into suitable intervals. Then one has, in effect, musical notation. The staff, invented in about the eleventh century, provides a scale for the vertical axis.[36] Despite Oresme's musical interests, it seems unlikely that he saw any connection between musical notation and graphs; if he had, he would surely have put scales on the axes of his graphs.

The usual 'Renaissance' gap in the development of science is evident with graphs, which appear to have had no use until 1600. Then their time came (surely it is the Wheel). Galileo's discovery of the uniform acceleration of free fall involved essential reference to a graph. At first, he drew a diagram of an inclined plane which pictured only spatial quantities. In such a diagram, there is no place for the time variable, making it difficult to reason about time. Galileo was misled by it into thinking that the speed of a falling body is proportional to the distance it has travelled from rest. Later, he drew almost the same diagram, but with one axis now representing *time*, and arrived at the correct conclusion: that speed is proportional to the time from rest, and hence distance travelled proportional to the square of the time.[37]

While Galileo is not thought to have read Oresme, the evidence is overwhelming that he did read certain printed books which incorporated Oresme's ideas. In particular, one can find antecedents of Galileo's graphical proof of the 'Merton mean speed theorem' (that in uniformly accelerated motion, the distance travelled is equal to the distance that would be travelled by a body with uniform velocity equal to half the maximum velocity of the original accelerated body[38]).

1.6 Tables

Academics, especially at exam time, become very aware of the importance of 'setting out', that is, the arrangement of pieces of information in a spatial pattern that allows the information to be grasped easily. Simple uses of space like indentation of new paragraphs, blank lines between sections, justification of type, headings in large type (or illuminated capitals) can reduce the reader's cognitive load remarkably.[39] The main principle is that a spatial division, such as a blank line, should correspond to a major division in the text's meaning. Similar considerations apply to tables of figures. While a table is not precisely a kind of diagram, it is not unlike the 'negative' of a tree diagram: blanks, rather than lines, represent the relations between parts.

A page from a book of accounts, such as that of Impyn's textbook,[40] is notable in several geometrical ways. It is not all text; in fact half of it is empty space. The text comes in aligned blocks, and (besides the sums of money) there are also important non-text items, like lines, crossings-out, and marginal numbers indexing where the entries in the journal have been transferred into the more permanent ledger. All of these things might be different; if they were, making the inferences about the state of the owner's finances would be much more difficult. (These inventions are not sixteenth century; they are all visible in fourteenth-century accounts.[41])

The neat tables of the accountants can give the impression that Renaissance statistical information is generally arranged to allow perspicuous inference about it. This is far from true. Nef records the experience of trying to get a picture of the increase of coal shipments in England in the period after 1550.[42] The Public Record Office contains hundreds of port books with entries about coal shipments, but separating them from those dealing with glass, salt and so on is an exercise for the modern researcher. That is, the records are a mass of items of information, but contain no suitable setting out of what it means; of statistical summaries, in modern language. The modern user of spreadsheets will notice that the account books, themselves much more sophisticated in these matters than the public records, contain places for 'totals', but none for 'averages'. And no bar graphs or pie charts, either.

Tables of compound interest were produced by the fourteenth century Italian merchants,[43] and for some time thereafter were regarded generally as trade secrets. Stevin produced the first printed tables.[44]

Astronomy was, as usual, rather in advance. The Alfonsine tables are ruled very neatly.[45] Astronomical tables differ from accounts in that they are projections onto the discrete realm, so to speak, of something continuous. Inference with them needs to keep in mind that they tabulate only a succession of points in a motion that is really continuous. Such inference is essential to Napier's invention of logarithms, about 1600. One can learn logarithms from a purely numerical point of view, as a set of rules about how to manipulate numerals, but that was not how Napier thought. Trained in spherical trigonometry for the calculation of astronomical tables, he actually defined logarithms in terms of moving points on continuous scales.[46]

The spatial organisation of text to facilitate a grasp of its meaning will shade off into considerations of punctuation, which improved markedly in late medieval times,[47] and such matters as decimals and algebraic notation in mathematics (to both of which Stevin made notable contributions). Interesting as these matters are, they would take us too far afield.

We now move on to diagrams whose geometry represents something which is literally geometrical itself. The material here is generally better known than are the more abstract kinds of diagrams just considered. Here, we will briefly survey the field, calling attention to the reasoning processes needed to interpret the pictures.

1.7 Scientific Illustrations

Drawings of machines are particularly interesting, from the point of view of reasoning, because the viewer has to infer how the machine *works.* To work is, among other things, to change over time, and neither time nor change can appear in the diagram. (In principle, it is possible to draw a series of diagrams representing the phases, as is often done in showing how a car piston works, or as the Bayeux tapestry shows the Battle of Hastings; pre-modern machine diagrams do not attempt this, and in any case, one must understand a single diagram of such a series reasonably well before being able to infer how it changes to the next.) Nor can the diagram explicitly show forces and their transmission, or the direction in which a part is intended to move. The difficulty of the exercise is shown by the fact that it is usually impossible to discover how a machine works from a photograph of it.

An added difficulty with interpreting diagrams of machines is that one has to infer their 3D structure from a 2D picture. Techniques of perspective drawing were useful, after their invention, but are neither necessary nor sufficient for a clear diagram. They are not necessary, since there are other ways of including all the essential information, like drawing sections, or flat diagrams that one imagines folded. They are not sufficient, since a perspective drawing that is of the outside of a machine lacks the working parts entirely, while a perspective projection of everything will normally be too cluttered to understand.

Ancient and medieval machine diagrams have been unfairly disparaged because the conventions by which they solve these problems differ from the modern ones. The few surviving ancient diagrams of pulleys, war machines and the like, are reasonably easy to interpret, though they are neither exactly perspective nor plan views: each part is shown from its own most natural viewpoint.[48] For the modern viewer, the convention is initially confusing, but not hard to learn. It simply requires some mental origami to rotate the parts suitably. Islamic and medieval drawings use the same convention,[49] and it is only replaced by more modern perspective-based techniques in the fifteenth century.[50]

Medical illustrations do not have the problem of the time dimension (until one begins to think of the heart as a pump). But a body is much harder to understand by looking at it than a machine is, being more complex, and largely invisible when in working order. So medical illustrations have serious problems in showing internals, and simplifying enough to show the main structure, or separate different kinds of structure. Here the convention of cut away diagrams was essential. The perfection of Leonardo and Vesalius[51] should not blind us from perceiving that the essentials are present in Guido da Vigevano's *Anathomia*, of the 1340s.[52] There seems to be some connection with anatomical models.[53]

Printing obviously has something to do with the spread of scientific illustrations. But the excitement over Gutenberg (or Koster, or whoever) can make one forget the fact (which of course one knows, when reminded) that he did not invent printing itself, but only printing with movable type (an invention which itself needs—indeed consists in—a basic act of spatial imagination).

Movable type is only useful for printing text, especially in alphabetic languages. Printing of pictures came first, and was reasonably common in Europe from about 1400.[54] Again, the path to the perfection of the art in Dürer is one of evolution, not revolution.

Edgerton, in arguing for the importance of scientific illustration in the Scientific Revolution, usefully juxtaposes some European machine diagrams with Chinese copies of them. At first glance, the copies are reasonable, but it is soon clear that the Chinese artist has misunderstood the diagrams at exactly the places where inference is required. Where a rope carrying tension is shown in a cut away section, and a continuation is shown in another part of the diagram, for example, the Chinese copy does not have the two parts collinear.[55] Mahoney, replying to Edgerton's thesis, points to the crucial role of algebra in the work of, for example, Huygens and Newton.[56] True as that is, Mahoney's examples are from a different period. The later phase of the Scientific Revolution is indeed algebraic, but the earlier one is diagrammatic.

Modern research on cross-cultural psychology has tended to confirm that there are major differences in how cultures perceive pictures, although research has concentrated on cultures that differ from the Western more than does the Chinese.[57]

1.8 Plan and Elevation

A building poses different drafting problems to a machine. It typically has more detail, but it is conveniently hollow. More importantly, most buildings are approximately rectangular prisms, so most of the information on their 3D shape can be incorporated in three 2D diagrams, their 'plan' and two 'elevations'.

Not much is known about ancient plans, though there are such things as scale drawings of Greek temples, and construction lines can sometimes be seen on the actual buildings.[58] Not a great deal is known about medieval building plans either, but there is enough to show that reasoning with plans was a normal part of cathedral building. There is a suggestion that making inferences about a building from plan and elevation was part of the famous 'secrets' of the masons.[59] It is interesting that the Rheims Palimpsest, of about 1250, which is one of the very few surviving early plans, sketches one half of the elevation in detail, and the other only roughly. The artist knows about using symmetry for inference.[60]

The Renaissance classic on architecture is Alberti's *On Building*. It well explains the difference between a plan and a picture:

> The difference between the drawings of a painter and those of the architect is this: the former takes pains to emphasize the relief of objects in paintings with shading and diminishing lines and angles; the architect rejects shading, but takes his projections from the ground plan, and, without altering the lines and by maintaining the true angles, reveals the extent and shape of each elevation and side—he is the one who desires his work to be judged not by deceptive appearance but according to certain calculated standards.[61]

That is, the inference in the case of plans is performed consciously.

The requirements of plan and elevation are, it must be emphasised, opposite in a way Alberti does not mention to those of a sketch which 'looks like' its object. The plan and elevation of a cube are both just squares, which are not very informative about the three-dimensional structure. To get a sketch of a cube that does convey the shape at a glance one should take a view from a 'generic' angle; that is, one that bears no special relationship to the edges of the cube.[62] It is notable that the stylised pictures of buildings and cities in the backgrounds of Byzantine and early medieval paintings take such a generic view (without being very expert in getting the angles right).[63]

1.9 Maps and Projections

A map, like an anatomical diagram, will simplify, select and label to facilitate inference (whether true or false).[64]

Medieval *mappaemundi* are more like diagrams, in the sense of the London Underground map, than maps drawn to scale, or according to a definite projection.[65] Some of them are combined with Wheels of Fortune, to illustrate conditions on earth,[66] and various other purposes can dominate the purely spatial information. Matthew Paris's itinerary map of the journey to the Holy Land is spatial, but essentially one-dimensional: it draws a road with icons of towns indicating the stages of the journey.[67]

Conceiving of a map as unlike a simple picture suggests the idea of adding something to it to indicate such spatial relations as direction. Latitude and longitude were reasonably familiar to the Arabs and medievals, but more in connection with the time differences between places than in drawing maps. Grids on maps are first found in diagrams of the planetary motions against the star background.[68]

The idea of a projection is especially important for inference from maps. The maker of a world map must understand some way of projecting round onto flat, and the user must understand it well enough to infer back from flat to round. Roger Bacon explains the need for a projection:

> Since these climates [i.e., zones] and the famous cities in them cannot be clearly understood by mere words, our sense must be aided by a figure. In the first place, then, I shall give a drawing of this quarter with its climates, and I shall mark the famous cities in their localities by their distance from the equinoctial circle, which is called the latitude of the city or region; and by the distance from the west or east, which is called the longitude of the region.[69]

There follows a scheme of projection where the spacing of parallels decreases systematically towards the Pole. Projections became much better understood with the rediscovery of Ptolemy's work in the fifteenth century,[70] and in the sixteenth, Mercator discovered his famous projection. Apart from aggrandising the circumpolar powers, its advantage is that of allowing the direction between any two places to be inferred directly from the map.

The inferring of directions is also the point of the fourteenth-century portolan charts, which were practical aids for navigation. These include accurate maps of the Mediterranean coast, criss-crossed by many rhumb lines, which one follows with the eye to work out the directions between the important locations. Some also have compass roses and scales.[71] Navigation involves difficult geometrical reasoning, as one must mentally co-ordinate three different spaces: that of a chart, that of an instrument, and that of the surrounding actual space. A great deal of effort was expended on the subject, in view of the costs of mistakes.

Local maps for legal and administrative purposes were occasionally found in the middle ages, but seem not to have been the normal thing until after 1500.[72] Hence, it seems that in general the ability to read maps and plans belonged only to certain specialised professions before about 1500, but became more general thereafter. Alberti describes how to construct a plan of Rome using an odometer and triangulation; surveying as a standard practice is also common after 1500, but not before.[73]

1.10 Instruments

Diagrams are not necessarily drawn on paper. For purposes of use, it may be better to inscribe them on something more durable, like metal. For making the inferences, it may be useful to include moving parts. The most impressive of the common instruments at the time was the astrolabe, useful for measuring anything from the position at sea to heights of towers. It is an analogue computing device, in the same sense as a slide rule is: it represents various continuously varying quantities by lengths on the instrument, and makes inferences back from the instrument to the quantities represented.[74]

One can have diagrams of the universe that incorporate the time dimension by having moving parts. There is a single ancient example, the Antikythera mechanism,[75] and various medieval 'equatories'.[76] If the mechanism is automated, it will become more and more like a clock. The earliest mechanical clocks of which there are reasonably complete descriptions are the complicated mid-fourteenth century planetary models of Richard of Wallingford and Giovanni de Dondi, which also happen to tell the time. The image of the universe as a clock, which we think of as so much an emblem of the Scientific Revolution, is an idea of Oresme[77] (though the phrase *mundi machina* is already in Sacrobosco's *Sphere*[78]).

Clocks provide the public with extensive training in spatial reasoning, since they pose a difficult three-way co-ordination problem. The circular space of the dial, actual (linear) time, and the numerals must be mentally identified. The addition of a minute hand makes the problem even worse, since the dial must be read as two spaces, one for each hand. No wonder it takes forever to teach a child to tell the time.[79]

All these diagrams are circular, as is one of Oughtred's original slide rules of 1632, and Galileo's 'geometrical and military compass', which is an analogue calculating device more than a compass.[80] 'Mathematics' did not have the suggestion of an abstract subject divorced from reality that it has today, but was much more closely associated with practical skills with a variety of instruments.[81]

Figure 5. *Stevin's Wreath of Spheres.*

1.11 Forces

A central place in the medieval applications of geometry was held by the science of weights, as classically expounded by Jordanus Nemorarius in his thirteenth-century treatise. His derivation of the law of the lever is interesting in that, unlike Archimedes, he relies on 'virtual displacements': he demonstrates that weights are in equilibrium by considering the movements they would undergo if they were not. The movements with which he calculates exist only in the imagination.[82]

It is the science of weights that includes one of the earliest actual discoveries of the Scientific Revolution, a successful application of reasoning with a diagram. In 1586 Stevin printed the 'Wreath of Spheres' diagram (fig. 5) as the title page of his *Elements of the Art of Weighing*.[83] The text which decorates it says, approximately, 'Wonderful, but not incomprehensible', which it is. It is clear that the circle of balls does not tend to rotate either clockwise or anticlockwise. But the balls hanging below the horizontal line are in equilibrium. One could cut them in the middle, and allow them to hang down, or even remove them altogether, without disturbing the balls resting on the two inclined planes. These upper balls, then, are at rest. The numbers of balls on each side of the apex are in inverse proportion to the sine of the angles at which the planes are inclined. One has derived, therefore, the law of the inclined plane, or, equivalently, the resolution of forces into components.[84] From the point of view of this paper, it is important that this, probably the only significant discovery of the early Scientific Revolution concerning forces, is expressed as pure geometry.

There seem to be no medieval or early modern diagrams that represent forces by arrows (even in the sections on stresses in beams in Galileo's *Two New Sciences* where one is almost forced to imagine arrows[85]). Nevertheless, someone who can design a flying buttress must have some mental representation of the direction of stresses;[86] likewise Brunelleschi in understanding that the hemispherical dome of the Cathedral of Florence could be constructed without scaffolding, a feat impossible with an arch.[87] Cannon and crossbows redirect forces, but it may have been possible to design and use them without imagining the forces. On the other hand, it is surely impossible to tack into the wind[88] without keeping constantly in mind the relations between the wind direction and the angle of the sails.

1.12 Euclid's Geometry

The history of geometry, in the formal sense, is a well-worn topic, and here we may just select a few aspects that are especially relevant to the thesis of the article.

As is well known, Euclid's reasoning cannot be carried out purely in terms of propositions, but relies essentially on the diagrams.[89] Kant is only the most famous of many who have been greatly impressed by the indispensability of 'constructions' in geometrical reasoning.[90] In Book I, proposition 1 Euclid draws a line, and then two circles with centres on the ends of the line and with radius equal to the length of the line, as in fig. 6. He next considers the points

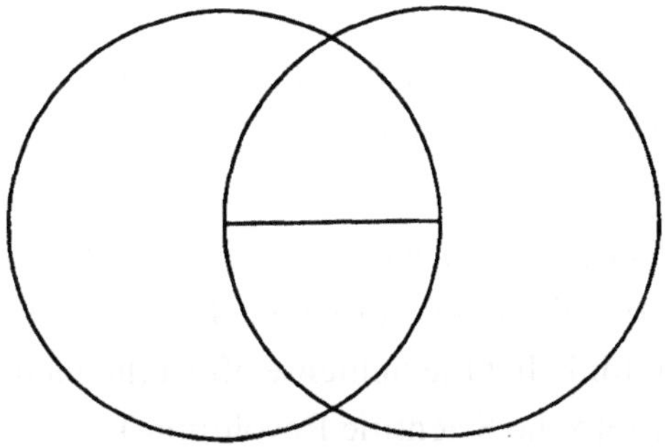

Figure 6. *Euclid I.1.*

where the circles interest. He does not use any axiom to justify the fact that they do intersect, but simply infers that from the diagram.

The intrinsic importance of whatever is in the early propositions of Euclid is magnified a thousandfold, as far as the history of ideas is concerned, by the central place the *Elements* has played in Western education since about 1200.[91] The pedagogical profession, to its lasting credit, held firm on Euclid (at least until around 1960), despite the immense consumer resistance that has left us the phrase *pons asinorum* (Euclid I.5).[92] A whole civilisation followed Euclid across that narrow bridge, into a bright new land of expanded horizons of the intellect.

The transferability of the skills learned with Euclid is of course an issue for debate, as it is when any modern mathematics educator demands more money on the grounds that 'mathematics teaches you to think'. Nevertheless, those involved seem to have had little doubt about the usefulness of geometrical training. Ramelli, for example, in his profusely illustrated book of 1588 on ingenious machines, praises Euclid at length; a modern commentator cannot understand the point of this, when Ramelli does not actually use any Euclidean theorems.[93] The answer must be, as it is to the modern question about the usefulness of training in abstract mathematics, that the more intelligent the pupil, the more transferable his skills, while even the most intelligent of pupils cannot be expected to recapitulate the history of mathematical discovery on his own. (Interestingly, the case for the usefulness of training in logic is much weaker.[94] While it is not impossible that training in formal logic should prove useful—as happened in the 1940s during the development of computers—there seems no reason to think that the immense effort the medievals put into formal logic had any input into the Scientific Revolution.)

Nevertheless, while there is a certain amount of agreement that the West's possession of Euclid was important,[95] there is room for confusion on what that means. For Euclid represents two very different things: logical rigour, and geometry. Philosophers emphasise the ideal of rigorous proof, according to which Euclid fulfils the goal set for all science by Aristotle's *Posterior Analytics*, of a set of theorems deduced from self-evident axioms. That is the aspect of Euclid that Matteo Ricci saw as lacking in Chinese mathematics:

> Nothing pleased the Chinese as much as the volume on the Elements of Euclid. This perhaps was due to the fact that no people esteem mathematics as highly as the Chinese, despite their method of teaching, in which they propose all kinds of propositions but without demonstrations. The result of such a system is that anyone is free to exercise his imagination relative to mathematics without offering a definitive proof of anything.[96]

That is, Chinese geometry lacks the discipline of rigorous inference. Modern studies have found that indeed mathematics in traditions that do not descend from the Greeks, though often excellent at developing complicated algorithms, is almost entirely lacking in proof.[97] Non-western mathematics resembles modern computer programming much more than it resembles modern mathematics. But, though the ideal of rigour is exciting, it is not obviously useful for anything outside geometry. For all the talk about axioms and certainty, no other science was put on an axiomatic basis, nor were there even any serious attempts. Not even the sciences closest to geometry became axiomatised: Ptolemy's *Almagest* has no self-evident axioms, and there is hardly a proof in number theory between Diophantus and Fermat, let alone an axiom.

Still less are there any medieval or Renaissance attempts like Spinoza's to develop ethics, for example, *more geometrico.*

On the other hand, Euclid represents geometry, understood as reasoning about spatial magnitudes and shapes. Here, the evidence is clearer for the usefulness of geometrical training. There was a medieval tradition of 'practical geometry' that depends on Roman surveying techniques and pre-dates the influence of the translations of Euclid.[98] Nevertheless, it was already consciously about inference: 'Practical [geometry] is that which is done by certain instruments and by inferring (*coniciendo*) proportionally one [distance] from others'.[99] When Euclid did become available, the practical geometries incorporated some of his ideas. Thus, even those interested only in the practical applications of geometry were encouraged to take some interest in reasoning. The widely-discussed medieval split between theory and practice, while genuine enough in general, is probably at its narrowest in geometry.[100] This is important, if the view is taken that a necessary condition for a scientific revolution is a feedback loop between theoretical science and technological innovation.

A sign of how deeply Euclid penetrated is the non-trivial use made of it by Bartolus, the foremost medieval authority in law, a discipline legendary for its innumeracy. Bartolus wrote a book applying Euclid to the division of lands affected by flooded rivers; it remains one of the few legal books with geometrical diagrams.[101] He also indulges in some simple optical reasoning: 'I saw Titius in the mirror at the barber shop, and recognised him' is not completely certain evidence that I saw Titius, because an image in a mirror is smaller than the thing, and so harder for the eye to discern.[102]

Yet, there are several respects in which Euclid is unsatisfactory as a training in spatial reasoning. For one thing, the proofs are difficult, so difficult that struggling through them can make learning the actual geometry very slow going. Secondly, the geometry of the *Elements* is almost all two-dimensional. It is true that Euclid knew some three-dimensional geometry, and something about how to represent it in two dimensions, as is clear from the diagram of parallelepipeds (fig. 7) from the oldest manuscript.[103]

But three-dimensional geometry occupies only small portions of the text, and they are in the little-read later books. This is a pity, as the difficulties of applying geometry to optics or astronomy or architecture lie mainly in coping with the third dimension. Finally, there is no motion; Euclid does not train the reader in what happens when shapes move, and avoids methods of proof that

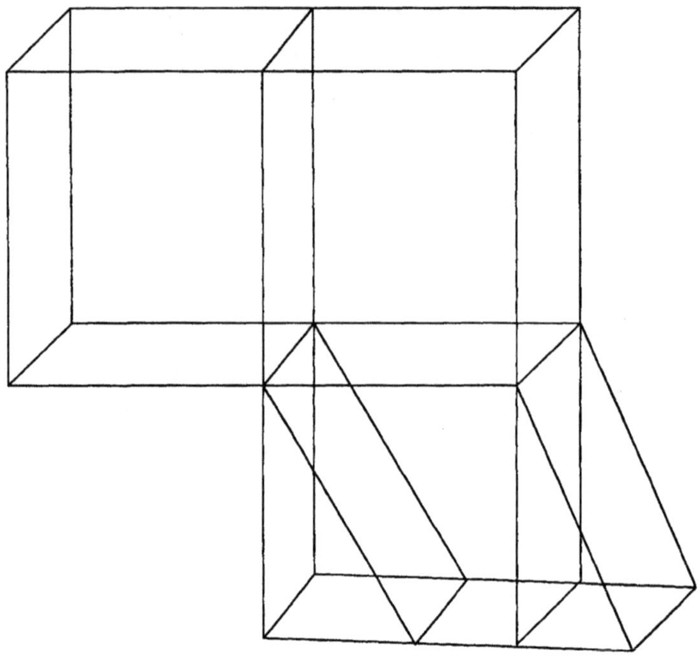

Figure 7. *Euclid XI.31,* from *The 'Heiberg' Manuscript.*

involve, for example, superimposing one figure on another. 'Geometry is occupied with immobile magnitude', Hugh of St Victor says, but adds, 'astronomy with mobile'.[104] Astronomy supplied all the other lacks in Euclid, too.

1.13 Astronomy

The standard introduction to astronomy in the university curriculum from about 1300 to 1600 was one of the best-written textbooks ever, the *Sphere* of Sacrobosco. The *Sphere* was the book on which Copernicus received his training on celestial orbs and their revolutions. It explains the celestial spheres, the zodiac, equinoxes and solstices, eccentrics and epicycles (very briefly) in a way that is always concise, clear and to the point. It expects of its readers a considerable willingness to imagine in three dimensions, but rewards average persistence with genuine understanding. Its strengths are illustrated by the two definitions of a sphere, which open the first chapter:

> A sphere is thus described by Euclid: A sphere is the transit of the circumference of a half-circle upon a fixed diameter until it revolves back to its original position. That is, a sphere is such a round and solid body as is described by the revolution of a semi-circular arc.
>
> By Theodosius a sphere is described thus: A sphere is a solid body contained within a single surface, in the middle of which there is a point from which all straight lines drawn to the circumference are equal, and that point is called the 'center of the sphere'.[105]

A diagram on paper is no use here: the reader must construct one in his imagination. Anyone who has done so and understood that the two definitions are equivalent has learned something substantial about modelling in the imagination.

Some of the later portions of the book are helped by illustrations. The edition of Venice, 1485, prints an eclipse diagram in three colours. Many of the sixteenth-century printed editions had sheets of volvelles to be cut out and pasted in.[106]

The capacity to visualise the universe from different points of view was not restricted to professional astronomers. The round earth itself required some work in the imagination for everyone: Madeville's *Travels* explains that the inhabitants of the southern hemisphere are upside down, to us, but seem to themselves to be the right way up.[107] (As far as is known, no other civilisation was asked to contemplate such a thing of other peoples.) There is something similar in the *Divine Comedy*,[108] a work generally heavy with geometry—its last thirteen lines contain a geometer, a circle, an image, the (faculty of) phantasy, and astronomical revolutions.[109] Dante imagines looking down from the sphere of the fixed stars and seeing the northern hemisphere from Cadiz to Asia.[110] We might think of it as a NASA's-eye view,[111] and see a connection between it and a good deal of later speculation about what is 'out there';[112] in any case, it is exactly the view shown in the map in Ptolemy's *Cosmographia* of 1482, which uses some interesting techniques to make the spherical shape of the earth obvious.[113] Edgerton remarks, 'Only by this kind of three-dimensional mind's-eye imagining could [Columbus] convince himself, as well as others, that it would be possible to reach the East by sailing west'.[114] It would be possible to mount a large thesis concerning the effect on Western thought of the 'what if?'

style of counterfactual imagining, implicating everyone from the Pre-Socratics[115] to science fiction, and no doubt such a thesis would be as defensible as ones of comparable size on religion and the rise of capitalism and the like. 'Here's fine revolution, an we had the trick to see't' (*Hamlet* V.i.89).

Since we are speaking of revolutions, it is pertinent to observe that the use of this astronomical metaphor to describe any major change is a medieval idea.[116]

So, when we speak of the 'heritage of Greek geometry', we should keep in mind not only Euclid, but Parmenides (or his contemporary), who first concluded the earth was round from a leap of geometrical imagination,[117] and Hipparchus, who fitted the numerical data of the Greeks and Babylonians to a geometrical scheme.[118] Astronomy has had a wide sphere of influence.

1.14 Optics and Perspective

Perspective in painting is an even more well-worn topic than geometry, and again we select just those aspects most relevant to spatial inference.

Geometrical optics, including the theory of vision, was, as is well-known, one of the best developed of medieval sciences.[119] The books of Alhazen, Bacon, Grosseteste, Witelo and Pecham are as scientifically sophisticated as anything the medievals produced. Optics include possibly *the* high point of medieval science, Theodoric of Freiberg's explanation of the rainbow (which depends essentially on a diagram).[120]

Books on perspective still sometimes read as if perspective sprang fully formed from the head of Brunelleschi.[121] This is ridiculous from the point of view of geometrical reasoning, and also from the point of view of illusionistic art. The rules of perspective are a marvellous discovery, but they are of the same sort as a number of earlier and simpler ones. (Giotto was 'the inventor and discoverer of many methods which had been buried for about six hundred years', according to Ghiberti.[122] The best one or two surviving antique perspective paintings do seem to justify the claims that the ancients knew some geometrical rules;[123] Giotto should have had little trouble reverse engineering them.) Consider the view of a ceiling with rafters in fig. 8.

Five lines in different directions, all converging, are more than a coincidence; the artist clearly knows something about how to represent three dimensions in two. The artist in question did not have the benefit of

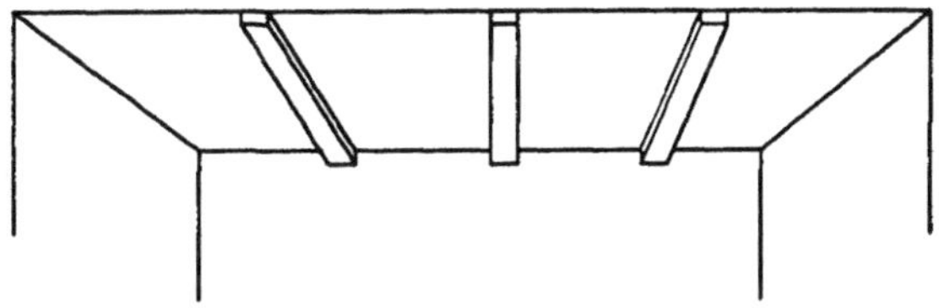

Figure 8. *Duccio's 'Perspective'.*

Brunelleschi's and Alberti's tuition, as he lived a century earlier. The view is from Duccio's *Maesta* of 1308-11;[124] he repeats it a number of times, as does Giotto, who on one ceiling has eleven converging lines, all in slightly different directions.[125] Duccio also has a complicated array of boxes seen from a generic angle, with the angles correct. Giotto draws the circular rims of jars seen from an angle as ellipses.[126] In these cases, the ellipses are rather flat, and could as well be lozenge shapes. This is not the case, however, with the quite wide ellipses Giotto uses to represent circles in his *coretti* in the Arena Chapel, perhaps the first genuine *trompe l'oeil*, at least since antiquity.[127]

The history of ellipses makes an interesting microcosm of the thesis of this paper. Ellipses are, in theory, fully treated in Apollonius' *Conics*, but as that work is almost impossible to read, they had only a marginal role in geometry in the tradition of Euclid's *Elements*.[128] They have little part in practical geometry either, which is based on straight rules and strings, and compasses. But in ancient painting, it was well-known that a shield or wheel seen side-on should be drawn as an ellipse, and theorem 36 of Euclid's *Optics* says that a wheel is seen sometimes as a circle, sometimes as an ellipse.[129] The ellipse belongs more to 'perspective'[130] than to formal geometry until its scientific apotheosis in Kepler's planetary theory. The method of constructing an ellipse with a string stretched between two pins was mentioned by Anthemius of Tralles, the architect of Hagia Sophia, and investigated by Kepler. Kepler opens his *New Astronomy* with a lament that is hard to find suitably prepared readers, as hardly anyone works through Apollonius, so that few can understand diagrams well.[131]

The most obvious way in which Giotto's and Duccio's 'perspective' differs from the real thing, as in the fifteenth-century masters, is that the earlier painters are not prepared to make the lines orthogonal to the picture plane long, with the result that they cannot have their perspective scheme covering the whole painting. They do not attempt real depth for the whole scene: there is, for example, always a blank wall preventing the rafters from continuing into the picture away from the viewer, covering the vanishing point (the 'perspective fig-leaf').

One of the things necessary to make this step—arguably the main one, in view of the effort Alberti and his contemporaries make in explaining it—is a conscious, general method of foreshortening. Lines parallel to the picture plane must recede into the distance, with the distances between equally spaced real things (for example the edges of tiles) diminishing in the picture in geometrical proportion. A method of accomplishing this appears in Pietro Lorenzetti's *Birth of the Virgin* of 1342, which Kemp calls 'the *tour de force* of fourteenth-century perspective'. It uses simply similar triangles, in much the same way as Alberti later describes; the construction lines can be traced incised in the wall in some parts.[132] Some of the painters' methods of geometrical construction on walls are described in Cennini's handbook of the late fourteenth century.[133]

The step from a collection of special tricks to a method of organising a whole picture requires the addition of geometrical theory. The part played by theoretical geometry and optics in the invention of perspective is not as clear as it might be, because neither Brunelleschi's two original perspective panels, nor his method of constructing them, has survived. It has been argued with at least some show of reason that he did not know any rules.[134] But the oldest relevant written evidence, the treatises of Alberti and Ghiberti, are fully based on geometrical theory. Book I of Alberti's *On Painting* is entirely an exposition of geometry, though punctuated by claims that he speaks 'as a painter', not as a mathematician. All that means is that his lines have a finite width, unlike the abstract widthless lines of the mathematicians. In fact, he goes so far as to retain barely relevant theory from Euclid, such as the axion that 'all right angles are equal'.[135] The point is to reason about what to draw in a painting:

> Now, since we have said that the picture is a cross-section of the pyramid we ought to investigate what importance this cross-section has for us. Since we have these knowns, we now have new principles with which to reason about the plane from which we have said the pyramid issues.[136]

What the reader is supposed to reason with is, in effect, an imagined three-dimensional diagram, containing a pyramid of visual rays and a cross-section of it, the plane of the painting. There follows a set of recipes on what lines to draw on the picture plane. Alberti promises that he possesses demonstrations of the recipes from reasoning about the visual pyramid, but omits them 'for the sake of brevity'.[137] The connection of early perspective with optical theory is even clearer in Ghiberti's *Third Commentary*, which is a series of extracts from the optical writings, especially Alhazen.[138] Though Ghiberti did not write the *Commentary* until the 1450s, when he was old, he claims to have studied optics when young and still learning the art of painting.[139]

The various Renaissance treatises on perspective provide the best examples of genuinely applied mathematics of the time. They are highly theoretical, but universally recognised as successful in practice.

An interesting application of optical reasoning in the Scientific Revolution itself is Galileo's reinvention of the telescope. He claims that the Dutch discovered it by accident, but that he, 'incited by the news mentioned above, discovered the same by means of reasoning'; that is, by reasoning as to what combination of convex and concave glasses would give a clear magnified image.[140]

1.15 The Mathematical Revolution

The thematic survey of geometrical reasoning just undertaken has perhaps obscured the broad chronological outlines of the Mathematical Revolution which preceded the Scientific one. Let us briefly review the chronology, emphasising how much was geometry, in the broad sense. To begin with, the ancient inheritance was extremely geometrical: the three most successful ancient sciences (not counting geometry itself) were astronomy, optics and statics, which consisted of great super-structures of geometry anchored in only few and easily acquired observations.[141] For later developments, a table (fig. 9) will provide the most perspicuous representation, as time is linear. Something that

	PURE GEOMETRY	OPTICS	ASTRONOMY	OTHER SPATIAL	NUMERICAL
1100	Latin Euclid, Archimedes			Music with staff Compass, quadrant	
1200	Euclid in curriculum Fibonacci's geometry	*Perspectives* Glass mirrors	Alfonsine tables Sacrobosco's *Sphere*	Villard's sketchbook Cathedral plans Jordanus on weights	Arabic numerals Fibonacci
1300		Theodoric on rainbow Spectacles Giotto's "perspective"	Wallingford Chaucer's *Astrolabe*	Portolan charts Oresme on graphs Merton "intension of forms" Anatomical diagrams Vigevano: machine drawings Compound interest tables	Double-entry Insurance Abacus schools Dice calculations
1400		Rules of perspective		Leonardo's sketches Henry "the Navigator"	
1500	Italian, English Euclids	Telescope	Copernicus Brahe	Dürer's illustrations Vesalius' illustrations Mercator's projection Stevin's "wreath of spheres"	Solution of cubic Decimals Viète's algebra Logarithms

Figure 9. *Table of developments.*

is clearer from the table than it would be from a mass of text is the clustering of applications of mathematics in the fourteenth century. The Scientific Revolution thus inherited a tradition of applying mathematics already two hundred years old. In some sense, it did not inherit much else of use—there was, for example, no comparable tradition of controlled experimentation. A glance at Grant's *Source Book of Medieval Science* shows that almost all medieval and Renaissance science is included in the above table, except for some depressing nonsense on matters like alchemy and herbs.

A snapshot of the state of mathematics after the Mathematical Revolution, but at the very beginning of the Scientific Revolution, as usually calculated, can be found in Billingsley's Euclid, the first English translation, of 1570 (a book otherwise notable for its pop-up figures of 3D geometry[142]). It has a preface by John Dee, 'specifying the chief Mathematicall Sciences, what they are, and wherunto commodious'. He arranges in a tree the two principal sciences, arithmetic and geometry, and some thirty derivative sciences. There are 'vulgar' arithmetic and 'vulgar' geometry, the latter divided into eleven sciences, concerned with surveying, in one, two and three dimensions, on both land and sea. Then there are nineteen miscellaneous sciences, most with a distinctly geometrical aspect. Of these, perspective, astronomy, music, cosmography, astrology, statics and navigation are clear enough, but the remaining eleven have names now unrecognisable. 'Anthropographie' is about the proportions in the human body; 'Zographie' is something like the modern theory of rendering in computer graphics; 'Trochilike' studies circular motions, simple and compound; 'Hydragogie' 'demonstrateth the possible leading of Water by Natures Law, and by artificiall helpe, from any head' and 'Pneumatithmie' 'demonstrateth by close hollow Geometrical figures (Regular and Irregular) the straunge properties (in motion or stay) or the Water, Ayre, Smoke and Fire'. Some of these are plainly more commodious unto nascent capitalism than others, but the total picture is of a suite of mathematical, mostly geometrical, sciences, in common and successful use, with investors queuing.[143] One of the mathematical sciences that certainly did attract serious money was navigation.[144] Military engineering was another steady earner.[145] Renaissance universities, for all the obloquy heaped on them, were also reliable sources of money for research into geometry, and especially astronomy (though not for algebra or non-medical experimental research).[146]

The evidence is, then, that when all due allowance has been made for the religious, hermetic, scholastic, military, astrological, ancient, alchemical and mercantile roots of the Scientific Revolution, the true compost in which those roots struck and grew strong was mathematics, and especially applied geometry.

2. THE VIEW FROM THE INSIDE

'One must consider the affection which is produced in the soul, and in that part of the body which contains the soul—the affection, the lasting state of which we call memory—as a kind of picture': Aristotle[147]

'There are no such things as mental pictures': Ryle[148]

We live at the end of a period which, perhaps more than any other, has hidden the pictorial life of the mind from intellectual view. Philosophy in the mid-century regarded 'sense data' as fictions, arrived at by (bad) inference and suitable for disposal as an undergraduate exercise.[149] It was seriously maintained that all inner representation was propositional.[150] (That is philosophy in the English-speaking tradition, continental philosophy being, if anything, even more word-oriented.) Psychology in the behaviourist decades was not much concerned with any inner life, pictorial or not, and 'imagery' tended to be associated with the Gestalt school, labelled 'unscientific'. Introspection, despite being reproducible with reasonable consistency, was ruled out as a source of experimental data. Frege, Russell and Hilbert, followed by Turing and the computer scientists and Artificial Intelligentsia, imposed on the learned world a view of inference as the manipulation of uninterpreted symbols according to formal rules. No room for pictures there. Even mathematics, once centred on geometry, maintained hardly any formal role for pictures, holding that geometrical intuition is unreliable. Galton was surprised to find even at the end of the nineteenth century that scientists were claiming to think in symbols, not images, and supposed that science had atrophied the imagination.[151] It is true that in the physical sciences there has been a kind of tradition of remarks on the need for mental geometrical intuition, but it is one of those 'traditions' that presents itself as a novelty every time it is reinvented.[152] At about the time Ryle was denying the reality of mental pictures, Einstein was saying:

> The words or the language, as they are written or spoken, do not seem to play any role in my mechanism of thought. The psychical entities which seem to serve as elements in thought are certain signs and more or less clear images which can be 'voluntarily' reproduced and combined ... The above mentioned elements are, in my case, of visual and some of muscular type. Conventional words or other signs have to be sought for laboriously only in a secondary stage, when the mentioned associative play is sufficiently established and can be reproduced at will.[153]

Generally, though, the image was left to intellectual marginals: diagrammatic inference to the engineers, with their slide rules, flow charts and circuit diagrams, and mental images to the Freudians and their dream fantasies.

The result is that the late-millennial intellectual has several layers of defence against accepting the medieval unselfconsciousness about using the imagination as a tool for doing science. First, we doubt the existence of the inner life at all. Or, we regard it as reached only by a chain of inference, not open to view. Or, we regard it as vague, and primarily emotional, not precise and scientific. Or, we presume the 'stream of consciousness', if there is one, is a flow of words, the preserve of psychiatrists and novelists. Or, if we do take the visual side of the imagination seriously, we think of it as something like a muse, 'inspiring' artists and poets in a manner too sublime to analyse. We have every excuse for misunderstanding.

At the leading edge of science, this picture is no longer true. Two developments especially have made the difference: psychological experiments on mental images, and scientific visualisation by computer. The psychological work is especially relevant. Around 1970, Shepard and Metzler found that the time subjects took to decide whether one 3D figure could be rotated to fit in the same space as another was proportional to the angle required, suggesting that the subjects were actually performing the rotation in some kind of mental space.[154] Since then, similar techniques have been used to investigate how subjects imagine themselves in an environment while mentally searching it,[155] and how they construct a mental model from a description of a scene.[156] It is found generally that reasoning about space is done by means of mental models, not via chains of propositions.[157] In view of what was said above about the importance of reasoning in three dimensions, it is significant that 3D imagining is found to be remarkably powerful: one can, for example, mentally scan across an imagined space from *any* viewpoint: the space is mentally encoded in 3D, but can be 'displayed' from any viewpoint in 2D.[158]

The other development that has given images scientific respectability is the very recent availability of computer-intensive visualisation tools, used for such purposes as understanding complex fluid flows.[159] The images are not themselves mental, but of course the point of computer visualisation is to create pictures that lead to understanding, that is, cause suitable mental representations. To some extent, the two themes are connected by research in education, which finds that mental visualisation is a necessary skill for learning in the sciences.[160] Anyone with some sympathy for the idea that ontogeny recapitulates phylogeny will be ready to look for some parallel process in the history of science.

These developments are yet to flow through into general intellectual consciousness, but progress has been sufficient to assert, at least, that Aristotle was more right than Ryle, and that there are no longer any barriers in principle to taking literally what the medievals were saying about mental images.

These studies also encourage us to add some further subtleties to the photo/line-drawing/diagram distinction drawn at the beginning of the paper. In order to *infer* something, whether from a diagram or from propositions, those entities must be represented internally, in the mind, soul or brain. In recent years some understanding has been reached of how this is done: put crudely, there is something like a picture inside; a kind of mental image which one can inspect. The image is, however, much more like a diagram than a photograph, in that it leaves out some things, while emphasising and labelling others.[161] So it is not quite correct to distinguish sharply between an image and a 'model' or metaphor.[162] Because the image is labelled, it contains information about, and can replicate (some of) the structure of its object; and hence support inference about the object. Speaking in an older idiom, Albertus Magnus says that memory is the storehouse not of images alone, but also of the *intentiones* drawn from them by the estimative power, with the image including the *intentio* within itself.[163] (Conversely, of course, from a purely physical point of view, photos, line drawings and diagrams are all just marks on paper. So to distinguish even them one must at least implicitly refer to the internal representation that they cause.)

2.1 The History Of Mental Images: Phantasms, Memory Theatres And Visions

We are now ready to see with new eyes the image-laden, text-poor world of the late medieval at his devotions. It is impossible not to speculate on the richness of the inner imaginal life of those fortunate to have viewed Fra Angelico or Giotto when newly painted. Fortunately, it is not necessary to be content with speculation, since those who did master text (surely a sample biased the wrong way) are eager to tell us in detail all about what it looked like inside.

The ancient legacy of discussion on mental images was moderately rich—richer, certainly, that the available ancient store of actual diagrams, of perspective pictures, or of psychological writings generally. 'There is no thinking without an image'[164] is one of Aristotle's most quoted sayings, and his discussion of imagination posits physical entities like pictures in the sensory apparatus.[165] Talk of 'phantasms', or mental images, is common currency in Stoic and Neoplatonist discussion of perception.[166] If Augustine is not the discoverer of the inner life here to the same extent as he is with autobiography, he certainly talked enthusiastically of inner images. He writes (the portion in italics is quoted in Aquinas' *Summa Theologiae*):

> The bodily sight cannot exist without the spiritual sight, because at the very moment when the bodily sense is touched by a body, there is made in the soul a something which is not this but is like it. If this were not produced, there would be no sense able to perceive those things which lie outside. *For the body does not sense, but the soul through the body, which it uses as a messenger for reproducing within itself what is announced from without.*[167]

He applies the expression 'the mind's eye' (*oculus mentis*) to a kind of intellectual vision.[168] The 'phantasm' occurs in the famous passage of Augustine where he anticipates Descartes' *Cogito, ergo sum*:

> But without any delusive representations of dreams and phantasms, I am most certain that I am ... For if I am deceived, I am.[169]

The western scholastics, following Avicenna, produced a very elaborate, and widely-known, theory of the 'inward wits'. It involved five internal faculties, including the 'imagination', which stored images, and the 'phantasy' (English 'fancy'), which recombined them.[170] There was also a baroque population of entities like the impressed and expressed species in which the various faculties dealt, but the most crucial one was still the 'phantasm'. The startlingly physical view of mental images that the scholastics took is preserved in later discussions of the effect of a mother's imagination on the foetus:

> Now from many instances it is clear that the imagination of the parents has an extraordinary power to modify and inflect the tempering and the formative power. So, if it is extremely strong and intent, sometimes the figure of what is thought, even if quite alien, is induced in the foetus. Thus it sometimes happens that a black child is born of two white parents, because the mother was turning over in her imagination an Ethiopian which she had depicted to herself in the bedchamber ... Given that the imagination can exercise such force, it is difficult to explain it. One opinion is this: The mother apprehends with intent thought, say, the form of an Ethiopian, whose image she imprints on the spirits which are carried back into the passages of the brain. The spirits then act as a vehicle to carry the image to the place of conception, where it

> imbues the material that the foetus is made from with the black colour that later appears. It so modifies the formative power inherent in the semen that what should make the foetus like the parents now makes it like the image, and so what happens is that the foetus degenerates into the form of an Ethiopian.[171]

(Before laughing about quaint views on the efficacy of mental images, one should perhaps recall that imagining walking increases the heart-rate.[172])

There is a dark side to regarding images as real things, which the soul somehow 'has' or 'receives'. It is that one can start brooding over their source, and suspecting their veracity. From Augustine again:

> Whatever we perceive by the body, even when not present to the senses, may be present to the imagination, as when we are asleep or angry: yet we cannot discern by the senses, whether what we perceive be the sensible object, or the deceptive image thereof.[173]

One will be particularly worried if one believes in dark forces whose mission is to deceive, like devils or witches.

> Both in bodily sights and the images of bodies which appear in the spirit, good spirits instruct and bad ones deceive.[174]

The result of these speculations in the fevered imaginations of the witch inquisitors makes an alarming and depressing story, an illuminating case study on the effects to which mistakes in abstract thinking can lead.[175] (Though on the other side the optical writer Witelo wrote a book *On the Nature of Demons*, attributing them to optical mistakes in bad light.[176]) Less harmful, but in principle much the same rationally, were beliefs in the efficacy of the imagination in Renaissance writings on magic and 'fascination'.[177]

After that, the visual imagery (*sic*) in Shakespeare's sonnets reads less as the excesses of the poetical imagination (*sic*) than as sober science:

> Mine eye hath played the painter and hath stell'd
> Thy beauty's form in table of my heart
> My body is the frame wherin 'tis held

And pérspective it is best painter's art.
For through the painter must you see his skill
To find where your true image pictured lies,
Which in my bosom's shop is hanging still *Sonnet 24*

Since I left you mine eye is in my mind
And that which governs me to go about
Doth part his function and is partly blind,
Seems seeing, but effectually is out,
For it no form delivers to the heart
Of bird, of flower, or shape, which it doth latch:
Of his quick object hath the mind no part
Nor his own vision holds what it doth catch. *Sonnet 113*[178]

For some more Shakespearean science:

> A foolish extravagant spirit, full of forms, figures, shapes, objects, ideas, apprehensions, motions, revolutions. These are begat in the ventricle of memory ...[179]

The references here to perspective and the theory of the internal senses are, it will be observed, not just decoration. They concentrate on perception as inference, with the possibility of that inference being faulty.

Conversely, talk about the 'imagination' of poets still meant something literal. Philip Sidney's *Apology for Poetry* explains that poetry is better than both philosophy and history, since the pictures it induces in the imagination have more force than the dry abstractions of philosophy, while still having a generality that the particular facts of history lack. He means by the 'imagination', as usual, a faculty of visualising by recombining images.[180] Sidney occupies a place towards the end of a long tradition of connecting the activity of the poet with the exercise of the (literal) faculty of imagination.[181]

So much for the vivid content of the medieval and Renaissance imagination. What was its purpose? In the first instance, it was for memory. Before memory was exported from the mind to written records, and the art of memory decayed, the geometrical resources of the visual imagination were used as a way of ordering any complicated body of ideas that had to be committed to memory, such as a long speech. The invention of the art was ascribed to Simonides of Ceos, who, leaving a banquet just before the dining hall collapsed, was able to identify the mangled bodies of the diners from the places where they lay; for he found he possessed a mental image of where the diners had been sitting.[182] From the start, a mental image was recognised as a structured entity, in which the relationship between the parts was useful for drawing conclusions. Simonides' idea was taken up in the later antique and medieval 'art of memory' that was the subject of Yates' famous book.[183] Cicero explains:

> The most complete pictures are formed in our minds of the things that have been conveyed to them and imprinted on them by the senses, but the keenest of all our senses is the sense of sight, and consequently perceptions received by the ears or by reflexion can be most easily retained if they are also conveyed to our minds by the mediation of the eyes.[184]

'Keenest' is perhaps not quite right; smells, for example, are 'keen', and one can recognise remarkable numbers of them. But the 'space' of smells does not seem to have a natural structure, whereas it is the geometry of images that makes them so structured, and hence useful for representing the structure of anything else. This is how the actual art works, as explained classically in the *Rhetoric to Herennius*: the orator imagines a building, with rooms, arches, statues. He places in them objects that will stimulate recall of the matter to be remembered. A ram with huge testicles, for example, will suggest testimony. Then while delivering the speech, he mentally visits the places in the correct order, thus recalling the speech. The medievals revived the art, applying it to the many texts that were memorised by all educated people, notably the Psalms. The illuminated capitals and marginal grotesques in medieval books are not just decoration. They are to enhance the visual memory of the page.[185]

Memory was of course closely connected with education, especially training 'by rote'. One of the architects of medieval pedagogy was Hugh of Saint Victor, inventor of possibly the largest diagram of the middle ages. It is his Mystical Ark of Noah, intended to organise the whole of knowledge and allow its recall. None of the surviving manuscripts attempt to picture it, as it is obviously too big to draw, and only fits in the imagination. It has all the patriarchs, popes and so on, also a map of the world, the vices, virtues, seasons and so on and on. It is organised with ladders, wheels and trees. Everything is in it.[186]

By what one may call the *Cutty Sark* phenomenon, the art of memory reached its most perfect form when it was already superseded, after the invention of printing. Father Ricci amazed the Chinese with this piece of Western technology, as with many others, using a vast memory theatre to achieve recall of Chinese characters.[187] Even more remarkable were the vast museum pieces of Giulio Camillo and Robert Fludd. Camillo's description of what he is doing is too overgrown with Hermetic and occult accretions to understand, but it is interesting in that it was actually constructed, in wood (though on what scale is now impossible to tell). It also attracted a brief description by a not entirely sympathetic visitor, who found in the idea of it something still of note:

> The work is of wood, marked with many images, and full of little boxes; there are various orders and grades in it. He gives a place to each individual figure and ornament, and he showed me such a mass of papers that ... He calls this theatre of his by many names, saying now that it is a built or constructed mind or soul, and now that it is a windowed one. He pretends that all things that the human mind can conceive and which we cannot see with the corporeal eye, after being collected together by diligent meditation may be expressed by certain corporeal signs in such a way that the beholder may at once perceive with his eyes everything that is otherwise hidden in the depths of the human mind. And it is because of this corporeal looking that he calls it a theatre.[188]

A mind realised in hardware, with all human knowledge in it, arranged in a natural way, hence content-addressable and immediately accessible! It is a dream yet to be realised by the AI and database communities. The modern equivalent is Lenat's CYC project, which aims to achieve artificial intelligence by having teams of typists enter all commonsense knowledge;[189] so far it is in the same state as Camillo's theatre appeared to his patron, the king of France: money goes in, and promises come back out. But one effect of the CYC project has been to make clear that AI confirms what the medievals presumed: to make sense of experience, one must know (that is, remember) a lot.

The unique aspect of Camillo's plan is that the organisation is topographic, in the most literal possible sense. The reader can no doubt visualise the place in his local library where the books on his favourite subject are kept, and finds unsettling the librarians' periodic rearrangements.

The second use of mental images, perhaps the most widely applied in medieval and Renaissance times, was to encourage meditation during prayer. The majority of mental pictures, like the majority of physical pictures, were painted to assist the soul. The medieval soul is not, according to itself, full of voices, but of forms. If modern introspection reveals neuroses and a stream of words, the medieval tended rather to find sins and visions. If the soul of a medieval was touched by God, it did not experience a voice from God so much as a 'vision'. (Some of the visions are very diagrammatic, too, especially those of Joachim of Fiori and Hildegard of Bingen, regarded as having considerable meaning.[190]) The imagination was made much of by the School of Saint Victor, in the twelfth century.[191] Visions may be for saints, but

> anyone can enter into the interior of his conscience and meditate in his mind's eye on Christ's wound, so that he conforms to Christ's sufferings.[192]

Each of Saint Ignatius of Loyola's *Spiritual Exercises* begins with a 'prelude' along the lines of:

> The second prelude is to form a mental image of the scene and to see in my imagination the road from Nazareth to Bethlehem. I will consider its length and breadth, and whether it is level or winding through valleys and over hills.[193]

Teresa of Avila's Interior Castle is more like a picture gallery than a library; though she is concerned to distinguish the true visions that adorn it from mere works of the imagination.[194]

The typical late medieval, then, was trained to exercise his imagination from his first youth.

2.2. The Imagination as a Tool of Scientific Visualisation

It has been argued several times above that various scientific thinkers must have been reasoning spatially in their imagination, particularly in cases like astronomy and perspective that involve heavy use of three dimensions. This ought to be obvious, but the ludicrous misunderstandings possible are illustrated by C.S. Lewis. Lewis would be expected to understand mental images, since, according to his student Ken Tynan, he was usually able to quote from a page, given the bay number in his room, the shelf number, how many books from the left, and the page number. Yet he comments on Albert the Great:

> I do not understand why *boni imaginativi* should tend, as he says they do, to be good at mathematics. Can this mean that paper was too precious to be wasted and you geometrised, as far as possible, with figures merely held before the mind's eye? But I doubt it; there was always sand.[195]

No-one can reason in the sand, since his brain is not there. You might as well say that someone who reads music needs no internal sense of rhythm, on the grounds that the notes are all there on the page. Or that a reader of words does not need to know their meanings, since the meanings are all in the dictionary. *One can only reason with internal representations.*

We now come to the evidence for saying that those who reasoned with the imagination were in general consciously aware of doing so, and saw the imagination as a tool of scientific visualisation.

Plato sometimes uses mental diagrams to represent abstract relations of proportion,[196] and Aristotle has the memory forming a kind of scale model of magnitudes, both spatial and temporal,[197] but generally the earliest writers do not much emphasise the role of the imagination in mathematics. Proclus, however, in the only developed ancient philosophy of mathematics, holds that the imagination is where geometry is done. For the usual Platonist reasons, he thinks physical diagrams are unsuitable: 'the circle in sensible things is inferior in precision, infected with straightness, and falls short of the purity of immaterial circles'. On the other hand, the pure understanding cannot do geometry either, since its concepts are simple, or 'wrapped up', and there is only one of each kind, so that it cannot deal with circles of different sizes. So the understanding projects images 'distinctly and individually on the screen of the imagination', which provides a kind of 'intelligible matter' for them.[198] The passage is an influential one. Kepler is enthusiastic,[199] while A. Piccolomini claims that Proclus' placing of the mathematicals in the imagination explains the certainty of mathematics.[200]

Alhazen remarks that visual rays are imaginary (*lineae imaginabiles* in the Latin);[201] 'lyne ymagined' is also Chaucer's phrase for meridians of longitude and the ecliptic.[202] The point that a diagram is a device to create a three-dimensional model in the imagination is made by a thirteenth-century Islamic writer:

> In drawing ... I have not aimed for completeness. My purpose was to present an arrangement so it can be understood in the whole and in detail. One realizes that there is obscurity in the representation of solid bodies, but in the imagination one can fit one thing to another, view it from any angle, dissect it, and thus assemble it step by step. All the drawings that I have made are simple, so that they give a clear picture.[203]

Hence, the fact that medieval scientific and technological diagrams are sometimes hard to interpret for us does not show that those diagrams failed to support good scientific inference. Living in a post-perspective world, we have it easy.[204]

There may be a case for connecting later medieval reasonings in the imagination with the use in philosophy of cases *secundum imaginationem*, which can only exist through God's absolute power.[205] The work on 'physics' of the 'Merton School', it has been pointed out, is purely 'in the head': it considers only imaginary cases, and confronts real experience only via 'well-known' (that is, remembered) facts such as that a spinning top occupies the same place.[206] Heytesbury says at one point that he is proceeding only *secundum imaginationem*: cases like acceleration to infinity and diminution to zero quantity are not physically possible, but are imaginable and should be considered.[207] It has been maintained that this procedure separates the medieval scientific methodology from the modern, experimental one. It could just as well be maintained that medieval 'physics' is really mathematics. In the modern theory of differential equations, one will certainly want to consider various limiting cases, and for the same reasons as the medievals.

Still, the imaginative and the philosophical methods of reasoning are not entirely compatible. Henry of Ghent, about 1300, calls some opponents

> those of whom the Commentator says that in them the imaginative virtue dominates over the cognitive virtue, and so, he says, they do not believe

> demonstrations unless the imagination accompanies them, for they cannot believe that there is neither plenum nor vacuum nor time outside the world ... mathematical imaginations and what is outside the heaven seem to them infinite ... Therefore such people are melancholy and make the best mathematicians, but the worst metaphysicians, because they cannot extend their understanding beyond site and magnitude.[208]

It will come as no surprise to find Oresme the most explicit on reasoning in the imagination. He thought of his graphs not, in the first instance, as on paper, but in the imagination:

> Every measurable thing except numbers is imagined in the manner of continuous quantity. Therefore, for the mensuration of such a thing, it is necessary that points, lines and surfaces, or their properties, be imagined. For in them [i.e. the geometrical entities], as the Philosopher has it, measure or ratio is initially found, while in other things it is recognized by similarity as they are being referred by the intellect to them [i.e., to geometrical entities]. Although indivisible points, or lines, are non-existent, still it is necessary to feign them mathematically for the measures of things and for the understanding of their ratios. Therefore every intensity which can be acquired successively ought to be imagined by a straight line perpendicularly erected on some point of the space or subject of the intensible thing, e.g., a quality. For whatever ratio is found to exist between intensity and intensity, in relating intensities of the same kind, a similar ratio is found to exist between line and line, and vice versa.[209]

The imagination is better than paper for graphs in higher dimensions:

> By a like imagination the quality of a surface is imagined as a kind of body, whose longitude and latitude is the extent of the surface and whose depth is the intensity of the quality.

So what about the quality of a three-dimensional object? Do we need a fourth dimension to graph its intensity? No,

> because while a flowing point is imagined as causing a line, a line a surface, and a surface a body, it is not necessary, if a body is imagined flowing, that it causes a fourth type of quantity, but only a body.[210]

This can only mean that the flow in the imagination is what represents the fourth dimension. Oresme himself describes the forerunners of his idea, in an effort to excuse himself from the vice of novelty:

> It is sought whether a quality is to be imagined as a surface.
> It is argued on the negative ...
> I respond that the statement is true and could be confirmed by the writers on perspective like Witelo and Lincoln [Grosseteste], who in this manner imagine the intensity of light, and by Aristotle, who in the fourth [book] of the *Physics* imagines time by means of a line, and by the Commentator [Campanus] in the fifth [book] of this [commentary on Euclid's *Elements*], where he holds, in expounding ratios, that everything having the nature of a continuum can be imagined as a line, surface or body.[211]

The originals[212] do use geometrical magnitudes to represent non-geometrical magnitudes, but only the optical writers speak of anything 'imaginary'.

Aristotle had maintained that the form of a work of art, or a building, exists beforehand in the soul of the artist.[213] According to Alberti, architecture does not happen on the building site, or even on paper, but in the imagination:

> It is quite possible to project whole forms in the mind without recourse to the material, by designating and determining a fixed orientation and conjunction for the various lines and angles. Since that is the case, let lineaments be the precise and correct outline, conceived in the mind, made up of lines and angles, and perfected in the learned intellect and imagination.[214]

It is clear that the role of the imagination in both pure and applied mathematical contexts was taken for granted by the time of the Scientific Revolution.

2.3 Galileo's Thought Experiments

It has been a source of embarrassment for many historians of science that Galileo, when he is supposed to be founding modern science by performing experiments, is actually caught doing his experiments in thought. Examples are common; let us take a crucial one in his early work, *On Motion*:

> … a larger stone does not fall more swiftly than a smaller. Those who are surprised by this conclusion will also be surprised by the fact that a very large piece of wood can float on water, no less than a small piece. For the reasoning is the same. Thus, if we imagine (*si mente conciperemus*) that the water on which a large piece of wood and a small piece of the same wood are afloat, is gradually made successively lighter, so that finally the water becomes lighter than the wood, and both pieces slowly begin to sink, who could every say that the large piece would sink first or more swiftly than the small piece?
>
> …
>
> Again, if we imagine, for example, a large piece of wax floating on water, and we mix this wax either with sand or some other heavier substances …
>
> I argue as follows in proving that bodies of the same material but of unequal volume move with the same speed. Suppose there are two bodies of the same material, the larger *a*, and the smaller *b*, and suppose, if it is possible, as asserted by our opponent, that *a* moves more swiftly than *b*. We have, then, two bodies of which one moves more swiftly. Therefore, according to our assumption, the combination of the two bodies will move more slowly than that part which by itself moved more swiftly than the other. If, then, *a* and *b* are combined, the combination will move more slowly than *a* alone. But the combination of *a* and *b* is larger than *a* is alone. Therefore, contrary to the assertion of our opponents, the larger body will move more slowly than the smaller.[215]

Where does the 'combination' of *a* and *b* take place? According to Aquinas, it is the 'phantasy or imagination' that is the organ of 'combining and dividing'.[216] Galileo's 'gradually' and 'successively' in this passage are significant: like Oresme with his 'flowing point', he regards the imagination as a (more or less) continuous medium. By movement through that continuous medium, he can perform his characteristic transformation to a limiting, ideal case, which is still physically meaningful, but inaccessible to real experiment.[217] Where the medievals used imaginary cases, like motion in the void, largely for critical purposes, Galileo uses them constructively, regarding them as simple versions of reality, to which complex real cases approximate.[218]

And is Galileo evasive about the imaginary nature of his experiments? Modest? Repentant? No, he is brazen:

> SIMPLICIO: So you have not made a hundred tests, or even one?
> SALVIATI: Without experiment, I am sure that the effect will happen as I tell you, because it must happen that way.[219]

As Kuhn remarks, 'Surely he did experiments, but he is even more noteworthy as the man who brought the medieval thought-experimental tradition to its highest form'.[220]

Experimenting in the imagination is a lot cheaper than in the lab, of course, and faster, like modern simulations by computer, but like them is exposed to the objection that it will not tell you how the real world is, since in thought, presumably, anything can happen.

This is the nub of the matter. Imagination, used as a tool for scientific reasoning by an expert like Galileo, is *not* infinitely plastic, and hence divorced from reality, but incorporates a lot of structure, parts of which restrict what can happen in other parts. This is why actual experiments on motion on an inclined plane can be relevant to what would happen with vertical motion in a vacuum, and the imagination can mediate between the two. Feynman, a modern physicist who was champion of imagination over mathematical formalism, explains that the point of the scientific imagination is the *constraints* it incorporates:

> The whole question of imagination in science is often misunderstood by people in other disciplines. They overlook the fact that whatever we are *allowed* to imagine in science must be *consistent with everything else we know* ... We can't allow ourselves to seriously imagine things which are obviously in contradiction to the known laws of nature. And so our kind of imagination is quite a difficult game.[221]

The question is, how does the imagination acquire that structure that mimics the world, and which allows it to be used as a bridge between real experiments and what would happen in counterfactual circumstances? One way or another, the imagination must be structured by the flux of ordinary experience. An indication of how it happens can be had by recalling Steven's Wreath of Spheres diagram, which is a thought experiment not unlike Galileo's. The fact that the spheres do not revolve in either direction, but hang in equilibrium, is a deliverance of the imagination, but is not a logical truth. It must, therefore, be a distillation of experience. It follows that the process of using the imagination to learn the results of thought experiments is a kind of reminiscence. That is what Aquinas says: the 'phantasy or imagination' is for the 'retention and conservation' of forms.[222] More to the point, it is what Galileo says. Salviati is encouraging Simplicio to realise that a stone released from a sling moves off tangentially:

> SALVIATI: The unraveling depends upon some data well known and believed by you just as much as by me, but because they do not strike you, I shall cause you to resolve the objection by merely recalling them.
> SIMPLICIO: I have frequently studied your manner of arguing, which gives me the impression that you lean toward Plato's opinion that *nostrum scire sit quoddam reminisci* [our knowing is a kind of reminiscence] ...
> SALVIATI: Well, then, what is its motion?
> SIMPLICIO: Let me think a moment here, for I have not formed a picture of it in my mind.
> SALVIATI: Listen to that, Sagredo; here is the *quoddam reminisci* in action, sure enough. Well, Simplicio, you are thinking a long time.
> SIMPLICIO: So far as I can see, the motion received on leaving the notch can only be along a straight line ...[223]

Two other arguments in the *Dialogue* involve Socratic questioning of Simplicio's 'memory'; they are both purely geometric arguments. In one of them, Simplicio is gradually forced to draw a diagram of the Copernican universe, led by purely geometrical considerations.[224]

Let us separate two questions that may occur on reading Galileo. Can one get a reasonably consistent answer by quizzing people's imagination on, for example, what happens when a body whirled on a string is released? And if so, is the answer the correct one, as Galileo suggests? Modern psychology has investigated and the answers are respectively 'yes' and 'sometimes'. Consistent patterns of expectation about motion are found on eliciting subjects' 'intuitive physics' or 'naive physics' by asking them to imagine what motion would occur in various circumstances. Some subjects give the correct answers, but a high proportion, even those educated in physics, persist in mistakes like expecting curved motion to continue in a curve when released.[225] The authors of these studies claim that all the different medieval errors about the motion of projectiles and circular impetus can be found among present-day college students. This suggests, among other things, that the medievals were doing their physics by the same kind of imaginative reasoning from experience as the psychologists are now studying.

2.4 Epilogue: Philosophy Moves Inside

The story would not be complete without at least a brief mention of the fact that a Philosophical Revolution occurred at the same time as the Scientific one, and that it too was driven by inference from mental images.

Kant claimed to have effected a 'Copernican revolution' in philosophy, meaning a fundamental change in viewpoint. He had in mind his claim that necessities, such as mathematical ones, previously thought to be in the world, were removed, in his philosophy, into the cognitive apparatus. Almost everyone has thought that such a reversal of point of view was indeed a revolution, but that its true Copernicus was Descartes. It is universally agreed that 'modern' philosophy began when the Cartesian *cogito* transferred the central question of philosophy from metaphysics to epistemology. It is the interior viewpoint that Descartes starts from which makes the problem of knowledge of the external world central. The point was argued at length by the modern Thomists, who saw Descartes as having diverted philosophy into several centuries of the 'way of ideas' and idealism by beginning with the certainty of mental images, and asking, 'How do we get out?'[226]

Descartes certainly had a well-trained imagination, and its training was on geometry. In the *Discourse*, he says of his early studies in geometry:

> [Geometry] is so closely tied to the examination of figures that it cannot exercise the intellect without greatly tiring the imagination.[227]

That is the opposite to the naive thought that imagining a few pictures should be easy, even if the logical reasoning is difficult. Not so difficult, though, that he will not recommend hard work with the imagination to others. Rule 14 of *Rules for the Direction of the Mind* is:

> The problem should be re-expressed in terms of the real extension of bodies and should be pictured in our imagination entirely by means of bare figures. Thus it will be perceived much more distinctly by our intellect.[228]

Among the figures suggested is a family tree. It is only with Rule 15 that permission is given for the diagrams to be drawn on paper.

On the question of whether training in mathematics is for the purpose of learning imaginative modelling or logical reasoning, Descartes prefers the latter, but concedes it is possible to do mathematics imaginatively:

> The fact that there are some people who are clever at Mathematics but less successful in subjects like Physics, is not due to any defect in their powers of reasoning, but is the result of their having done Mathematics not by reasoning but by imagining—everything they have accomplished has been by means of imagination. Now, in Physics there is no place for imagination, and this explains their signal lack of success in the subject.[229]

In the *Meditations*, Descartes examines the difference between the imagination and the 'pure understanding':

> When I imagine a triangle, for example, I do not merely understand that it is a figure bounded by three lines, but at the same time I also see the three lines with my mind's eye as if they were present before me; and this is what I call imagining. But if I want to think of a chiliagon, although I understand that it is a figure consisting of a thousand sides just as well as I understand the triangle to be a three-sided figure, I do not in the same way imagine the thousand sides or see them as if they were present before me ... But suppose I am dealing with a pentagon: I can of course understand the figure of a pentagon, just as I can the figure of a chiliagon, without the help of the imagination; but I can also imagine a pentagon, by applying my mind's eye to its five sides and the area contained within them. And in doing this I notice quite clearly that imagination requires a peculiar effort of mind which is not required for understanding ...[230]

Elsewhere, Descartes claims that a heptagon or octagon can only be imagined with difficulty, but that was he, 'who is a fairly imaginative man and has trained his mind in this field for some time', can do it 'reasonably distinctly'. In the same place he emphasises how like imagination perception is, the only difference being that the images are imprinted in one case without and in the other case with external objects.[231] Again, in defining his crucial term 'idea', Descartes distinguishes an idea from an image in the imagination, but the difference is not large:

> Thus it is not only the images depicted in the imagination that I call 'ideas'. Indeed, in so far as these images are in the corporeal imagination, that is, are depicted in some part of the brain, I do not call them 'ideas' at all; I call them 'ideas' only in so far as they give form to the mind itself, when it is directed towards that part of the brain.[232]

It is true that here, by agreeing with the scholastics that the imagination is 'corporeal', Descartes distinguishes some internal pictures from what is really in the 'ego'.[233] Nevertheless, it is clear that Descartes' 'idea' has a spatial content lacking in the modern 'concept', and that this survival of the medieval theory of the imagination is at the bottom of much of what moderns find bizarre in the seventeenth and eighteenth-century 'way of ideas' in philosophy.[234]

And it is no accident that Descartes' 'representative' theory of perception is essentially the same as Aristotle's theory of memory. According to Aristotle, one can regard a memory image, like any picture, either simply as a painted thing or as a likeness.[235] It is in the latter mode that memory is a kind of inference, from image to thing pictured. Indeed, Aristotle emphasises that recollecting is a sort of inference, hence, he thinks, peculiar to humans:

> For when a man is recollecting he infers that he has seen or heard or experienced something of the sort before, and the process is a kind of search.[236]

Aristotle says that memory is inference from internal pictures. Descartes' Copernican revolution in philosophy is the claim that all perception is inference from internal pictures.

School of Mathematics, University of New South Wales

NOTES

[1] Emphasised in J.V. Field, *Kepler's Geometrical Cosmology* (London, 1988).

[2] M.G. Winkler and A. van Helden, 'Representing the heavens: Galileo's visual astronomy', *Isis* 83 (1992) 195-217.

[3] Galileo, *The Assayer*, in *The Controversy on the Comets of 1618*, trans. S. Drake and C.D. O'Malley (Philadelphia, 1960) pp. 237-8.

[4] E.g., J. le Goff, *The Medieval Imagination*, trans. A. Goldhammer (Chicago, 1988); C.S. Lewis, *The Discarded Image* (Cambridge, 1964).

[5] J.H. Larkin and H.A. Simon, 'Why a diagram is (sometimes) worth ten thousand words', *Cognitive Science* 11 (1987) 65-99; N.H. Narayanan (ed.), *AAAI Spring Symposium on Reasoning with Diagrammatic Representations* (Stanford, CA, 1992).

[6] See J.C. Frakes, *The Fate of Fortune in the Early Middle Ages* (Leiden, 1988); F.P. Pickering, *Literature and Art in the Middle Ages* (London, 1970) ch. 3.

[7] Villard de Honnecourt, *The Sketchbook of Villard de Honnecourt*, ed. T. Bowie (Bloomington, Ind, 1959) plate 64; cf. R. Bechmann, *Villard de Honnecourt: la penseé technique au XIIIe siècle et sa communication* (Paris, 1991).

[8] D. Londey, 'Apuleius and the square of opposition', *Phronesis* 29 (1984) 165-73.

[9] D. Gilman, 'A new perspective on pictorial representation', *Australasian Journal of Philosophy* 70 (1992) 174-86.

[10] The two main sources on medieval diagrams are J.E. Murdoch, *Album of Science: Antiquity and the Middle Ages* (N.Y., 1984) and M. Evans, 'The geometry of the mind', *Architectural Association Quarterly* 12:4 (1980) 32-55, which relies considerably on A.C. Esmeijer, *Divina quaternitas: A Preliminary Study in the Method and Application of Visual Exegesis* (Amsterdam, 1978).

[11] Murdoch, ch. 6; Evans, section 5; S.Y. Edgerton, *The Heritage of Giotto's Geometry: Art and Science on the Eve of the Scientific Revolution* (Ithaca, N.Y., 1991) p. 30.

[12] Evans, section 5.3; L. Means, 'The vulnerability of volvelles in manuscript codices', *Manuscripta* 35 (1991) 43-54.

[13] Evans, section 7.1; *The Figurae of Joachim of Fiore*, ed. M. Reeves and B. Hirsch-Reich (Oxford, 1972).

[14] M. Carruthers, *The Book of Memory* (Cambridge, 1990) p. 252.

[15] Illustrations in several media in L. Lee, G. Seddon and F. Stephens, *Stained Glass* (London, 1976) pp. 36-7; see A. Watson, *The Early Iconography of the Tree of Jesse* (Oxford/London, 1934); A. Watson, 'The *Speculum virginum* with special reference to the Tree of Jesse', *Speculum* 3 (1928) 445-69; M.W. Evans, *Medieval Drawings* (Feltham, N.Y., 1969) plate 69.

[16] Seneca, *De beneficiis* III.xxviii.2; Pliny, *Natural History* XXXV.ii.6.

[17] M.T. Clanchy, *From Memory to Written Record* (London, 1979) plate XIII.

[18] E.H. Wilkins, 'The trees of the 'genealogia deorum'', *Modern Philology* 23 (1925-6) 61-5; Murdoch, *Album of Science*, illustration 41.

[19] Evans, *Medieval Drawings*, plate 71; Murdoch, *Album of Science*, illustrations pp. 37-9.

[20] Quoted in H. Child, *Heraldic Design* (London, 1965) p. 113.

[21] J. Huizinga, *The Waning of the Middle Ages* (Harmondsworth, 1965) pp. 194-5.

[22] S.M. Stigler, 'Stigler's law of eponymy', *Transactions of the New York Academy of Sciences,* 2nd series 39 (1980) 147-57.

[23] N. Kretzmann et al. (eds), *The Cambridge History of Later Medieval Philosophy* (Cambridge, 1982) p. 129.

[24] L. Pacioli, *Summa de arithmetica* (Venice, 1494) fol. 82r, reproduced in J.B. Geijsbeek, *Ancient Double-Entry Bookkeeping* (Denver, 1914, repr. Osaka, 1975) p. 26.

[25] H. Gruber, 'Darwin's 'Tree of nature' and other images of wide scope', in J. Wechsler (ed.), *On Aesthetics in Science* (Cambridge, Mass, 1978) 121-40.

[26] S. Mac Lane, *Categories for the Working Mathematician* (N.Y., 1971) introduction; M.A. Arbib and E.G. Manes, *Arrows, Structures and Functors* (N.Y., 1975) ch. 1; cf. G. Birkhoff, *Lattice Theory* (3rd edn, Providence, RI, 1967).

[27] A.S. Maida, article 'Frames', in *Encyclopedia of Artificial Intelligence*, ed. S.C. Shapiro (2nd edn, N.Y., 1992) vol. I pp. 493-507.

[28] L. Euler, *Opera Omnia* 3rd series vol. 11 p. 233, trans. in *Letters on Different Subjects of Natural Philosophy* (N.Y., 1833, repr. N.Y., 1975) p. 341.

[29] A.-T. Vandermonde, 'Remarques sur les problèmes de situation', *Histoire de l'Académie des Sciences* (1771) 566-74, trans. in N.L. Biggs, E.K. Lloyd and R.J. Wilson, *Graph Theory 1736-1936* (Oxford, 1976) 22-6.

[30] See J.V. Fleming, *From Bonaventure to Bellini: An Essay in Franciscan Exegesis* (Princeton, 1982) pp. 3-32.

[31] Dante's vision is depicted in Nardo di Cione, *The Inferno*, fresco, Santa Maria Novella, Florence (reproduced in R. Hughes, *Heaven and Hell in Western Art* (London, 1968) p. 158).

[32] F. Yates, *Theatre of the World* (London, 1969).

[33] S.K. Heninger, *The Cosmographical Glass: Renaissance Diagrams of the Universe* (San Marino, CA, 1977); F. Saxl, 'Microcosm and microcosm in medieval pictures', in *Lectures* (London, 1957); Evans, *Medieval Drawings*, plate 81; R. Fludd, *Utriusque cosmi ... historia* (Oppenheim, 1617-21) discussed in R.S. Westman, 'Nature, art and psyche: Jung, Pauli and the Kepler-Fludd polemic', in B. Vickers (ed.), *Occult and Scientific Mentalities in the Renaissance* (Cambridge, 1984) 177-229.

[34] *Nicole Oresme and the Medieval Geometry of Qualities and Motions*, ed. and trans. M. Clagett (Madison, Wisc, 1968); Oresme, *Quaestiones super geometriam Euclidis*, ed. H.L.L. Busard (Leiden, 1961).

[35] [Nicole Oresme], *Perutilis tractatus de latitudinibus formarum* (Padua, 1486).

[36] R. Rastall, *The Notation of Western Music* (London, 1983) ch. 2; C. Parrish, *The Notation of Medieval Music* (London, 1957, repr. N.Y., 1978) plate VII.

[37] M.S. Mahoney, 'Diagrams and dynamics: Mathematical perspectives on Edgerton's thesis', in J.W. Shirley and F.D. Hoeniger (eds), *Science and the Arts in the Renaissance* (Washington, 1985) 198-220.

[38] M. Clagett, *Nicole Oresme and the Medieval Geometry of Qualities and Motions* (Madison, 1968) pp. 104-6.

[39] Murdoch, *Album of Science*, ch. 2.

[40] G.P. Di Bianchi and J. Christoffels Ympyn, *A notable and very excellente woorke: expressyng and declaryng the maner and forme how to kepe a boke of accoptes or reconynges* (London, 1547, ed. B.S. Yamey and O. Koiima, Kyoto, 1975) plate V.

[41] A. Martinelli, 'The ledger of Cristianus Lomellinus and Dominicus de Garibaldo, stewards of the city of Genoa (1340-41)', *Abacus* 19 (1983) 83-118, plates 1-4.

[42] J.U. Nef, *Cultural Foundations of Industrial Civilization* (Cambridge, 1958) pp. 11-4.

[43] F.B. Pegolotti, *La Pratica della Mercatura*, ed A. Evans (Cambridge, Mass, 1936) pp. 301-2.

[44] S. Stevin, *Tafelen van Interest* (Antwerp, 1582; Amsterdam, 1590).

[45] *Alfontij regis castelle illustrissimi celestium motuum tabule* (Venice, 1483).

[46] J. Napier, *Mirifici logarithmorum canonis descriptio* (Edinburgh, 1614) bk. I ch. 1.

[47] M.B. Parkes, *Pause and Effect: An Introduction to the History of Punctuation in the West* (Aldershot, 1992).

[48] K. Weitzmann, *Ancient Book Illumination* (Cambridge, Mass, 1959) figs 4-9.

[49] B. Gille, *Engineers of the Renaissance* (London, 1966) chs 1-3.

[50] P.J. Booker, *A History of Engineering Drawing* (London, 1963, repr. London, 1979); F.M. Feldhaus, *Geschichte des technischen Zeichnens* (Wilhelmshafen, 1959); F.D. Prager and G. Scaglia, *Mariano Taccolo and his Book De Ingeneis* (Cambridge, Mass, 1972); A.L. Matthies, 'Medieval treadwheels: artists' view of building construction', *Technology and Culture* 33 (1992) 510-47. Further in R.G. Mazzolini (ed.), *Non-Verbal Communication in Science Prior to 1900* (Florence, 1993).

[51] J. Ackerman, 'The involvement of artists in Renaissance science', in Shirley and Hoeniger, *Science and the Arts in the Renaissance*, 94-129; C. Truesdell, *Essays in the History of Mechanics* (Berlin, 1968) ch. 1.

[52] S.Y. Edgerton, 'The Renaissance development of the scientific illustration', in Shirley and Hoeniger, *Science and the Arts in the Renaissance*, 168-97, at pp. 174, 176.

[53] L.C. MacKinney, 'The beginning of western scientific anatomy', *Medical History* 6 (1962) 233-9.

[54] L. Febvre, *The Coming of the Book* trans. D. Gerard (London, 1976) pp. 45-9; A.M. Hind, *Introduction to a History of Woodcut* (1935, repr. N.Y., 1963) vol. 1 ch. 3; discussion in W.M. Ivins, *Prints and Visual Communication* (Cambridge, Mass, 1953) ch. 2.

[55] Edgerton, 'Renaissance development', at pp. 187, 191; see *Heritage* pp. 277, 279.

[56] M.S. Mahoney, 'Diagrams and dynamics: Mathematical perspectives on Edgerton's thesis', in Shirley and Hoeniger, *Science and the Arts in the Renaissance*, 198-220.

[57] R.J. Miller, 'Cross-cultural research in the perception of pictorial materials', *Psychological Bulletin* 80 (1973) 135-50; J.B. Deregowski, *Illusions, Patterns and Pictures: A Cross-Cultural Perspective* (London, 1980).

[58] L. Haselberger, 'The construction plans for the Temple of Apollo at Didyma', *Scientific American* 253:6 (1985) 126-32.

[59] J. Gimpel, *The Cathedral Builders*, trans. T. Waugh (London, 1988) p. 101-2.

[60] Gimpel, p. 117.

[61] Alberti, *On the Art of Building in Ten Books* ed. and trans. J. Rykwert, N. Leach and R. Tavernor (Cambridge, Mass, 1988) p. 34.

[62] C.H. Thompson, *Fundamentals of Pipe Drafting* (N.Y., 1958) p. 18.

[63] An example with the angles right in Ambrogio Lorenzetti, *Allegory of Good Government*, Palazzo Pubblico, Siena (reproduced in J.G. Links, *Townscape Painting and Drawing* (London, 1972) pp. 14-5).

[64] M. Monmonier, *How to Lie with Maps* (Chicago, 1991).

[65] D. Woodward, 'Medieval *mappaemundi*', ch. 18 of *The History of Cartography*, vol. 1, ed. J.B. Harley and D. Woodward (Chicago, 1987).

[66] *History of Cartography*, p. 339; E. Kitzinger, 'World map and Fortune's wheel: A medieval mosaic floor in Turin', *Proceedings of the American Philosophical Society* 117 (1973) 344-73.

[67] *History of Cartography*, plate 38.

[68] B. Eastwood, 'Plinian astronomical diagrams in the early Middle Ages,' in E. Grant and J.E. Murdoch (eds), *Mathematics and its Applications to Science and Natural Philosophy in the Middle Ages* (Cambridge, 1987) 141-72.

[69] *The Opus Majus of Roger Bacon*, trans. R.B. Burke (1928, repr. N.Y., 1962) vol. 1 p. 315; see *History of Cartography*, p. 322.

[70] S.Y. Edgerton, *The Renaissance Rediscovery of Linear Perspective* (N.Y., 1975) ch. 7.

[71] T. Campbell, 'Portolan charts from the late thirteenth century to 1500', ch. 19 of *The History of Cartography*.

[72] P.D.A. Harvey, 'Local and regional cartography in medieval Europe', ch. 20 of *The History of Cartography*.

[73] J. Pinto, 'Origins and development of the ichnographic city plan', *Journal of the Society of Architectural Historians* 35 (1976) 35-50.

[74] J.D. North, 'The astrolabe', *Scientific American* 230:1 (Jan. 1974) 96-106; R.B. Thomson, *Jordanus de Nemore and the Mathematics of Astrolabes* (Toronto, 1978); Chaucer, *Chaucer on the Astrolabe: with the original illustrations* (2nd ed., Oxford, 1931); see J.D. North, 'Coordinates and categories: The graphical representation of functions in medieval astronomy', in Grant and Murdoch, *Mathematics and its Applications*, 173-88.

[75] D.J. de Solla Price, *Gears from the Greeks: the Antikythera Mechanism* (N.Y., 1975).

[76] *The Equatorie of the Planetis*, ed. D.J. Price (Cambridge, 1955); E. Poulle, *Equatoires et horlogerie planetaire du XIIIe au XVIe siècle* (2 vols, Geneva, 1980).

[77] *Nicole Oresme and the Kinematics of Circular Motion: Tractatus de commensurabilitate vel incommensurabilitate motuum celi*, ed. and trans. E. Grant (Madison, Wisc, 1971) p. 295; N. Oresme, *Le livre du ciel et du monde*, ed. A.D. Menut and A.J. Denomy (Madison, Wisc, 1968) p. 288; cf. L. Thorndike, *History of Magic and Experimental Science*, vol. 4 (N.Y., 1934) p. 169; see also D. de Solla Price, 'Automata and the origins of mechanism and mechanistic philosophy', *Technology and Culture* 5 (1964) 9-23.

[78] Ch. 1, Thorndike, *Sphere*, p. 119; earlier refs in A. Funkenstein, *Theology and the Scientific Imagination* (Princeton, 1986) p. 317 n. 52.

[79] Cf. E.L. Edwardes, *Weight Driven Chamber Clocks of the Middle Ages and Renaissance* (Altrincham, 1965) pp. 60-2, 78-80.

[80] Galileo, *Operations of the Geometric and Military Compass,* trans. S. Drake (Washington, DC, 1978); S.A. Bedini, 'The instruments of Galileo Galilei', in E. McMullin, *Galileo: Man of Science* (N.Y. 1967) 256-92.

[81] J.A. Bennett, 'The mechanics' philosophy and the mechanical philosophy', *History of Science* 24 (1986) 1-28; D.J. Bryden, *Napier's Bones* (London, 1992).

[82] E.A. Moody, *The Medieval Science of Weights* (Madison, Wisc, 1960) especially p. 139.

[83] S. Stevin, *De Beghinselen der Weeghconst* (Leyden, 1586).

[84] H.A. Simon, *The Sciences of the Artificial* (1st ed., Cambridge, Mass, 1969) pp. 1-2; R. Laymon, 'Thought experiments by Stevin, Mach and Gouy: Thought experiments as ideal limits and as semantic domains', in T. Horowitz and G.J. Massey (eds.), *Thought Experiments in Science and Philosophy* (Pittsburgh, 1991) 167-91; J.R. Brown, *The Laboratory of the Mind: Thought Experiments in the Natural Science* (N.Y., 1991) pp. 3-6.

[85] Galileo, *Dialogues Concerning Two New Sciences*, Second day, in *Opere*, ed. Favaro, vol. 8 p. 159, trans. S. Drake (Madison, Wisc, 1974) p. 117.

[86] J. Ackerman, 'Ars sine scientia nihil est', *Art Bulletin* 31 (1949) 84-111.

[87] H. Saalman, *Filippo Brunelleschi: The Cupola of Santa Maria del Fiore* (London, 1980) ch. 3; S. Sanpaolesi, 'Ipotesi sulle conoscenze matematiche, statiche e mecchaniche del Brunelleschi', *Belle Arti* 2 (1951) 25-54.

[88] J.H. Pryor, *Geography, Technology and War* (Cambridge, 1988) pp. 32-5, 40-2.

[89] T.L. Heath, trans., *The Thirteen Books of Euclid's Elements* (N.Y. 1956) vol. 1 p. 242.

[90] Kant, *Untersuchung über die Deutlichkeit der Gründsätze der natürlichen Theologie und der Morale* First Reflection §1 (Berlin Academy ed.) vol. II, pp. 276-8, Kant, *Selected Pre-Critical Writings* trans. G.B. Kerferd and D.W. Walford (Manchester, 1968) pp. 6-8; *Critique of Pure Reason*, B 741; see J. Franklin, 'Artifice and the natural world: Mathematics, logic, technology', in *Cambridge History of Eighteenth Century Philosophy*, ed. K. Haakonssen, to appear.

[91] See B.L. Ullman, 'Geometry in the medieval quadrivium', *Studi di bibliografia e di storia in onore di Tammaro de Marinis*, IV (Verona, 1964) pp. 263-85; L.R. Shelby, 'Geometry', ch. 8 of *The Seven Liberal Arts in the Middle Ages*, ed. D.L. Wagner (Bloomington, Ind, 1983).

[92] T.L. Heath, trans, *The Thirteen Books of Euclid's Elements* (N.Y., 1956) vol. 1 pp. 415-6.

[93] E.S. Ferguson, 'The mind's eye: Non-verbal thought in technology', *Science* 197 (1977) 827-36, at p. 833.

[94] See A. Arnauld, *The Art of Thinking*, trans. J. Dickoff and P. James (Indianapolis, 1964) p. 21.

[95] J. Needham and Wang Ling, *Science and Civilization in China* vol. 3 sections 19-25 (Cambridge, 1959) pp. 150-68; but see J. Needham, *The Grand Titration: Science and Society in East and West* (Toronto, 1969) pp. 41-51.

[96] *China in the Sixteenth Century: The Journals of Matthew Ricci,* trans. L.J. Gallagher (N.Y., 1953) p. 476.

[97] G.G. Joseph, *The Crest of the Peacock: Non-European Roots of Mathematics* (Harmondsworth, 1991); see review by J. Franklin in *Metascience* NS 2 (1993) 97-8; Li Yan and Du Shiran, *Chinese Mathematics: A Concise History*, trans. J.N. Crossley and A.W.-C. Lun (Oxford, 1987) pp. 194-5.

[98] S.K. Victor, *Practical Geometry in the High Middle Ages* (Philadelphia, 1972); R. Shelby, 'The geometrical knowledge of the medieval master masons', *Speculum* 47 (1972) 395-421; cf. O.A.W. Dilke, *The Roman Land Surveyors* (Newton Abbot, 1971).

[99] Hugh of St Victor, *Practica geometriae,* in *Opera propaedeutica*, ed. R. Baron (Notre Dame, Ind, 1966) p. 16; cf. Hugh of St Victor, *Didascalicon*, bk. 2 ch. 13 (trans. J. Taylor, N.Y., 1961, p. 70).

[100] G. Beaujouan, 'Réflexions sur les rapports entre théorie et pratique au moyen âge', in J. Murdoch and E. Sylla (eds.), *The Cultural Context of Medieval Learning* (Dordrecht, 1975) 437-84; more fully in G. Beaujouan, *Par raison de nombres: L'art du calcul et les savoirs scientifiques médiévaux.* (Aldershot, 1991).

[101] Bartolus, *De fluminibus seu Tiberiadis* (Rome, 1483); *Tyberiadis* (Bologna, 1576); *La Tiberiade di Bartole da Sasoferato del modo di dividere l'alluuione, l'isole and l'aluei* (Rome, 1579); see J. van Maanen, 'Teaching geometry to 11 year old 'medieval lawyers', *Mathematical Gazette* 76 (1992) 37-45.

[102] Bartolus, *De testimoniis, Opera Omnia* (Venice, 1615) vol. 10 fol. 160v.

[103] Euclid, *Elements* bk XI prop. 31, from Ms. Vat. gr. 190 (Ninth century) vol. 2 fol. 207v. The page may be viewed in the Vatican Library exhibit at:
http://sunsite.unc.edu/expo/vatican.exhibit/exhibit/d-mathematics/images/math22.jpg
The corresponding diagram in Health, vol. 3 p. 337, is somewhat adjusted. Other 3D figures from the same manuscript are in Murdoch, *Album of Science*, illustrations 117, 122.

[104] *Didascalicon,* bk. 2 ch. 14 (Taylor, p. 70).

[105] L. Thorndike, *The Sphere of Sacrobosco and its Commentators* (Chicago, 1949) p. 118.

[106] O. Gingerich, 'Sacrobosco as a textbook', *J. for the History of Astronomy* 19 (1988) 269-73.

[107] *Mandeville's Travels*, ch. 20.

[108] Dante, *Inferno* canto XXXIV lines 100-13.

[109] Dante, *Paradiso* canto XXXIII lines 133-45; see T.E. Hart, 'Geometric metaphor and proportional design in Dante's *Commedia*', in G. di Scipio and A. Scaglione (eds), *The Divine Comedy and the Encyclopedia of Arts and Sciences* (Amsterdam, 1988) 95-146.

[110] Dante, *Paradiso* XXVII, 81-3; see Lewis, *Discarded Image*, ch. 5.

[111] See H. Nemerov, 'The backward look', in D.J. Enright, *Oxford Book of Contemporary Verse, 1945-1980*, (Oxford, 1980) p. 118.

[112] K.S. Guthke, *The Last Frontier—Imagining Other Worlds, from the Copernican Revolution to Science Fiction*, trans. H. Atkins (Ithaca, N.Y., 1990).

[113] Reproduced in, e.g. Edgerton, *Heritage*, p. 152.

[114] Edgerton, pp. 151-2; cf. V.J. Flint, *The Imaginative Landscape of Christopher Columbus* (Princeton, 1992); S.D. Westrem (ed.), *Discovering New Worlds: Essays on Medieval Exploration and Imagination* (N.Y., 1991).

[115] N. Rescher, 'Thought experimentation in Presocratic philosophy', in Horowitz and Massey, *Thought Experiments*, 31-41.

[116] *OED*, s.v. 'revolution'.

[117] L. Tarán, *Parmenides* (Princeton, 1965) pp. 296-8; A.H. Coxon, *The Fragments of Parmenides* (Assen, 1986) p. 229; Aristotle, *On the Heavens* (II.xiv); O. Neugebauer, *A History of Ancient Mathematical Astronomy* (Berlin, 1975) pp. 109-12.

[118] A. Jones, 'The adaptation of Babylonian methods in Greek numerical astronomy', *Isis* 82 (1991) 441-53; G.J. Toomer, 'Hipparchus and Babylonian astronomy', in E. Leichty, M. DeJ. Ellis and P. Gerardi (eds), *A Scientific Humanist: Studies in Memory of Abraham Sachs* (Philadelphia, 1988) 353-62.

[119] D. C. Lindberg, *Theories of Vision from al-Kindi to Kepler* (Chicago, 1976).

[120] E. Grant, *A Source Book in Medieval Science* (Cambridge, Mass, 1974) pp. 435-41; Murdoch, *Album of Science*, illustration 138; W. Wallace, *The Scientific Methodology of Theodoric of Freiberg* (Fribourg, 1959).

[121] M. Kubovy, *The Psychology of Perspective and Renaissance Art* (Cambridge, 1986).

[122] Ghiberti, *Second Commentary*, selection in E.G. Holt, *A Documentary History of Art* (Garden City, N.Y., 1957) vol. 1, p. 154.

[123] C. Wright, *Perspective in Perspective* (London, 1983) p. 38; cf. A.D. Trendall and T.B.L. Webster, *Illustrations of Greek Drama* (London, 1971) III. 3,43; R. Tobin, 'Ancient perspective and Euclid's *Optics*', *Journal of the Warburg and Courtauld Institutes* 53 (1990) 14-41; K. Andersen, 'Ancient roots of linear perspective', in J.L. Berggren and B.R. Goldstein (eds), *From Ancient Omens to Statistical Mechanics* (Copenhagen, 1987) 75-89.

[124] Duccio, *Maesta: The Virgin and St John* in *L'opere completa di Duccio*, ed. G. Cattaneo and E. Baccheschi (Milan, 1972).

[125] Giotto, *Jesus before Caiaphas*, Scrovegni Chapel, Padua (in A. Martindale and E. Buccheschi (eds), *The Complete Paintings of Giotto* (N.Y., 1966) no. 82); compare Giotto (?) *Innocent III Approving the Order*, Upper Church of St Francis, Assisi.

[126] Giotto, *The Wedding Feast of Cana*, Scrovegni Chapel, Padua (in *Complete Paintings*, plate XXVI); see also *The Wedding Feast of Cana* in the Upper Church of St Francis, Assisi.

[127] See Martindale and Baccheschi, *Complete Paintings*, pl. XXXVI. Also pictured in context in Edgerton, *Heritage*, p. 78.

[128] M. Clagett, 'Conic sections in the fourteenth century', in A. Maierù and A. Paravicini Bagliani (eds), *Studi sul xiv secolo in memoria di Anneliese Maier* (Rome, 1981) 179-217.

[129] Cf. Pappus, *Collections*, bk VI proposition 53; W. Knorr, 'When circles don't look like circles: an optical theorem in Euclid and Pappus', *Archive for History of Exact Sciences* 44 (1992) 287-329.

[130] E.g. Witelo's *Perspective*, book I propositions 112, 115 (in *Witelonis Perspectivae Liber Primus*, ed. and trans. S. Unguru, Wroclaw, 1977, pp. 127, 133); cf. Weitzmann, *Ancient Book Illumination*, Fig. 6.

[131] Kepler, *New Astronomy* Introduction (trans. W.H. Donahue, Cambridge, 1992) p. 45.

[132] M. Kemp, *The Science of Art* (New Haven, 1990) pp. 10-1.

[133] C. Cennini, *The Craftsman's Handbook*, trans. D.Y. Thompson (New Haven, Con, 1933, repr. N.Y., 1960) ch. 67.

[134] Kubovy, *The Psychology of Perspective*, ch. 2.

[135] Euclid, postulate 4; L. Alberti, *On Painting*, trans. J.R. Spencer (revised ed., London, 1966) p. 45.

[136] Alberti, p. 52

[137] P. 59; cf. J. Elkins, 'Renaissance perspectives', *J. of the History of Ideas* 53 (1992) 209-30.

[138] Original in *Lorenzo Ghibertis Denkwürdigkeiten,* ed. and trans. J. von Schlosser (Berlin, 1912); discussion in G. ten Doesschate, *De derde commentaar van Lorenzo Ghiberti in verband met de middeleeuwsche optiek* (Utrecht, 1940); G. Federici Vescovini, *Studi sulla prospettiva medievale* (Turin, 1987) especially chs 11-12, summarised in G. Federici Vescovini, 'La fortune de l'optique d'ibn al-Haitham', *Archives internationales d'histoire des sciences* 40 (1990) 220-38; briefly in Kemp, *Science of Art* pp. 26-7.

[139] Ghiberti, *Second Commentary*, in E.G. Holt, *Documentary History*, vol. 1 p. 156.

[140] Galileo, *The Assayer*, section XIII, in *The Controversy on the Comets of 1618,* trans S. Drake and C.D. O'Malley (Philadelphia, 1960) p. 213.

[141] T. Kuhn, 'Mathematical versus experimental traditions in the development of science', ch. 3 of *The Essential Tension* (Chicago, 1977).

[142] Murdoch, *Album of Science*, illustration 123.

[143] Euclid, *The Elements of Geometrie* trans. H. Billingsley (London, 1570, repr. Ann Arbor, 1967) Preface; cf. W.A. Wallace, *Galileo and His Sources* (Princeton, 1984) pp. 138, 145.

[144] J.W. Shirley, 'Science and navigation in Renaissance England', in Shirley and Hoeniger, *Science and the Arts in the Renaissance*, 74-93; cf. John Dee, *Perfect Arte of Navigation* (1577, repr. Amsterdam, 1968); M. Boas, *The Scientific Renaissance* (London, 1962) ch. 7.

[145] J.R. Hale, *Renaissance Fortification: Art or Engineering*? (London, 1977).

[146] J. Gascoigne, 'A reappraisal of the role of the universities in the Scientific Revolution', in D.C. Lindberg and R.S. Westman (eds), *Reappraisals of the Scientific Revolution* (Cambridge, 1990) 207-60.

[147] Aristotle, *On Memory and Reminiscence* 450a27-30.

[148] G. Ryle, *The Concept of Mind* (London, 1949) p. 254.

[149] J.L. Austin, *Sense and Sensibilia* (Oxford, 1962) ch. 3; D.M. Armstrong, *Perception and the Physical World* (London, 1961) ch. 2.

[150] Armstrong, *Perception and the Physical World*, ch. 9; Z. Pylyshyn, 'What the mind's eye tells the mind's brain: A critique of mental imagery', *Psychological Bulletin* 80 (1973) 1-24.

[151] E.S. Ferguson, 'The mind's eye: Non-verbal thought in technology', *Science* 197 (1977) 827-36.

[152] R.S. Root-Bernstein, 'Visual thinking: The art of imagining reality', *Transactions of the American Philosophical Society* 75 (1985) part 6: 50-67; A.I. Miller, 'Visualization lost and regained: The genesis of the quantum theory in the period 1913-1927', in J. Wechsler (ed), *On Aesthetics in Science* (Cambridge, Mass, 1978) pp. 72-102; D. Hilbert and S. Cohn-Vossen, *Geometry and the Imagination* (N.Y., 1952).

[153] J. Hadamard, *The Psychology of Invention in the Mathematical Field* (Princeton, 1945) pp. 142-3; cf. A.I. Miller, *Imagery in Scientific Thought: Creating 20th-Century Physics* (Boston, 1984); R. Penrose, *The Emperor's New Mind* (Oxford, 1989) pp. 548-50; J. Gleick, *Genius: Richard Feynman and Modern Physics* (London, 1992) pp. 131, 244-5.

[154] R.N. Shepard and J. Metzler, 'Mental rotation of three-dimensional objects', *Science* 171 (1971) 701-3; R.N. Shepard and L.A. Cooper, *Mental Images and Their Transformations* (Cambridge, Mass, 1982).

[155] N. Franklin and B. Tversky, 'Searching imagined environments', *J. of Experimental Psychology: General* 119 (1990) 63-76; B.J. Bryant, B. Tversky and N. Franklin, 'Internal and external frameworks for representing described scenes', *J. of Memory and Language* 31 (1992) 74-98.

[156] B. Tversky, 'Spatial mental models', *The Psychology of Learning and Motivation* 27 (1991) 109-45.

[157] R.M.J. Byrne and P.N. Johnson-Laird, 'Spatial reasoning', *J. of Memory and Language* 28 (1989) 564-75.

[158] S. Pinker, 'Mental imagery and the third dimension', *J. of Experimental Psychology: General* 109 (1980) 354-71.

[159] R.A. Earnshaw and N. Wiseman, *An Introductory Guide to Scientific Visualization* (N.Y., 1992); G.M. Nielson and B. Shriver (eds), *Visualization in Scientific Computing* (Los Alamitos, Ca, 1990); *IEEE Computer Graphics and Applications* 11 (3) (May, 1991) special issue on visualization; E.R. Tufte, *Envisioning Information* (Cheshire, Conn, 1990); W. Bown, 'New-wave mathematics', *New Scientist* 131 (3 Aug 1991) 31-5.

[160] Many refs in K. Rochford, A.P. Fairall, A. Irving and P. Hurly, 'Academic failure and spatial visualization handicap of undergraduate engineering students', *International J. of Applied Engineering Education* 5 (1989) 741-9; A.J. Bishop, 'Review of research on visualization in mathematics education', *Focus on Learning Problems in Mathematics* 11 (1-2)(Win-Spr, 1989) 7-16.

[161] M. Tye, *The Imagery Debate* (Cambridge, Mass, 1991); T.P. McNamara, 'Memory's view of space', *The Psychology of Learning and Motivation* 27 (1991) 147-86.

[162] W.H. Leatherdale, *The Role of Analogy, Model and Metaphor in Science* (Amsterdam, 1974) pp. 117, 200.

[163] Albertus Magnus, *De bono,* Tract. IV q. 2 art. 2 ad 13, in *Opera Omnia* ed. H. Kühle *et al.* (1951) at p. 251, described in Yates, *Art of Memory* p. 64; trans. in Carruthers, *Book of Memory*, p. 279.

[164] De Anima 431a17, cf. 432a8.

[165] 427b28-429a9; see D.W. Modrak, *Aristotle: The Power of Perception* (Chicago, 1987) ch. 4; M.W. Bundy, *The Theory of Imagination in Classical and Medieval Thought* (Urbana, Ill, 1926) ch. 3; M. Schofield, 'Aristotle on the imagination', in J. Barnes, M. Schofield and R. Sorabji (eds), *Articles on Aristotle: Psychology and Aesthetics* (London, 1979) pp. 103-32; M.V. Wedin, *Mind and Imagination in Aristotle* (Yale, 1988) chs 2-3; R. Lefebvre, 'Aristote, l'imagination et le phénomène', *Phronesis* 37 (1992) 22-45.

[166] Bundy, chs 4, 6, 7; Liddell and Scott, *Greek-English Lexicon*, s.v. φαντασία, φαντασιόω; cf. Epicurus in Diogenes Laertius, *Lives of Eminent Philosophers* bk X.33; A. Sheppard, 'Phantasia and mental images: Neoplatonist interpretations of *De Anima 3.3'*, *Oxford Studies in Ancient Philosophy* supplementary vol, 1991, pp. 165-73; E.K. Emilsson, *Plotinus on Sense-Perception* (Cambridge, 1988) pp. 107-12; G. Watson, *Phantasia in Classical Thought* (Galway, 1988).

[167] Augustine, *De Genesi ad Litteram*, bk XII, 24, quoted in Aquinas, *Summa Theologiae*, part I q. 84 art. 6.

[168] Augustine, *De Trinitate* XI, ch. 8.

[169] *City of God*, XI ch. 26.

[170] A Kenny, *Aquinas* (Oxford, 1980) ch. 3; A. Kenny, 'Intellect and imagination in Aquinas', in *Aquinas: A Collection of Critical Essays* (Garden City, N.Y., 1969) 273-96; H.A. Wolfson, 'The internal senses', *Harvard Theological Review* 28 (1935) 69-133; E.R. Harvey, *The Inward Wits* (London, 1975); Bundy, ch. 9; E.P. Mahoney, 'Sense, intellect and imagination in Albert, Thomas and Siger', ch. 30 of *The Cambridge History of Later Medieval Philosophy*, ed. N. Kretzmann *et al.* (Cambridge, 1982); N.H. Steneck, 'Albert on the psychology of sense perception', in J.A. Weisheipl (ed), *Albert Magnus and the Sciences* (Toronto, 1980) 263-90; *On the Properties of Things: John Trevisa's Translation of Bartholomaeus Anglicus De Proprietatibus Rerum*, bk 3 ch. x-xi, ed. M.C. Seymour (Oxford, 1975) pp. 98-9; *Phantasia—imaginatio: Vo Colloquio internazionale del Lessico intelletuale europeo, Rome, 1986* ed. M. Fattori and M. Bianchi (Rome, 1988); refs to Avicenna's originals in D.L. Black, *Logic and Aristotle's Rhetoric and Poetics in Medieval Arabic Philosophy* (Leiden, 1990) p. 202 n. 66; antecedents of Avicenna in R.J. Hankinson, 'Galen's anatomy of the soul', *Phronesis* 36 (1991) 197-233.

[171] Collegium Conimbricensis, *Commentarii in duos libros de Generatione et Corruptione* (Conimbricae, 1597) Lib. I cap. 4 q. 30 art. 2, quoted in E. Gilson, *Index Scolastico-Cartésien* (Paris, 1912) p. 140; further refs on the force of ideas on foetuses in B. Hansen, *Nicole Oresme and the Marvels of Nature* (Toronto, 1985) p. 346 n. 131 and L. Demaitre and A.A. Travill, 'Human embryology and development in the works of Albertus Magnus', in Weisheipl, *Albertus Magnus*, pp. 405-40, at p. 435 n. 118.

[172] J. Decety, M. Jeannerod, M. Germain and J. Pastene, 'Vegetative response during imagined movement is proportional to mental effort', *Behavioural Brain Research* 42 (1991) 1-5.

[173] Augustine, *Quaestiones 83*, q.9, quoted in Aquinas, *Summa Theologiae* I q. 84 a. 6.

[174] Augustine, *De Genesi ad Litteram*, XII, 15, cf. *City of God*, XVIII, ch. 18; Thomas Aquinas, *Quaestiones disputatae de malo*, q. 16 art. 11.

[175] H. Kramer and J. Sprenger, *Malleus Maleficarum*, trans. M. Summers (London, 1928, repr. N.Y., 1971) pp. 58-9, 119, 125; cf. James VI and I, *Daemonologie* (Edinburgh, 1597, repr. Edinburgh, 1966) pp. 79-80.

[176] *Nicole Oresme and the Medieval Geometry of Qualities and Motions*, ed. and trans. M. Clagett (Madison, 1968) pp. 484-5.

[177] References in *Cambridge History of Renaissance Philosophy*, ed. C.B. Schmitt (Cambridge, 1988) p. 288.

[178] Cf. Dante, *Purgatorio* canto XVII lines 22-5.

[179] Shakespeare, *Love's Labour's Lost* IV.ii.66-69.

[180] Sidney, *Apology for Poetry*, ed. G. Shepherd (London, 1965) pp. 107-9; cf. Aristotle, *Poetics* 1451a36-b11; on the connection between 'imagination' in poetry and older ideas see generally J.M. Cocking, *Imagination: A Study in the History of Ideas* (London, 1991); M. Warnock, *Imagination* (London, 1976); D. Kelly, *Medieval Imagination; Rhetoric and the Poetry of Courtly Love* (Madison, 1978).

[181] Black, *Logic and Aristotle's Rhetoric and Poetics*, chs 6-7.

[182] Cicero, *De oratore* II.1xxxvi.351-4.

[183] F. Yates, *The Art of Memory* (London, 1966, 2nd ed., 1992); confirmed in H. Blum, *Die Antike Mnemotechnik* (Hildesheim, 1969); Carruthers, *Book of Memory* and J. Coleman, *Ancient and Medieval Memories* (Cambridge, 1992).

[184] Cicero, *De oratore* II.lxxxvii.357.

[185] Carruthers, *Book of Memory*, pp. 243-7.

[186] *Ibid.*, pp. 231-9.

[187] J.D. Spence, *The Memory Palace of Matteo Ricci* (N.Y., 1984).

[188] Viglius to Erasmus, quoted in Yates, Art of Memory, pp. 131-2.

[189] D. Lenat and E. Feigenbaum, 'On the thresholds of knowledge', *Artificial Intelligence* 47 (1991) 185-230.

[190] Evans, 'Geometry of the mind', section 6.4; pictures of Hildegard's visions in C. Singer, *From Magic to Science* (London, 1928) ch. 6.

[191] Richard of Saint Victor, *Benjamin Minor* (in Migne, *Patrologia Latina* vol 196 cols 1-64; Middle English translation in *Deonise hid Diuinite and Other Treatises on Contemplative Prayer*, ed. P. Hodgson (Early English Text Society, no. 231, Oxford, 1955) pp. 12-46.

[192] Peter of Limoges, *De oculo morali* cap. VII, quoted in D.L. Clark, 'Optics for preachers: The *De oculo morali* of Peter of Limoges', *Michigan Academician* 9 (1977) 329-43, at p. 338; see *The Book of Margery Kempe*, ch. 14 (ed. S.B. Meech and H.E. Allen, Early English Text Society, no.212, Oxford, 1940, pp. 29-30); *The Cloud of Unknowing*, ch. 65 (ed. P. Hodgson, Early English Text Society, no. 218, Oxford, 1944, pp. 117-8); A.J. Minnis, 'Langland's Ymaginatif and late-medieval theories of imagination', *Comparative Criticism: A Yearbook* 3 (1981) 71-103; Lydgate, 'The fifteen joys and sorrows of Mary', discussed in P. De Wit, *The Visual Experience of Fifteenth-Century English Readers* (D. Phil. dissertation, Oxford University, 1977) pp. 24-8.

[193] *The Spiritual Exercises of St. Igantius*, second week, first day, second contemplation, (trans. A. Mottola, N.Y., 1964, p. 71; cf. p. 54, etc.); cf. D.C. Steinmetz, 'Luther and Loyala: An exploration of the human imagination as an instrument of spiritual nurture and theological reform', *Interpretation* 47 (1993) 5-14.

[194] Teresa of Avila, *The Interior Castle*, especially I.i.4 and VI.ix.6.

[195] Lewis, *Discarded Image*, p. 163.

[196] Plato, *Republic* 509-11; R.S. Brumbaugh, *Plato's Mathematical Imagination* (Bloomington, Ind, 1954) ch. 3.

[197] Aristotle, *De memoria* 452b7-453a4.

[198] Proclus, *Commentary on the First Book of Euclid's Elements* trans. G.R. Morrow (Princeton, 1970) pp. 41-5; cf. Aristotle, *Metaphysics* 1036a9-12 and S. Gaukroger, 'Aristotle on intelligible matter', *Phronesis* 25 (1980) 187-97; J.F. Jones, 'Intelligible matter and geometry in Aristotle', *Apeiron* 17 (1983) 94-102.

[199] *Keplers Gesammelte Werke*, vol VI, ed. M. Caspar (Munich, 1940) pp. 218-21.

[200] A Piccolomini, *Peripateticae de anima disputationes* (Venice, 1575) fol. 95r, quoted in *Cambridge History of Renaissance Philosophy*, p. 694.

[201] Alhazen, *Opticae thesaurus* (Basel, 1572) p. 15.

[202] Chaucer, *Treatise on the Astrolabe*, Part II, section 39 and Part I section 21.

[203] Al-Jazari, *The Book of Knowledge of Ingenious Mechanical Devices*, ed. and trans. D.R.H. Hill (Dordrecht, 1974) p. 192.

[204] Cf. Edgerton, *Heritage*, p. 32.

[205] J. Murdoch, 'From social into intellectual factors: An aspect of the unitary character of late medieval learning', in Murdoch and Sylla, *The Cultural Context of Medieval Learning*, 271-348, at pp. 292, 297; E. Sylla, 'Mathematical physics and the imagination in the work of the Oxford Calculators', in Grant and Murdoch, *Mathematics and its Applications*, pp. 69-101.

[206] P. King, 'Medieval thought-experiments: The metamethodology of medieval science', in Horowitz and Massey, *Thought Experiments*, 43-64.

[207] Heytesbury, *Regulae solvendi sophismata*, in *Tractatus Gulielmi Hentisberi de sensu composito et diviso ...* (Venice, 1494) fol. 43v; E. Sylla, 'The Oxford Calculators', ch. 27 of *The Cambridge History of Later Medieval Philosophy* ed. N. Kretzmann *et al.*, at pp. 557-8; C. Wilson, *William Heytesbury: Medieval Logic and the Rise of Medieval Physics* (Madison, 1960) pp. 24-5; Funkenstein, *Theology and the Scientific Imagination*, pp. 164-74.

[208] *Quodlibeta Magistri Henrici Goethals a Gandavo*, quodlibet II, q. 9 (Paris, 1518, fol. 36r) quoted in A.G. Molland, 'Colonizing the world for mathematics: The diversity of medieval strategies', in Grant and Murdoch, *Mathematics and its Applications*, 45-66, at p. 60.

[209] *Nicole Oresme and the Medieval Geometry of Qualities and Motions*, ed. and trans. M. Clagett (Madison, 1968) pp. 165-7.

[210] Oresme, *Quaestiones super geometriam Euclidis*, q. 10 conclusio 3 (ed. Busard, p. 27).

[211] Oresme, *Questions on the Geometry of Euclid*, q. 11, quoted in Clagett, p. 537.

[212] Discussed in Clagett, pp. 50-4.

[213] Aristotle, *Metaphysics* 1032b32.

[214] Alberti, *On the Art of Building*, bk 1 ch. 1, Rykwert, Leach and Tavernor, p. 7.

[215] Galileo, *On Motion* ch. 8 (*Opere*, ed. A. Favaro, vol. 1, pp. 263-5, trans. I.E. Drabkin, Madison, 1960, pp. 27-9); A. Koyré, 'Galileo's treatise 'De motu gravium': the use and abuse of imaginary experiment', in A. Koyré, *Metaphysics and Measurement* (London, 1968) pp. 44-48.

[216] Aquinas, *Summa theologiae*, bk. I q. 78 art. 4.

[217] On which see M. Clavelin, 'Conceptual and technical aspects of the Galilean geometrization of the motion of heavy bodies', in W.R. Shea (ed.), *Nature Mathematized* (Dordrecht, 1983) 23-50.

[218] Funkenstein, *Theology and the Scientific Imagination*, pp. 174-9.

[219] Galileo, *Dialogue Concerning the Two Chief World Systems*, second day (*Opere*, vol. 7 p. 172; trans. S. Drake, Berkeley, 1953, p. 145).

[220] Kuhn, *The Essential Tension*, p. 42.

[221] R.P. Feynman, R.B. Leighton and M. Sands, *The Feynman Lectures on Physics* (Reading, Mass, 1963-5) II-20-10.

[222] Aquinas, *Summa theologiae*, bk. I q. 78 art. 4.

[223] *Dialogue*, second day (*Opere*, vol. 7 pp. 217-9, Drake, pp. 190-1); cf. W.R. Shea, *Galileo's Intellectual Revolution* (London, 1972) pp. 154-5; see R. Sorenson, *Thought Experiments* (N.Y., 1992) pp. 88-92.

[224] *Dialogue*, third day (*Opere*, p. 403, Drake, p. 376; and *Opere*, pp. 350-2, Drake, pp. 322-5).

[225] M. McCloskey, 'Intuitive physics', *Scientific American* 248:4 (Apr. 1983) 114-22; M. McCloskey and D. Kohl, 'The curvilinear impetus principle and its role in interacting with moving objects', *J. of Experimental Psychology: Learning, Memory and Cognition* 9 (1983) 146-

56; M.K. Kaiser, M. McCloskey and D.R. Proffitt, 'Development of intuitive theories of motion-curvilinear motion in the absence of external forces', *Developmental Psychology* 22 (1986) 67-71; cf. L.B. Flick, 'Interaction of intuitive physics with computer-simulated physics', *J. of Research in Science Teaching* 27 (1990) 219-31.

226 In immense detail in C. Fabro, *God in Exile*, trans. and ed. A. Gibson (Westminster, Md, 1968).

227 Descartes, *Discourse on the Method* part 2 (*Oeuvres de Descartes*, ed. C. Adam and P. Tannery, revised ed., Paris, 1964-76, vol. VI p. 17; *Philosophical Writings of Descartes*, trans. J. Cottingham, R. Stoothoff and D. Murdoch, vol. 1 p. 119).

228 Descartes, *Rules for the Direction of the Mind*, (AT X 438; CSM 1 56); see E. Pastini, 'Mathesis und Phantasie: Die Rolle der Einbildungskraft im Unfeld der Descartesschen *Regulae*', *Studio Leibnitiana* 42 (1992) 159-76.

229 Descartes, *Entretien avec Burman*, AT V 176-7, *Descartes' Conversation with Burman*, trans. J. Cottingham (Oxford, 1976) par. 79.

230 Descartes, *Meditations on First Philosophy*, sixth meditation (AT VII 72; CSM 2 50-1); cf. *Objections and Replies*, Fifth set of objections (AT VII 330-2; CSM 2 229-30).

231 *Burman*, AT V 162-3, Cottingham, par. 42.

232 Descartes, *Objections and Replies*, Second set of replies (AT VII 160-1; CSM 2 113).

233 E. Gilson, *Index Scolastico-Cartésien* (Paris, 1912) pp. 137-40; J.H. Roy, *L'imagination chez Descartes* (Paris, 1944).

234 A. Kenny, *Descartes* (N.Y., 1968) pp. 105-10; J. Franklin, 'Achievements and fallacies in Hume's account of infinite divisibility', *Hume Studies* 29 (1994) 85-101.

235 Aristotle, *On Memory and Reminiscence* 450b23.

236 *Ibid.*, 453a10-13.

JOHN SUTTON

BODY, MIND, AND ORDER: LOCAL MEMORY AND THE CONTROL OF MENTAL REPRESENTATIONS IN MEDIEVAL AND RENAISSANCE SCIENCES OF SELF

1. NEITHER WORD NOR IMAGE: CONFUSION AND COGNITION IN HISTORY

Historical cognitive science works between two projects. One is the analysis of other and older theories of mind, of how they relate to and differ from current approaches, and of what forgotten or neglected explananda they bring into focus. The other, relating to cognitive practices rather than theories, is the task of working out how such views about mind and self reflect or partly cause different historical forms of mental activity. The delicate equilibrium to be maintained is between allowing for the plasticity in human cognition which anthropological and historical data can suggest, and yet remaining not just aware of but embedded in the diversity of approaches in contemporary theories of mind, in order to make the history effective and utilisable in the growing interdisciplinary environment. In specific domains, such as visual perception, dreams, emotion, inductive reasoning, or (as here) memory and learning, the shifting interdependencies of cognition and culture can be traced from two directions. Firstly, tensions can be addressed in many periods between social or moral norms and theoretical commitments concerning body, brain, and mind; then, more self-consciously present-centred inquiry can employ polemically, within cognitive science, the extra breadth, context-sensitivity, and attention to discontinuity which historical work requires.

This paper, then, is a tentative step, at a very general level, towards the proposal of one set of analytical devices for historical cognitive science. The domain is an area of problems about memory and personal identity which cross levels between philosophy of mind, 'psychology', neurophysiology and medicine. The case, crudely, is that embroiled with these problems have often been a series of related conceptual dichotomies or, better, continua between

G. Freeland and A. Corones (eds.), 1543 and All That, 117 – 150

order and confusion, independence and blending, distinctness and dissolution. Attitudes towards these constructions, having real enough effects in particular contexts, colour theoretical debates about self, memory, and brain not only at levels which are obviously metaphorical and rhetorical.

Visual, graphic, and textual supplements to human cognition changed form rapidly in the 'proto-scientific revolution', and in turn altered needs for and capacities of visualising and imagining. Martin Kemp and Jim Franklin in this volume confirm the complexity of relations in the period between the external media of pictorial and linguistic representation, and the cognitive styles of learning and reasoning which evolved with new modes of illustration and instrumentation. Kemp warns against the temptation to seek a single general theory of the parallels between mental representation or information-processing and external technology-dependent representation across different domains for this or any period. But still open are a set of strategies which I apply here to familiar enough material on the case of remembering, rather than visualising or imagining.

As Kemp notes, in any one period common aesthetic and rhetorical metaphors may be found in sciences which otherwise call diverse models into play. He cites repeated invocations of cosmic symmetry designed by the orderly divine artisan, a symmetry impossible to disturb in any part 'without producing confusion in all the other parts':[1] theoretical visualisation of astronomical and anatomical bodies was less bound by the limitations or peculiarities of existing instruments or illustrations than by metaphors and analogies embedded in the Renaissance aesthetics and ethics of decorum.[2] In the case of the branch of natural philosophy dealing in mental models of memory and mind (rather than of planets or body parts), the constraints of various social or moral conceptions of order and decorum permeated theory construction even more thoroughly.

Despite Renaissance rhetoric of cognitive order, and the hatred of 'evil mixture' with its anarchic psychological and social consequences, a contrasting and conflicting attraction to confusions and dissolutions also had its powers. In social, metaphysical, cognitive, and physiological domains, the perils and beauties of mixtures induced both fear and fascination. This is old news, at least within sixteenth and seventeenth century social, political, and literary histories: order/chaos dualisms structure influential accounts of 'Renaissance thought'.[3] Renewed attention to breakdowns of such dichotomies is a feature of recent interdisciplinary history.[4] But their implications and complications have been less studied in relation to memory and self. I want to bring to bear two lines of thought, familiar in other contexts, which seek to bridge gaps between minds and cultures. First is the treatment of theoretical models of memory as specimens of the way cultural norms and artefacts can permeate ('proto')scientific views of inner processes. Second is the application of this analysis to the particular area of psychological control over one's own body, brain, and mind.

Metaphors and models for memory and mental representation can signal the projection inside of external aids. Overtly at least, medieval and Renaissance theorists agreed that such models had to allow for, or even guarantee, some conception of cognitive order and discipline. In the case of memory, this entailed both that individual representations or traces should be independent or isolable, not mixed up or interfering with others, and that some sense be given to the experience of and desire for active control over remembering and associative mental processes. In section 2 I address these requirements as they appear in the various forms of the arts and techniques of place memory. Moral psychologies of memory were always linked to, and could come into tension with, alternative philosophical or physiological approaches which couldn't ensure in advance the subject's control over distinct items in memory.

This leads to the second point of departure. In a wonderful essay on Donne, Elaine Scarry has argued that Renaissance theory of the human body, the animation and entering of the body by science, religion, poetry, and language in general, is often an 'interiorisation of the artefactual', driven by a desire to revise the body in order to render it susceptible to voluntary control.[5] I apply this suggestion to the longer tradition of improving or bypassing 'natural memory' by deliberately internalising artefactual models. In the arduous processes of 'self-fashioning', boundaries between inner and outer, between bodily or psychological contents and cultural or technological items, could be shifted.[6] Coexisting inside theories and individuals were often two conflicting attitudes. On the one hand, fear, disgust, and loathing of the confusions attendant on physical and cognitive mixture could fuel concerns to find guarantees of immunity from melding, inducing theoretical rage for order. On the other hand, seductions of and fascination with blending and fusing, desires to dissolve dully independent mental items or selves into new combinations, could keep up residual dissatisfactions with over-rigid impositions of order.

This, anyway, is the broad framework, the schema behind this over-general analysis.

There is a final, more present-centred, motivation for examining these topics. Modern cognitive scientists have tended to think of pictorial and linguistic media as opposing and mutually exclusive candidates for being the medium of mental representation.[7] But in the recent resurgence of 'new connectionist' approaches to memory and mind (in the form of parallel distributed processing models and neural networks),[8] the suggestion has arisen that there might be a fundamentally different form of representation, neither pictorial nor linguistic, neither word nor image.[9] A brief digression on this possibility may illuminate the later analysis.

The key to such 'distributed representations' is the idea of 'superpositional storage'. Representations or traces are not kept passively in separate boxes or at distinct addresses, waiting in cold storage for an active executive to pull them out for processing. Instead, many traces are overlaid on or in the same physical (sub)system, not as distinct explicit items, but as dispositions for the reconstruction of patterns of activation across the system. Any one trace is 'stored' across many parts of the network, and any one part of the network is involved in the 'storage' and reconstruction of many traces. Since only one explicit pattern can be active in a network at one time, an activity pattern not explicitly present does not, in one sense, exist anywhere: it is only there, along with all the other implicit representations, as a disposition for the re-evocation of that explicit pattern.[10]

Distributed representation is interesting not only as the fad causing current turmoil in cognitive science. It provides direct connections with the analyses of metaphors of order and confusion. As a direct consequence of superpositional storage, distributed representations, overlapping in implicit representational space, tend to interfere one with another, to blend and mix, all of them affecting all ongoing processing and being in turn affected by the changing state of the system as new traces are laid down and old ones altered or activated. In the contemporary context, optimists take such patterns of interference to promise provocative modelling of phenomena of generalisation, blending effects, and prototype extraction in human learning and memory.[11] Critics argue that realistically-scaled networks will be unable to distinguish (reproduce) any of the superposed representations, to achieve even the degree of order in remembering which humans do, but will suffer catastrophic interference as overlaid patterns are obliterated and forgotten.[12] These debates, then, are in part about the explananda, about what is most characteristic of the way human memory works, and so deal inevitably in rhetoric and assumptions about order and confusion as well as in the technical details of the particular models available. Linking them with parallel historical debates is one way to broaden the scope of current discussions to include attention to the issues of cognitive architecture, mental control, and subjectivity which are already implicated in the more technical controversies.

If there is anything in the possibility that distributed representations are indeed different from both logico-linguistic and pictorial-imagistic forms of representation, then there should be hints of this alternative in the disagreements and unresolved tensions of other models of memory. I have argued elsewhere that this is the case for seventeenth and eighteenth century theories of memory, mental representation, and personal identity.[13] Here I seek to do the same for earlier periods, using medieval and Renaissance scholars' research on memory arts, models of memory which clearly do not explicitly sanction distributed representation. The modern debates can play no further direct part in this paper, but the need to trace these implications and issues underneath debates over memory and cognition applies no less firmly to our own sciences.

The last set of connections, in section 4, expands beyond memory again to hint at simultaneous theoretical movements or pressures away from and towards confusion across other domains in which issues of blending and distinctness arise. Developing this approach more fully would allow the exploitation of memory's boundary-blurring connections, across physiology and medicine, providing the matter which both dreams and reason trawl, and reaching up to play central roles in attitudes towards problems of personal identity and psychological conflict, and, further, towards the bridge between individual psychology and social relations. This section is necessarily even more sketchy than the rest, but at least gestures towards the range of historical issues about cognition and culture which memory and its attendant confusions can open up.

2. LOCAL MEMORY AND COGNITIVE DISCIPLINE

In the sprawling traditions of place memory, imagistic and textual models for internal storage functioned in complementary rather than opposed fashions, united in that both fulfilled theorists' desires for order in memory. Both mental models of mental representation were uneasily tied to medical traditions: the need to discover or impose inner discipline on the memory is all the more urgent if roving, nimble animal spirits are the fickle medium of mental control.[14] But the venerable physiological spirits were less pivotal in Renaissance neurophilosophy than they would become in the seventeenth century.[15] Here, then, rather than addressing the physiology of memory directly, I tour backwards through norms of local memory, recalling the complexity of the metaphorical associations of the memory art which Renaissance natural philosophers knew well. As Kemp suggests, references by leading scientists to the importance of external aids for strengthening the memory are 'unlikely to have been casual'.[16] The moral and normative constraints on theories and practices of remembering which are obvious in these contexts do not disappear from later or proto-scientific models.

The following sparse analyses of conceptual foundations of ancient, medieval, and Renaissance arts of memory rely almost entirely on the exciting and complicated materials provided by Frances Yates and Mary Carruthers.[17] These historians have teased out the details of readings, misreadings, and applications of Aristotelian associationism, faculty neuropsychology, the rules for places and rules for images found in 'Tully's' *Ad Herennium,* complicated by Cicero and Quintilian and filtered through the Arabs, the medieval ethics of memory, and Renaissance occultism and Neoplatonism. But for the history of theories of mental representation, it is useful to extract influential ideas from this complex of traditions. Such a survey will not cleanly assign all of its results to the particular, often distinct elements of the conglomerate of memory practices and theories which self-conscious moderns would soon find intensely alien. But it can seek to push on the historicising of a naturalistic cognitive science by examining shifts in, motivations for, and attractions of belief in what the historians tend to see only as 'certain enduring requirements of human recollection'.[18]

2.1 Localist Models: the Independence of Stored Items

Ideas encoded as images or notae in or on the various places of the memory systems must be independent of each other, and must map individually onto the places. John Willis advises the lacing of every idea in order 'always provided, there be but one idea assigned to one place'.[19] This is why strict division of material is a precondition for successful encoding: the items must be isolated, kept distinct one from another.[20] Division guarantees that memory will be, in John of Salisbury's words, 'a sure and reliable place of safe-deposit for perceptions'.[21] Because each idea is in principle independent of all others, yet stands in a fixed, ordered relation to them, it is possible, in executing a search procedure, arbitrarily to access any item, proceeding in any direction, skipping some material or moving around at random.[22] In this lasting local model, then, no two ideas can be in the same place: they must be 'actually different and separate one from another'.[23] It is this isolating of ideas at the time of encoding which wards off error in remembering: as Carruthers argues, ancient and medieval memory theorists considered mistakes as due not to distortions at the time of recollection, but to failures to make images 'sufficiently distinct from one another' in the first place.[24] Each memory address in the place system is loaded with only one item, and the system's capacity must not be overloaded by the number of items stored.[25]

Further support for this principle of the independence and isolation of memory items one from another was garnered from the medical traditions of faculty neurophysiology. Memory as an entire faculty was distinct from the other internal senses (perception, phantasy, common sense, and sometimes reason), and this distinctness was grounded in the localisation of memory in the posterior ventricle.[26] Now a view that memory capacities are, in general, localised in certain parts of the brain does not entail a localist view of the storage of *particular* memories: memory could, globally speaking, be localised, while memories, within the local areas, are still distributed.[27] But there was an easy slippage to the notion that individual memories in turn must be located separately, 'placid bi rewe along in ye heed, and ech in his propre celle'.[28]

2.2 Rigid Order and Memory as Inner Writing

The second general feature of the local art of memory, repeatedly invoked, is that 'order most secures the memory'.[29] Items are rigidly fixed in the soul by the use of mental grids, alphabetical systems, lists, plans and so on, on which they could be independently placed.[30] Fixed ordering of items in memory is possible just because of a prior ordering of reusable memory places.[31] It is up to the prudent soul trained in the art of memory to extract, manipulate, and combine (by juxtaposition rather than fusion) the items stored.

These features of independence (of images or items) and order (of places or addresses), as Carruthers recognises, make medieval memory systems exemplars of random access memories. Mental bins or cells remain unchanged as an executive system[32] arbitrarily manipulates, extracts, or replaces independent items arbitrarily stored therein.[33] Renaissance and baroque memory places could be vast and convoluted theatres, wheels, and layered circles: but the point of all such inner edifices was to ensure the total separation of more and more distinct loci within.

A more specific model which preserved these features often enhanced the practitioners' faith in the power of their trained memories to ward off fears of loss or obliteration. The writing of memories into areas as if on wax is a mechanism for fixity, so that contents are held more firmly.[34] Theorists thought of the medium of mental representation as both pictorial and linguistic: to use David Krell's felicitous labels for the classes of memory metaphor, the *typographic* imprinting of contents as on a wax block slid easily into the *iconographic* copying of originals into memory images, which in turn slipped into the *engrammatological* inscription of contents in a language of thought.[35] The textual model for mental representation seemed to ensure the required semantic stability of local items, which retained their content across different contexts, even when moved around the place system or the book of memory.[36]

These metaphors provide metaphysical points of entry for a homuncular soul, separate from the distinct imprints, pictures, or writings which it can somehow interpret, decode, or read. Such a central executive is a moral as well as a psychological necessity, since the idea of order in memory is linked with the requirement that discipline be imposed on one's memories. Rigid order is in some contexts a theoretical discovery about the nature of memory and elsewhere an imposition on memory, to be maintained on pain of falling into confusion.

Pictorial and linguistic representations were not mutually exclusive candidates for mental representation. Both clear and distinct images and inner writing will ward off confusion. Discipline is aided and exemplified in treating the brain as a 'book and volume' in which clear current contents can, in principle, wipe away trivial observation or unwanted 'pressures past'.[37] The normative requirement is clear in such a system: only the resolute, disciplined, prudent, trained soul will be able thus to obliterate the improper and retain in unmixed independent form the morally appropriate contents alone.

Although, as Carruthers convincingly argues, the arts of memory and the associated cognitive practices were not simply displaced by written records, the need to supplement memory by (external) writing, a need which the figure of memory as writing would in theory render unnecessary, did become more prominent. The art of memory is useful, John Willis remarks, only when you can't get contents down on paper, downloading memories into external words, 'the most happy keepers of any thing in memorie'.[38] In an odd chapter, Willis even includes in his art the use of (not just a linguistic representation but) the object as its own memorial symbol: placing a book by the door as a reminder is a surer guarantee that we will remember to take it than imaging or inscribing it on a memory place.[39] Hamlet has to write the ghost's commandment on external 'tables' as well as on the table of his memory.[40]

In other traditions of moral memory the downloading of inner script into an external medium met a stronger ethical demand. Janet Coleman's account of the 'blanched' memory of Cistercian tradition shows that Neoplatonist needs for the past to be purified were met by purging 'filthy traces' through the scriptural word.[41] We are again in the domain of mental control as technology of the self: the moral life is the disciplining, ideally the obliterating, of potentially corrupting inner items.

2.3 Will, Discipline, and the Moral Control of Memory

The possibility of such discipline over one's own representations requires curious division, the separation of a self from its memories. Ideals of powerful executive control sat well with local memory, for independent ordered items in their places were already passive, waiting for the active executive to hunt them out. Reminiscence or recollection was ethically prior to recognition and associative memory.[42] Similar models of control occur in the well-documented field of political metaphors for cognition. In Kenelm Digby's mid-seventeenth century localist model, the cognitive agent (the will, brain, or fancy) can light on, pull out, and move around passive atomic items: when it has trouble in recollecting something,

> ... it shaketh again the liquid medium they all floate in, and rooseth every species lurking in remotest corners, and runneth over the whole beaderoule of them; and continueth this inquisition and motion, till eyther it be satisfyed with retriving at length what it required, or that it be grown weary with tossing about the multitude of litle inhabitants in its numerous empire, and so giveth over the search, unwillingly and displeasedly.[43]

Not only does this cognitive agent, prone to boredom and petulance, have the power to scan and search through its liquid empire: it also recognises when it has retrieved the required memory. Wholly distinct from the stored items in its dark cells, the processor of local memories itself needs remarkable capacities for recollection:[44] a set of tasks and techniques must thus follow to enhance its chances of success.

Morality must be branded in the memory. In guaranteeing that one's traces are 'lofty'[45] rather than filthy, in ensuring that the prudent soul can order the confusion of the body and the traces it conceals, students of local memory could find solace in the regimen suggested by their psychophysiological theories. Differing views of the relation between natural and artificial memory all left room for learned improvement. Easily, though not always, assimilated to the distinction between corporeal and spiritual or intellectual memory, this natural/artificial distinction rendered problematic any attempt to specify what memory was 'really' like. 'Natural' memory as conceptualised in any theory was itself already cultural, shaped as a lack or incapacity to be remedied by particular artifices.

Artificial memory, in various systems, could be characterised simply as the imposition of discipline on natural memory.[46] The independence of stored items and the rigid order of places had always to be struggled for, through conscious cultivation and practice. Memory techniques were cognitive tools which, in different periods and traditions, were intended to complement, improve, strengthen, or supplant and bypass entirely the natural memory.[47] While some saw natural human capacities as merely needing some honing, others took natural memory to be irretrievably weak, arguing for example that artificial divisions are required between sets of connected material because 'the memory is lazy and rejoices in brevity'.[48]

The historians of memory have theorised this artificial/natural distinction variously. Carruthers sees it as confirmation of the modern cognitive psychological finding that (natural) short-term memory is limited to around seven independent items, so that memory techniques are ways of compressing more information into a single such local image.[49] Yates, with a wider historical extension into Renaissance arts, argued that increasing trust in the potential of human memory showed a pattern, typical of the 'Renaissance psyche', of new confidence in human capacities to express the divine and grasp the intelligible world.[50] Here I want to probe what it reveals about the local memory theorists' concerns, rather than their confidence, about the sources and degrees of cognitive order and control.

The medieval arts took as their domain the sadly imperfect capacities of human memory after the fall. Artifice is needed because of corruption and bodily confusion of memory from its prelapsarian state.[51] It is as a concession to the fallen soul, entranced by corporeality and particularity, that images are used.[52] But for sin, humans like angels would have no need of memory.[53] After Eden order must be fought for. With the fragility of natural memory apparent, space is carved out for the imposition of a range of techniques on the self and its murky contents. Even if Hermetic and Neoplatonist strands of Renaissance thought did, as Yates argues, hint at a more positive view of the potential strength of controlled recollection, this still required long immersion in occult arts as intricate as Bruno's.[54] The darker vision persisted in many religious contexts.

The weaknesses attributed to natural memory derived from its physical sources and mechanisms. In any individual it depended on elemental constitution, astrological endowment, and physiological fortunes. The irretrievably bodily character of natural memory was one source of difficulty for moral memory. Confusion was taken to be natural to 'natural' memory: one way of thinking of the artificial techniques was as a creation or substitution of an external memory, albeit one internalised as places and images, for the initially weak internal one.[55]

Defenders and critics of the different arts of memory alike made claims to order and unity. Theorists in both Aristotelian and Neoplatonic strands of the place techniques stressed the capacity of their art to unify the chaos of particulars which would otherwise swamp and overload the memory. Whether aiming at oneness with the One,[56] or merely enhanced abilities in perceiving and extracting forms and prototypes across their particular instantiations,[57] rhetorical or propagandist statements promised some kind of cognitive order to overcome or exclude associative excess and undesirable mixture.

2.4 Against Confusion and Multiplicity

Joseph Glanvill criticised the theory that memories were distributed motions or shifting patterns of animal spirits. If this were true, any chance mental process occurring as we tried to remember something would 'put all the other Images into a disorderly floating, and so raise a little Chaos of confusion, where Nature requires the exactest order'.[58] After Descartes, the need to combat distributed animal spirits models of corporeal memory was pronounced.[59] In attacking them, Glanvill reveals the ease with which the order previously imposed on natural memory by artifice could, as the arts of memory declined, be projected inside to become the explanandum for any psychology of memory. Yet there is a slippage still between the *assumption* that memory is 'capable of regularity' by nature (a 'uniformity' which all theories would then have descriptively to guarantee), and the *wish* that memories '*should* so orderly keep their Cells without any alteration of their site or posture'[60] when properly controlled (when a theory of memory would be normative, prescribing how we can avoid ataxy and impose order to retune the mental disharmony left us after the Fall).[61]

Medieval and Renaissance writers were less urgently concerned to deny or ward off interference. But they did seek recipes for maintaining the independence of stored items and the rigid order of places. Without a trained and prudent memory, local memory theorists warned, we would be lost in the murky forests of memory, wandering amidst 'a mass of unrelated and disordered material'.[62] The arts of memory exist to help us avoid confusion, though they may only succeed, as the *Ad Herennium* had warned, if confusion is already avoided in the initial establishment of independent places. In laying down images on images, those who fail to make proper divisions will proceed 'without order and in a confused form'.[63] Carruthers summarises the perceived dangers of not using a safe technique, of 'relying on simple chance to fish what one wants out from the murky pool of one's undifferentiated and disorganised memory': only rigorous discipline ensures the power or control to move around among one's own internally stored items without confusion.[64]

Analysis of particular kinds of confusion reveals both assumptions about and fears of possible forms of memory breakdown. Within a local model, the first problem is of restricting the number of images to the number of places or addresses: otherwise memory will be crushed beneath a weight of images.[65] Seventeenth century localist theories had to contend with the problem of finding room in the brain for the whole variety of independent items remembered.[66] In the earlier period, the fear is more that overloading local places will break down the distinctness of the stored items, with the chaos of interference bound to result. Albertus Magnus described how 'confusion is engendered' when too many heaped up images 'break up in the soul and do not remain, just as a great number of waves break up in water'.[67] Carruthers comments that this concern is not that of the *Ad Herennium*, which advised only against the initial imprinting of crowded places. But there were ancient precedents for worry over how persisting ingredients in cognitive mixtures in an internal environment could retain their original identity and avoid being dispersed.[68] The danger is that the carefully constructed order of local items may break down, and interference ensue.[69]

One strategy against this was to seek the liberation of memory from the confusing body. Hamlet's wish to inscribe the ghost's commandment 'all alone' in the book of his memory 'unmix'd with baser matter',[70] links not only to a retreat from trivial contents but also to a desired decorporealising of memory, a shaking off of the dirt added to thoughts by the body.[71] In its more grandiose Neoplatonic forms the inner art was a discipline of slowly amending the corrupting embodiment caused by the Fall, drawing away from the 'punishments of matter'.[72] Even where moral memory was a moral physiology of balance, *hexis,* and proportion,[73] the fragile equilibrium of bodily *krasis,* proper blending, in the Aristotelian metaphysics of mixtures had to be keenly monitored.[74] The arts of memory were tied to humoral physiology, and advice abounded on the appropriate non-naturals required to retain sufficiently rigid order.[75]

The strength of the requirements of order and discipline in these various traditions, the constant keeping of confusion at bay, perhaps suggests the depth of the internal tensions in the models. As well as bodily intrusions into mnemonic order, local memory theories were complicated by the ancient method of using emotionally charged images to aid retrieval, images which if not treated prudently might induce exactly the indiscipline which was meant to be stamped out.[76] Images of violence, seduction, blood, mud, uses of zodiacs and bestiaries could leave the imagination dangerously free.[77] Reactions against the dangers of indiscipline were most pronounced in the Ramist memory theorists who proposed, in contrast, imageless dialectical systems to intensify the principle of order.[78] Even though many did recognise the importance of powerful context-based and emotive associations, every participant in any debate about the arts of memory had to claim for their own system a route to true order, while banishing confusion to the opposition.[79]

Dichotomous fixations with order left no space for exploring the possibilities and complications of different forms of order. Ideas about the achievement of recollective order by the movement of images rather than by their static fixture seem not to have been developed.[80] While, as Carruthers' readings show, the entire powers of imagination were harnessed, through images of violence, dismemberment, and titillation, for moral or religious ends, they were seen only as auxiliary to quests for order. Bruno, following Lull, sought to introduce a dynamic element into the arts of memory, giving the images some activity as they are animated within.[81] But his more complex conceptions of order as folds, rather than simple chains or isolated rooms, was still subordinated to the dualistic task of passing beyond the convolutions to an angelical unity without disturbance, body, or multiplicity.[82]

3. MENTAL BRACHYGRAPHY

Both pictorial and linguistic mental representations, I have argued, are easily assimilated to localist models. Images and semantic items can have clear boundaries, which may be taken to map in simple isomorphism onto the equally clear boundaries of whatever it is they represent. On any general atomistic theory of representation, images or items have meaning in their own right, independent of their place in a general system of representations and their relations with other distinct representations within that system.[83] This picture is easily projected inside, onto a language of thought or a system of mental images in which single items have their semantic properties in isolation. Complex semantic items will then be generated by the combination or juxtaposition of arbitrary strings of these individual contents, preserving all original semantic information by preserving the exact syntactic form of individual representations.

In alternative, anti-atomist accounts, items are not initially entirely distinct, they have causal influence even when they are not explicitly 'there', and they combine not by juxtaposition but by blending, fusion, and interference.[84] Information, then, is inevitably lost, altered, added, or distorted: representations have to be (re)constructed, since, being stored only superpositionally, they do not endure independently to be simply reproduced. The compression and transformation involved between encoding and retrieval forces attention to the methods or mechanisms of 'decoding', which here means only (re)separation from the mix. An important change in recent connectionism is attention to non-linear processes in these transformations, which has also influenced recent theory on the order/confusion polarity inevitably called into rhetorical play here.[85] But there are hints too of relevantly similar notions of compression in historical theories of representation and memory.

In surveying the arts of local memory, I have pointed to lines of internal tension. Theorists were desperate to guarantee order, but were often aware both that order is not 'naturally' to be found, and that some of the best techniques for remembering, exploiting the noise and excess information carried in emotional images, intrinsically tend towards interference and confusion. The strongest historical inklings of distributed models would come in neurophilosophical and psychophysiological contexts. But even at the psychological level, it is possible to piece together, even from writers overtly hostile to interference, suggestions of a non-local form of mental representation which doesn't rely on the independent storage of distinct items.[86] I'll suggest this first with a seventeenth century example.

The problem with theories which take representations to be patterns of motions in the fleeting animal spirits, complains Henry More, is that they 'force a great deal of preposterous confusion' on the memory.[87] Retained representations, if left alone to be acted on by 'the bare laws of matter', would become 'strangely depraved, if not obliterated'.[88] More argues that only an immaterial soul can keep 'intire and unconfused images of things without'.[89] It takes something incorporeal, free of the 'foulness and coursness of *Matter*', to maintain many local items each 'yet distinctly represented'.[90] The brain does, however, have a subordinate role as the soul's instrument:

> ... she might make an occasional use of some private marks she impresses in the Brain; which haply may be nothing at all like the things it would *remember*, nor of any considerable magnitude nor proportion to them ...[91]

These marks, More reasons,

> ... must be a kind of *Brachygraphie*, some small dots here and there standing for the recovering to Memory a series of things that would fill, it may be, many sheets of paper to write them at large.[92]

With cognitive order here guaranteed by the spiritual 'inward Center', there need be no resemblance between physical traces and what they represent. Instead, there can be a compression of complex contents into condensed form. Similarly, Wittgenstein would argue against mental images by describing marks on paper which are necessary for someone to reproduce a passage of text, and yet which are not a rendering, translation, or simple storage of the text.[93] Both More and Wittgenstein take these suggestions to rule out any materialist account of memory.[94] But what is of interest here is their acknowledgment of the possibility of squashings or transformations of contents which improve on common external methods of transferring contents into other symbolisms. If mental brachygraphy can encode many different 'things' within a restricted physical system, each part of the system must be involved in the 'storage' of many items, items which are no longer independent. Though More is sure the soul controls her own condensed marks, he has allowed confusion to re-enter the scheme. There is nothing *within* the physical system to maintain or guarantee the identity of any of the ingredients in such a mix.

Some ancient arts of memory were projections inside of the Greek use of shorthand symbols, or *notae*, although the shorthand ideal of retaining a definite symbol for every single item in 'memory for words' was considered too cumbersome.[95] More useful was that aspect of 'memory for things' (for the gist) which compressed many items into a few images or metaphors.[96] Carruthers takes this process in its medieval form to be a simple substitution which condenses 'large amounts of material into single markers', with the resultant units still each independent local items.[97] But when focussing on the somatic side of the medieval models, she acknowledges within the Aristotelian associationist tradition a more dynamic kind of compression, in which the composition, construction, breaking and rejoining of condensed images, processes uncertainly controlled by our cogitative and deliberative powers, could result in shifting, fracturing bodily traces of ambiguous identity.[98]

External systems for selecting and compressing bodies of material in brachygraphy and tachygraphy were theoretically comprehensible, since translation rules were constructed by humans and consensually accepted. It would be useful to examine relations between the development and theory of shorthand and stenography in the late Renaissance, and their possible roles as models for forms of mental representation.[99] Mental brachygraphy would be more mysterious, hieroglyphics and tracings in a more reckless algebra, in which the mapping of representations onto items represented could always be many/many rather than one/one. This might seem useful in later theories of memory, seeking to do without a soul which already knows the structure of the task domain:[100] but it would discomfort even anti-dualists who retain the wish for prior control over their own representations.

The need to bring passions, contexts, and diverse associations into play even within some local models shows the need for what we would call a content-addressable memory to supplement the hard-earned random access one. It was, perhaps, only the power of the metaphors and theoretical images of external order (rooms, cells, theatres, books) which kept up confidence in the internal preservation of single isolated items.[101] This rendered problematic the grounding of mental representation, brachygraphic or not, in unstable matter. But increasing attention to the possibility of incorporeal memory, which would culminate in Descartes' undeveloped and vain attempt to set an intellectual memory free from the confusion which his own physiology of memory allowed,[102] never overturned the dominant awareness among memory theorists of how much would be lost without body and brain.[103] In thinking of how remembering related to patterns of motions in internal environments, the wish to impose greater control and regularity on fluid innards contended with the acknowledgment that it was also interesting to construct internal mixtures from which might emerge unknown blends of altered ingredients.

4. BODY, MIND, AND MIXTURE

The remainder of the paper allows the body some space in these schemes, briefly exposing some of the innards. The movements of submission and resistance to bodily confusion could, in some contexts, be simultaneous. Put another way, desire to retain autonomy and strict identity while mixed with matter could coexist with desire to be immersed in or merged with different corporeal blends. The approach to personal identity at work here is not the quest for criteria for unity and continuity of self,[104] but description of undersides, abnormalities, and excesses of the normative concept, pathologies of self which may throw light on what's missing from explicit philosophical theory.[105] The idea that there are attractions in loss of self as well as in the clear definition and separation (from matter, mortality, or society) of an autonomous self has become alien only to some modern philosophy of personal identity. But not only has it always featured in certain forms of mysticism, where the ecstasis of unravelling the soul can be a sought dissolution of dull identity rather than a discovery of the more perfect identity of the personal soul: it was a central issue in Renaissance philosophy of mind. In struggles over the Averroist doctrine of 'one mind', the will to lose memory and self in merging with the single immaterial soul was one draw to the idea of impersonal immortality as opposed to Thomistic personal survival.[106] In the early sixteenth century immortality controversies around Pomponazzi and his uses of Alexander of Aphrodisias, a key issue both in the interpretation of Aristotle and in the quest for truth was whether the intellect was 'unmixed' or was, as in certain medical traditions, simply the best-proportioned mixture of elements and qualities. If the latter, immortality seemed untenable, for all material mixtures will dissolve.[107]

Such naturalistic theories of mind threatened not only absolute immortality, but also the control of mental representations and bodily processes which, I have argued, local memory theorists sought. Without immaterial substance to order material mixtures from outside, the difficult task approached by philosophers like Pomponazzi and later Hobbes, as well as by medical theorists, was to construct a regimen for the immanent maintenance of psychological and bodily balance: there could be no single unified central executive rigorously ordering multiple cognitive contents and keeping the body in place. It's in these Renaissance medical and literary traditions, often closely connected,[108] that modern history and philosophy of the human body helps to display strands of the fragility and disharmony of the decorum and discipline officially sought for, in, and between selves.

4.1 Secrets Within[109]

When Vesalius attacked ventricular faculty neuropsychology in the *Fabrica,* preferring to avoid theory of higher cognitive function, he used charged metaphors to describe and discredit traditional views.[110] Discussing the passage from the third to the fourth ventricle, he denies the existence of a controllable process by which fantasies may enter the seat of memory, which may then return its 'captives',[111] collected in 'the prison of memory', to the alleged seat of reason in the middle ventricle.[112] The language of confinement and domination is indeed appropriate to place memory schemes,[113] and provides one point of contact between memory and medicine. Imaginary anatomies, surviving from medieval through Renaissance metaphoric physiology, pictured nested internal organs as treasures, secrets to be kept safe, ordered, and in their 'natural' places.[114] In all the medieval models and schemata which structured medical theory, pathologies and confusions were understood not through the common architectural and artificial metaphors, but through analogies from animal and vegetal worlds in which improper matter intrudes across the seals of the human body.[115] Just as in the case of artificial memory, the projection inside of comprehensible and cleanly-structured external artifice functioned to secure an image of stability and inner discipline which was otherwise threatened by the fluidity and proneness to disorder of brain and body parts.

Potential for internal confusion was, then, even harder to deny in general 'pre-modern' physiology than in the case of memory. The open and porous body of traditional humoral theory provided a language structuring bodily experience, which could thus be tumultuous and dramatic, with body parts having their own affective capacities as boundaries between inner and outer are continually permeated.[116] Belief in the interconvertibility of body fluids (blood, milk, fat, sperm, sweat, tears) and processes (lactation, menstruation, concoctions) made it hard to separate brain function from the active runny parts of the churning internal environment with its needs for purging, bleeding, and sealing.[117]

Restoring decorum by localising and isolating independent parts in a rigid order was unlikely to succeed in either general physiology or theories of memory. The breakdown of ventricular localisation made it desperately difficult for those still pushing localist models of memory to find physiological grounding.[118] But this led to no greater tolerance for confusion: the wish to isolate, separate, and thus control items in body and mind remained even when the impossibility of doing so cleanly was recognised. Later iatromechanical theorists still advised on guarding the purity of the imagination against 'dangerous traces' by 'strategems' like thinking of 'eternity, or some other solid thought' to impose order and fixity on the fleeting spirits by sheer external will.[119] And in physiology, knowledge gained through anatomical practice across late Renaissance Europe required the destruction, usually violent, of a previous organisation in order to divide and classify.[120] It is one of the tenacious cliches about the late Renaissance that recognition of the cost and disruption incurred in the production of new knowledge and order was more visibly, spectacularly on display than in later, more 'civilised' times.[121]

4.2 Crucifying the Self, Dissolving the Self

Those who give way to violent passions, Burton warned in pithy summary of Renaissance moral psychology, 'are torn in pieces, as Actaeon was with his dogs, and crucify their own souls'.[122] The astonishing *écorchés* (flayed) figures in baroque landscapes of Renaissance anatomists like Casserio, Spieghel, and Bucretius who exaggerate the illustrative techniques used in the Vesalian muscle-men and skeletons,[123] grotesque bodies often kindly holding back the folds of their own skin to reveal body secrets, are a symptom of wider fascination with self-inflicted violence.[124] Despite increasing current study, even in non-psychoanalytic psychology, of control exercised over one's own memories and mental representations,[125] there is an oddity in wishing to separate one's self or will entirely from one's memories comparable to that of disembowelling one's own innards, to intensify conflict in the quest for greater order.

Donne's claim to have 'cut up mine own anatomy, dissected myself' was already conventional.[126] The myths of Marsyas and Actaeon, repeatedly renarrated through the Renaissance, provided a range of motifs for understanding the processes of tearing self from self in which the agents of violence, for instance Actaeon's hounds, were often interpreted as internalised metaphor for one's own snarling, conflicting thoughts and desires.[127] Models of psychological division, literary or philosophical, required physiologies of self-mastery. Sweeping away 'anie filthinesse' in 'the secret closets and private chambers of thine heart' was a physical as well as theological duty.[128] But how can one dominate the cells, fibres, and spirits of one's own brain? The traditional *psychomachia* against rebellious passions never sat easily with physiological schemes. If distemper or 'lewed perturbations' have gripped our spirits, our 'intestine enemies' confusing reason,[129] it may be too late to tear the self 'that breeds all disquiet' from that other (true?) self 'that stilleth what the other hath raised'.[130]

This suggests a psychophysiological implementation or historical realisation of one strand of Stephen Greenblatt's schematic account of Renaissance self-creation. Greenblatt argued, around a series of literary and religious lives and case studies, that (i) Renaissance selves were fashioned in relation to some alien or hostile thing or power, that (ii) this alien could be constructed as chaotic or unformed, without order, that (iii) this chaotic alien could be internalised along with the authority which gains its identity from it, and that (iv) attacks on and suppression of such an alien could require excesses of power which threaten, efface, or dissolve the authoritative self which was meant to be defended.[131] My analysis of the memory arts starts at the third stage of this scheme, where the danger of confusion is always already internalised in 'natural' memory and requires the importation and imposition of external order as artificial memory to keep chaos at bay. The excess, frightening yet compulsive, sometimes produced in these exercises of theoretical and personal discipline encouraged Renaissance attention to suicide, dismemberment, and the other violations of self mentioned above. The effacement or loss of self which Greenblatt also mentions is the final topic here.

John Carey's account of Donne fixes on 'both [Donne's] urge to blend and the inescapable selfhood which prompted and frustrated it'.[132] Processes of blending, melding, merging into some new mixture in which the original ingredients are effaced, melting individuality away, could exert strong attractions at social, interpersonal, psychological, and physical levels. Symptoms, all familiar in their own domain, include dispute over whether or not angels make love by total interpenetration, mixing in the same place, horror of and fascination with hybrids and monstrous mixtures in biology and proto-anthropology, moral indignation at various kinds of hermaphroditism and 'confusion of sexes', and concerns to guarantee immunity from melding and evil mixture with people or things at alien places on the various hierarchies of 'degree'.

Between selves, the key question was whether particular contacts were merely the juxtapositions of impenetrable individuals who would remain unchanged by the process, or strong fusions in which restoration to a former state, retention of identity through the mix, was impossible. Irrevocable alteration could be both sought and resisted. In some contexts, notably religious, the merging would be into a greater Other which simply incorporated more into itself, as the sea will swallow a winedrop, but in more challenging cases, of true confusion, none of the ingredients would survive. It is Diana/Cynthia in Jonson's treatment of the Actaeon myth who punishes the corrupt and disguised masquers who 'mixe themselves with others of the court':[133] these self-lovers have not given themselves up, have not been drawn from themselves as Petrarch/Actaeon claimed to have been when transformed after seeing Laura/Diana.[134] Revengers in Jacobean theatre dissolve their identities in ambiguous ways: critics still battle over the extent to which disguised malcontent/rulers in plays like *Measure for Measure* and Marston's *The Malcontent* retain any control over the mixing process or emerge from reseparation with identities intact.[135]

Mixtures of sexes, lovers, mutual body parts, of friends, social roles, and constitutions can all seem more enticing than internal psychological blends between memories or mental representations. Donne's uses of words as ways in to other bodies as well as his own, his placings of towns, farms, instruments, and measures in interior locales, his fascinations with melting kisses or inner plaits and folds, are fine examples of the struggle to achieve control of recalcitrant matter by seeking to embrace and order material mixtures from the inside.[136] Yet, I have argued, related patterns operated in the context of memory, mind, and brain. The sheer difficulty of an immanent account of mixtures is one result of the inquiry: interference cannot, it has often appeared, be controlled without an external agent. Naturalistic theories of mind and self find it little easier genuinely now to acknowledge all the odd phenomena of confusion which a focus on such mixtures reveals. The over-homogenising of many historical problems, domains, and attitudes which this paper has risked is perhaps worthwhile if some of these strange fields have been productively addressed and connected.

I have suggested that it is possible to find hints, in medieval and Renaissance accounts and practices of representation, of a form of representation which is neither word nor image. Related perhaps to schemes for external brachygraphy, these quasi-distributed mental representations would work by condensation and compression. As models for the mind, use of both linguistic and pictorial representations showed how rigid boundaries between items could be retained, limiting data compression and rendering context effects unlikely, or at least extrinsic to the medium, by having exact semantic stability across instantiations enforced. Such atomistic models were the only ones which could be easily articulated explicitly, due to the shared assumption that 'order is what is needed', rather than compression, for successful representation.[137] It is certainly hard to find explicit historical evidence for this bypassing of the word/image dichotomy: acknowledgement of semantic instability as explanandum for theories of mind would have violated methodological presuppositions about order in cognition, cosmos, and culture.

But by enlisting evidence about attitudes to psychological control, I have tried to show that indirect evidence can be found. Alongside obvious historical dreams of a pristine interior, that realm within of true freedom untainted by spirits, traces, and this-worldly power, there was doubt about the ideal of an over-ordered atomistic inner space. In seeking to supplement the weakness of inner capacities with external representational aids, to help the active self with passive and derivative storage, theorists could almost confine the scope of the 'active' out of existence (natural memory requires, by its weak nature, artifice). Shaping the self by moulding its representations, seeking to keep representation transparent, perfect, clean, and well-bounded, could in effect render it opaque, elaborate, filthy, fuzzy. Tensions between ethics and physiology, or between containing the self and dissolving the self, are formed in specific periods, in which particular kinds of interference are resisted. Behind the truism that medieval, Renaissance, and seventeenth-century philosophers were concerned urgently with the construction and maintenance of order in the mind as well as in society and world, there are still many leads to pursue towards the cognitive and cultural importance of the catastrophic confusion they sought to avert.

School of History, Philosophy and Politics, Macquarie University

NOTES

1 Copernicus, *De revolutionibus orbium coelestium* (1543), quoted by Kemp in this volume.

2 Kemp, this volume. The chance to read in advance the papers by Martin Kemp, Jamie Kassler, and Jim Franklin gave me considerable help in constructing my case here. My thanks to Kassler, Franklin, and Doris McIlwain for useful comments.

[3] Classic literary/history of ideas studies include E.M.W. Tillyard, *The Elizabethan World Picture* (London, 1943), *passim*; R. Colie, *Paradoxia Epidemica* (Princeton, 1966); R. Grudin, *Shakespeare and Renaissance Contrariety* (Los Angeles, 1979).

[4] See Jonathan Dollimore's provocative review in *Radical Tragedy* 2nd ed. (Brighton, 1989), introduction (xi-lxviii); other strands of the new history can be approached through Michel Foucault, *The Order of Things* (London, 1970), e.g., pp. 71ff; M.M. Slaughter, *Universal Languages and Scientific Taxonomy in the Seventeenth Century* (Cambridge, 1982) pp. 1-11, 212-17; A. Fletcher and J. Stevenson (eds), *Order and Disorder in Early Modern England* (Cambridge, 1985); Otto Mayr, *Authority, Liberty, and Automatic Machinery in Early Modern Europe* (Baltimore, 1986) pp. 115-36 and 181-9; Stephen L. Collins, *From Divine Cosmos to Sovereign State* (Oxford, 1989) pp. 3-39; Annabel Patterson, 'The very name of the game: Theories of order and disorder', in T. Healy and J. Sawday (eds), *Literature and the English Civil War* (Cambridge, 1990) pp. 21-37; Joy Wiltenburg, *Disorderly Women and Female Power in the Street Literature of Early Modern England and Germany* (Charlottesville, 1992) chs 2 and 10.

[5] Elaine Scarry, 'Donne: 'but yet the body is his booke'', in Scarry (ed.), *Literature and the Body* (Baltimore, 1988) pp. 70-105, especially 95-6 and notes 19 and 22, pp. 101-2.

[6] Stephen Greenblatt, *Renaissance Self-Fashioning: From More to Shakespeare* (Chicago, 1980), especially p. 9 on the internalization of 'the alien' constructed as chaotic. See also section 4 below on Renaissance destructions and dissolutions of self.

[7] As, most notably, in debates between Kosslyn and Pylyshyn, discussed in Jim Franklin's paper in this volume. See also Michael Tye, *The Imagery Debate* (Cambridge, MA, 1991).

[8] This is the briefest and most informal account of the rudiments of distributed representation. Good introductions include Andy Clark, *Microcognition* (Cambridge, MA, 1989) ch. 5; William Bechtel and Adele Abrahamsen, *Connectionism and the Mind* (Oxford, 1991) ch. 2; Patricia Churchland and Terrence Sejnowski, *The Computational Brain* (Cambridge, MA, 1992) pp. 163ff.

[9] Tim van Gelder, 'What is the 'D' in 'PDP'? A survey of the concept of distribution', in W. Ramsey, S. Stich, and D. Rumelhart (eds), *Philosophy and Connectionist Theory* (Hillsdale, N.J., 1991) pp. 33-59; van Gelder, 'Defining 'distributed representation'', *Connection Science* 4 (1992) 175-91; John Haugeland, 'Representational genera', in *Philosophy and Connectionist Theory*, pp. 61-89; also relevant is Terence Horgan and John Tienson, 'Representations without rules', *Philosophical Topics* 17 (1989) 147-74.

[10] For this point see G.E. Hinton, J.L. McClelland, & D.E. Rumelhart, 'Distributed representation', in Rumelhart and McClelland (eds), *Parallel Distributed Processing*, vol.1 (Cambridge, MA, 1986) pp. 77-109, at p.80; and compare John Locke, Essay II.10.2: '... our *Ideas* are said to be in our Memories, when indeed, they are actually no where, but only there is an ability in the Mind, when it will, to revive them again'. I have argued extensively that parallels between the dispositional accounts of memory of contemporary and early modern theorists are not coincidental, but depend on shared neurophilosophical views of distributed representation which are independent from specific realisations in animal spirit patterns, vibratiuncles, or neural nets: John Sutton, *Connecting Memory Traces: Studies of Neurophilosophical Theories of Memory, Mental Representation, and Personal Identity from Descartes to New Connectionism*, Ph.D thesis, University of Sydney, 1993. The present arguments do not, however, depend on such a strong case.

[11] J.L. McClelland and D.E. Rumelhart, 'A distributed model of human learning and memory', in McClelland and Rumelhart (eds), *Parallel Distributed Processing*, vol.2 (Cambridge, MA, 1986) pp. 170-215.

[12] M. McCloskey and N.J. Cohen, 'Catastrophic interference in connectionist networks: The sequential learning problem', *Psychology of Learning and Motivation* 24 (1989) 109-65; Robert

M. French, 'Semi-distributed representations and catastrophic forgetting in connectionist networks', *Connection Science* 4 (1992) 365-77, with references to earlier discussions.

[13] Sutton, *Connecting Memory Traces*: ch. 2 on the Cartesian philosophy of the brain, ch. 3 on English responses to Descartes, ch. 4 on Locke, Hume, and associationism, and ch. 5 on Hartley and his critics.

[14] On animal spirits see my *Connecting Memory Traces*, and Sutton, 'The quick and nimble animal spirits: A case study in the elimination of psychophysiological constructs', forthcoming.

[15] D.P. Walker, 'Francis Bacon and *Spiritus*', reprinted in P. Gouk (ed.), *Music, Spirit, and Language in the Renaissance* (London, 1985) p. 126: as well as the other papers in this volume see also Walker, *Spiritual and Demonic Magic from Ficino to Campanella* (1958, reprinted Notre Dame, 1975), e.g., pp. 27-9, 189-92, 230-1; Walker, 'Medical spirits and God and the soul', in M. Fattori and M. Bianchi (eds), *Spiritus* (Rome, 1984) pp. 223-44. Walker praised Vesalius and other sixteenth-century theorists in whose thought *spiritus* 'rarely plays a conspicuous part', in contrast to Bacon and Descartes, for wasting little time on the animal spirits, apparently believing that there were no outstanding explananda for which spirits or other hypothetical entities were needed (Walker, 'Ficino's 'spiritus' and music', p. 150). Compare Notes 112 and 113 below on Vesalius.

[16] Kemp, this volume, on Vesalius, p. 21.

[17] Frances Yates, *The Art of Memory* (London, 1966); Mary Carruthers, *The Book of Memory* (Cambridge, 1990); also Janet Coleman, *Ancient and Medieval Memories* (Cambridge, 1992), which I haven't been able fully to benefit from. I give most primary source references to the quotations in Yates and Carruthers, hoping that the interest of their discussions in this new context will make up for my imperfect understanding of medieval contexts. A provocative summary of the arts of memory is Lina Bolzoni, 'The play of images: The art of memory from its origins to the seventeenth century', in P. Corsi (ed.), *The Enchanted Loom: Chapters in the History of Neuroscience* (Oxford, 1991) pp. 16-26. There is a different approach to early medieval memory in James Fentress and Chris Wickham, *Social Memory* (Oxford, 1992) pp. 11-5 and ch. 4.

[18] Carruthers, *The Book of Memory*, p. 130: compare Coleman, *Ancient and Medieval Memories*, pp. xiii-xviii and 600-14.

[19] John Willis, *The Art of Memory as it Dependeth upon Places and Ideas* (London, 1621; reprinted New York and Amsterdam, 1973) p. 52. This was a partial translation by the author of his *Mnemonica* (1618): I haven't seen the full 1661 translation. See Yates, *Art of Memory*, pp. 324-6 and n.40, p. 415.

[20] Division is stressed, for example, by Martianus Capella (Yates, *Art of Memory*, p. 64), and Hugh of St. Victor (Carruthers, *Book of Memory*, p. 83).

[21] In Carruthers, *Book of Memory*, p. 113. The same point is often made within the wax-block tradition of memory models in encouragements to imprint individual ideas deeply, separately from all others (Willis, *Art of Memory*, pp. 52-3).

[22] Aristotle, De Memoria, 452a; *Ad Herennium* (Yates, *Art of Memory*, p. 22); Hugh of St. Victor (Carruthers, *Book of Memory*, p. 82).

[23] This formulation is from Robert Hooke's localist model of memory, in his *Lectures of Light*, in *The Posthumous Works of Robert Hooke*, R. Waller (ed.) (London, 1705) p. 142. Compare Kenelm Digby, *Two Treatises* (Paris, 1644, reprinted New York, 1978) pp. 284-5. On the local models of Digby and Hooke see my *Connecting Memory Traces* (Note 10) ch. 3, and 'Inner discipline: Confusion and cognition in English responses to Descartes' theory of memory', forthcoming. I was encouraged to look to Digby, Glanvill, and Hooke by Jamie Kassler, who takes

a quite different approach in her *Inner Music: Hobbes, Hooke and North on Internal Character* (London, 1995), ch. 3.

[24] Carruthers, *Book of Memory*, p. 61; compare pp. 75-77 on the parallel with descriptions of error given by the Russian mnemonist Shereshevski (S.) in A.R. Luria, *The Mind of a Mnemonist* (New York, 1968). Carruthers uses this case study to throw light on medieval memory sensitively; but she fails to stress how alien from 'normal' twentieth-century remembering S.'s method was, or how inexplicable Luria found the case. Carruthers, like Coleman, wants to reveal radical differences between medieval and modern *practices*, as well as theories, of remembering. I would add that the mass of evidence for the centrality of local memory schemes also suggests intense awareness of the fragility of 'natural' memory: stored items need to be rigorously ordered, isolated one from another, by technique and control just because this is not how they would be without the intervention of an external trainer.

[25] In Willis' system, for example, there are 18 repositories, each with two rooms: we can thus safely remember 36 items without overcharging our memory (Willis, *Art of Memory*, p. 36).

[26] Walter Pagel, 'Medieval and Renaissance contributions to knowledge of the brain and its functions', in *The History and Philosophy of Knowledge of the Brain and its Functions* (Oxford, 1958) pp. 95-114, especially 97-103; Katherine Park, 'The organic soul', in C.B. Schmitt & Q. Skinner (eds), *The Cambridge History of Renaissance Philosophy* (Cambridge, 1988) pp. 464-84. There are useful lists of the diverse catalogues of the internal senses offered by a large number of Arab and medieval writers in G.W. Bruyn, 'The Seat of the Soul', in F.C. Rose & W.F. Bynum (eds), *Historical Aspects of the Neurosciences* (New York, 1982) pp. 55-81.

[27] Hinton, McClelland, and Rumelhart, 'Distributed Representation' (Note 10) p. 79. There is a genuine issue here, about whether or not memory is separate from any 'executive' which deposits and accesses memories: but this can, at least initially, be kept clear of the problem of the form of representation within memory.

[28] Reginald Pecock, *The Folewer to the Donet*, 30, in E. Ruth Harvey, *The Inward Wits: Psychological Theory in the Middle Ages and the Renaissance* (London, 1975) p. 61; compare her discussion of Nemesius, pp. 31-7.

[29] Fortunatianus (Carruthers, *Book of Memory*, p. 86; cf. p. 80).

[30] Carruthers, *Book of Memory*, pp. 92f, 107f, 129 on grids, alphabets, and lines; Yates, *Art of Memory*, pp. 87, 115-6 on lists and plans, ch. 15 on stages and rooms. More complex background places were entire image systems like the Bestiary (Carruthers, *Book of Memory*, pp. 126-7). Some such systems (internal and external) immediately raise the issue of whether or not there was a centre: Penelope Reed Doob argues, for example (*The Idea of the Labyrinth* (Ithaca, 1990) pp. 51-63), that most mazes and labyrinths before the twentieth century had a centre meant to order, through controlled art, the wanderer's confused error.

[31] *Ad Herennium* (Yates, *Art of Memory*, p. 23).

[32] Which could be called self, soul, reason, will, or fancy. The point here is that it is a cognitive agent separate from the representations in memory: it is something which has memories, rather than *being* (in part) its memories. Storage is distinct from processing, the knowledge base from the executive.

[33] Carruthers, *Book of Memory*, pp. 131-2 and 146 on bins and cells; p. 7 on random access memory. Carruthers here equates 'rigid order' with 'easily reconstructible order'. This way of putting it is guided by the lengthy advices given for finding one's way around one's own memory system. But these methods are not genuinely reconstructive, for after careful local encoding all the images are always already there, waiting: they have only to be found by an active, searching consciousness or subject, and do not, like distributed memory representations, have themselves

to be recreated anew each time. This is what gives local representations their characteristic context-independence.

[34] Martianus Capella (Yates, *Art of Memory* p. 64 and Carruthers, *Book of Memory*, p. 147). Carruthers is sensitive to the ways the vast contemporary literatures on textuality, orality, and *mentalités* can help with or be upset by specific historical problems.

[35] David Farrell Krell, *Of Memory, Reminiscence, and Writing: On the Verge* (Bloomington, 1990), introduction and *passim*. For further examples of the engrammatological programme see Carruthers, *Book of Memory*, pp. 8-10, 16, 60, 89, 111; Yates, *Art of Memory*, p. 22 (*Ad Herennium*) p. 64 (Martianus) p. 213 (Bruno) p. 262 (Dicson); Coleman, *Ancient and Medieval Memories*, pp. 157-68 (Anselm). For recent work on memory metaphors see also Note 101 below.

[36] On the 'mind as text' image fuelling contemporary cognitivist local models see David Kirsh, 'When is information explicitly represented?', in P. Hanson (ed.), *Information, Language, and Cognition* (Vancouver, 1990) pp. 340-65, especially pp. 350-60; Andy Clark, *Associative Engines: Connectionism, Concepts, and Representational change* (Cambridge, MA, 1993) pp. 8-15, 39-40, 115-29.

[37] Shakespeare, *Hamlet*, I.v.90-112.

[38] Willis, *Art of Memory*, preface (unpaginated): more strongly still, he says that writing 'doth for speed and certaintie go beyond any art of Memorie': also ch. 16, pp. 96-104 on how to 'despatch' memory items and leave internal places empty again. Bacon in *De augmentis scientiarum* II.5 argues that 'the great help to the memory is *writing*', which must supplement 'the natural and naked force of thought and memory' (in D.J. Herrmann and R. Chaffin (eds), *Memory in Historical Perspective* (New York, 1988) p. 167).

[39] Willis, *Art of Memory, ch. 15, pp. 93-5.*

[40] *Hamlet* I.v.98, 107.

[41] St Bernard, in Coleman, *Ancient and Medieval Memories*, ch. 11, especially pp. 182, 185, 191: compare her ch. 5 on Plotinus. Bernard's image of the memory as a stomach where dirty and bitter multiple particulars reside and ache invites analysis in terms of the retentive, evacuative, and contaminatory functions played by body memory in a dualistic scheme: for a suggestive method in a later historical context see Frank Whigham, 'Reading social conflict in the alimentary tract: more on the body in Renaissance drama', *ELH* 55 (1988) 333-50.

42 The privileging of autonomous active reminiscence could be sanctioned in Aristotle (*De Memoria* 451b-452a) and Augustine, *Confessions* 10.8 (on the power of the 'I' to drive away naughty candidate memories until 'what I wish for be unveiled, and appear in sight, out of its secret place'). Compare also Aquinas and Averroes, in Carruthers, *Book of Memory*, pp. 58-60.

[43] Kenelm Digby, *Two Treatises* (Note 23) pp. 285-6.

[44] This gives rise to the common charge of circularity against all trace theories of memory: if either the subject or some internal agent already knows which trace is the one it wants, the postulation of the trace can seem redundant, since there is no further need to bridge the temporal gulf between experience and recall. I argue elsewhere that such objections have force only against localist models: Sutton, *Connecting Memory Traces*, ch. 6, with discussion of the literature.

[45] Hugh of St Victor (Carruthers, *Book of Memory*, p. 71). Carruthers describes well the basis of the character-moulding tasks of memory training in the Aristotelian typography of the stamping of memories on inner wax.

[46] Yates, *Art of Memory*, p. 266 (William Perkins' Ramist art) and p. 298 (the opposed art and 'inner discipline' of Bruno); Carruthers, *Book of Memory*, pp. 46, 78.

[47] Carruthers, *Book of Memory*, pp. 70, 78; Yates, *Art of Memory*, pp. 30, 63-4, 70; Willis, *Art of Memory*, preface (the art does not hurt, but corroborates natural memory). Compare Mary Pardo, 'Memory, imagination, figuration: Leonardo da Vinci and the painter's mind', in S. Kuchler and W. Melion (eds), *Images of Memory* (Washington, 1991) pp. 47-73 and 212-24, for Leonardo's use of a 'sketch-pad memory' acting as 'a thesaurus of single impressions' to represent the dynamic action natural memory could not.

[48] Hugh of St Victor (Carruthers, *Book of Memory*, p. 83).

[49] Carruthers, *Book of Memory*, p. 84.

[50] Yates, *Art of Memory*, pp. 173-4, 227.

[51] Boncompagno (Yates, *Art of Memory*, p. 70).

[52] Aquinas (Yates, *Art of Memory*, p. 82).

[53] On the dispensability of memory in incorporeal and atemporal existence see Coleman, *Ancient and Medieval Memories*, pp. 62, 68, 72-3, 155. The idea that angels need no memory lasted: in Locke's *Essay* II.10.9, they 'constantly have in view the whole Scene of all their former actions', showing up by contrast the meagre human need to scramble for the past in the face of oblivion.

[54] Bruno's arts aimed to extract the deep 'order and series' from the 'inform chaos' of primordial elements and numbers. The point of using memorable forms is to achieve an ordered disposition of the memory items, 'necessary for the control of memory': this is tied directly to the demand for independence of the local items one from another, maintaining 'certain distinct intervals' which ensure the requisite form and order. See Yates, *Art of Memory*, pp. 213-4, 249; Yates, *Giordano Bruno and the Hermetic Tradition* (London, 1964), *passim.*

[55] Yates, *Art of Memory*, pp. 136, 206, 221.

[56] Yates, *Art of Memory*, p. 159 (Ficino & Camillo) pp. 221-2 and 250-1 and 304 (Bruno).

[57] On the general problem of universals in Aristotelian Renaissance philosophy of mind see H. Skulsky, 'Paduan epistemology and the doctrine of one mind', *Journal of the History of Philosophy* 6 (1968) 341-61.

58 Joseph Glanvill, *The Vanity of Dogmatizing* (London, 1661, reprinted Brighton, 1970) p. 36; also p. 39 (against the view that memories are 'stored with infinite variety of divers, yea contrary motions, which must need interfere, thwart, and obstruct on another', when 'there would be nothing within us, but Ataxy and disorder').

[59] Sutton, *Connecting Memory Traces* (Note 10) chs 2-3.

[60] Glanvill, *Vanity of Dogmatizing*, 36 (my emphasis).

[61] *Ibid.*, pp. 37-39 and 1-16, especially 4-5, and my 'Inner Discipline' (Note 23).

[62] Carruthers, *Book of Memory*, p. 62, defining a 'common metaphorical extension in Latin of the word *silva*'.

[63] Thomas of Waleys (Carruthers, *Book of Memory*, p. 103).

[64] Carruthers, *Book of Memory*, p. 7.

[65] Cicero, *De oratore* (Yates, *Art of Memory*, p. 34). This leads to advice like John Willis' (Note 25) about the necessity to keep the number of memorabilia down to avoid overcrowding.

[66] Robert Boyle, *Works* (ed. T. Birch [1772], Hildesheim, 1965), 4.454; Robert Hooke, *Lectures on Light* (Note 23) pp. 143-4; J.J. MacIntosh, 'Perception and imagination in Descartes, Boyle, and Hooke', *Canadian Journal of Philosophy* 13 (1983) 327-52. On the alternative Cartesian distributed model of memory, where items are not independent one from another but are dispositions for reconstruction of an explicit pattern of motions in the animal spirits, this problem

could be dismissed: see Nicolas Malebranche, *The Search After Truth* (Columbus, 1980) II.I.5.iii, p.107, and my *Connecting Memory Traces*, chs 2-3.

[67] De bono 11 (Carruthers, *Book of Memory*, p. 138).

[68] Aristotle, *De memoria* 2, 453ab; *De somno* 3, 456a-457a; *De somniis* 2-3, 459a-461b; with *De generatione et corruptione* I.10, 327a-328b; compare Plato, *Theaetetus* 191de, 194c-195a; Krell, *Of Memory, Reminiscence, and Writing* (Note 35) ch. 2.

[69] There was also basic advice on how to avoid mnemonic confusion and perturbation by choosing carefully the context of encoding (Yates, *Art of Memory*, pp. 79, 110).

[70] *Hamlet* I.v.102-4.

[71] Coleman, *Ancient and Medieval Memories*, chs 5 and 11.

[72] Bruno's follower Dicson in Yates, *Art of Memory*, p. 263; also p. 315 (Fludd's spiritual memorizing to 'overcome the confusions of Babel').

[73] Aquinas in Carruthers, *Book of Memory*, p. 68. For the ancient background compare T.J. Tracy, *Physiological Theory and the Doctrine of the Mean in Plato and Aristotle* (Chicago, 1969).

[74] Zabarella, *De Mistione* [1594], with H.H. Joachim, 'Aristotle's conception of chemical combination', *Journal of Philology* 29 (1904) 72-86; Harald A.T. Reiche, *Empedoclean Mixture, Eudoxan Astronomy, and Aristotle's Connate Pneuma* (Amsterdam, 1960), especially ch. 2 on immanent and transcendent models of mixture; Sutton, 'Confusion and mixture in Aristotle's psychology', in preparation. Yates claimed (*Art of Memory*, pp. 163, 169-170) that the idea of thinking of successful artificial memories as proper mixtures was new with Ficino and the Italian academies, but this seems unlikely, given the power of the notions of proportion and balance in ancient and medieval medical traditions: see for example R. Klibansky, E. Panofsky, F. Saxl, *Saturn and Melancholy* (London, 1964) Part One.

[75] The complex relations between the art of memory and physiology play only passing roles in the existing histories: see Yates, *Art of Memory*, p. 70 (Boncompagno on humours) p. 80 (Albertus on melancholy) p. 191 (Ramon Lull on the use of medicines and plasters); Carruthers, *Book of Memory*, pp. 47-51 (general physiology of memory).

[76] The use of exceptional images goes back to the weird examples of the art of memory used in the *Ad Herennium* to supplement nature with art: see Yates' outstanding commentary, *Art of Memory*, pp. 25-30.

[77] Yates, *Art of Memory*, pp. 112, 203, 242, 358-9; Carruthers, *Book of Memory*, p. 109 and ch. 4.

[78] Yates, *Art of Memory*, pp. 229-32 (Ramus) p. 266 (Perkins).

[79] For some reactions against confusion see Yates, *Art of Memory*, p. 133 (moral humanists) pp. 231, 268-70 (Ramists), 255 (Protestants).

[80] Thomas Bradwardine, in Carruthers' description (*Book of Memory*, pp. 132-4 on the need for movement in recall to 'glue' the order of images, 144, 149-150 on matrices).

[81] Yates, *Art of Memory*, chs 8 (Lull) pp. 9, 11, 13, and 14 (Bruno), especially pp. 203, 249.

[82] For the fold see Bruno, *The Secret of Seals* (in Yates, *Art of Memory*, p. 243) on 'The Field' which is the memory, 'the ample folds of which are to be worked upon by the art of places and of images'; compare Gilles Deleuze, *The Fold: Leibniz and the Baroque* (Minneapolis, 1993), e.g., pp. 23-4. For Bruno against multiplicity see Yates, *Art of Memory*, pp. 222, 250-1, 296 (also 365-72 on Bruno, Leibniz, and memory).

[83] For a recent defence of atomism in theory of meaning against a range of holisms see J.A. Fodor and E. Lepore, *Holism: A Shopper's Guide* (Oxford, 1992), especially ch. 1. Wolfgang Kemp has argued ('Visual narratives, memory, and the medieval *Esprit du System*', in Kuchler & Melion (eds), *Images of Memory* (Note 47) pp. 87-108, 226-229) for a stricter distinction than Yates made between ancient mnemonic image-based systems, and medieval 'schematic-systematic' systems using figure and diagram. But he acknowledges that the two coexisted from the thirteenth century at least: my claim then is that their differences mattered less than the shared conceptual commitment to atomism and order.

[84] Such an account can spring from some forms of 'holism' in the theory of meaning, from the chemical (as opposed to mechanical) metaphors within the associationist tradition, or from a connectionist theory of distributed representation.

[85] Michel Serres, *Hermes II: l'interference* (Paris, 1972); Serres, *Hermes: Literature, Science, Philosophy*, J.V. Harari and D.F. Bell (eds) (Baltimore, 1982); N. Katherine Hayles, *Chaos Bound: Orderly Disorder in Contemporary Literature and Science* (Ithaca, 1990); Hayles (ed.), *Chaos and Order: Complex Dynamics in Literature and Science* (Chicago, 1991); Alexander Argyros, *A Blessed Rage for Order* (Ann Arbor, 1991); Philip Kuberski, *Chaosmos* (New York, 1994).

[86] Here as often in this paper it would be natural to question the relation of Descartes to the diverging traditions of the art of memory and the medical psychophysiology of memory. His discussion, in a physiological context, of the uses of 'compact' 'abbreviated representations' as 'safeguards against lapses of memory' (*Rules for the Direction of the Mind*, rule 12, AT X.417, *The Philosophical Writings of Descartes*, J. Cottingham, R. Stoothoff, and D. Murdoch (trans.) vol.1 (Cambridge, 1985) p. 43), is clearly relevant. Descartes' early rejection of the art of memory (*Cogitationes Privatae*, AT X.230) is, however, more complicated than is often allowed, and his attitude does change as he comes to construct a new physiological system. See my *Connecting Memory Traces*, ch. 2, and the forthcoming edition of Descartes' *Treatise on Man* by Stephen Gaukroger and John Sutton. A good recent paper on these problems is Dennis Sepper, 'Imagination, 'ingenium', and memory art', in S. Voss (ed.), *Essays on the Philosophy and Science of René Descartes* (Cambridge, 1993) pp. 142-61; also Paolo Rossi, *Clavis Universalis* (Milan, 1960) pp. 145, 153-61; Yates, *Art of Memory*, pp. 359-60; MacIntosh, 'Perception and imagination...' (Note 66); Dalia Judowitz, *Subjectivity and Representation in Descartes* (Cambridge, 1988) pp. 25-32; Stephen Gaukroger, *Cartesian Logic* (Oxford, 1989) pp. 31-38, 46ff.; Krell, *Of Memory, Reminiscence, and Writing* (Note 35) ch. 2.

[87] Henry More, *The Immortality of the Soul, in A Collection of Several Philosophical Writings* ([2 vols., 1662] reprinted New York, 1978) vol.2, Book II ch. 2 paragraph 7 (II.2.7) p. 68.

[88] More, *Immortality*... II.7.16, p. 93; II.10.9, p. 105.

[89] More, *An Appendix to the foregoing Antidote against Atheism*, in *A Collection*... vol.1, ch. 10 paragraph 10, p.173.

[90] *Immortality*... II.10.2, p. 102; Appendix, p. 173.

[91] *Immortality*... II.7.16, p. 93.

[92] *Immortality*... II.11.4, p. 107.

[93] L. Wittgenstein, *Zettel* (Oxford, 1967), 612 (= *Remarks on the Philosophy of Psychology* vol.1 [Oxford, 1980], 908).

[94] 'The *Memory* is in the *Soul*, and not in the *Brain*' (More, *Immortality*... II.11.4, p.107). For relevant Wittgensteinian remarks see my *Connecting Memory Traces*, ch. 8, and David Stern, 'Models of memory: Wittgenstein and cognitive science', *Philosophical Psychology* 4 (1991)

203-18; also A.J. Cascardi, 'Remembering', *Review of Metaphysics* 38 (1984) 275-302. More, unlike Wittgenstein, thinks or hopes that a dualist theory can replace the materialist ones.

[95] Yates, *Art of Memory*, pp. 30-1, 39, 55; Carruthers, *Book of Memory*, p. 124 on the medieval ars notataria and its links with the arts of memory.

[96] Yates, *Art of Memory*, pp. 76-7 (Albertus Magnus, *De Bono*).

[97] Carruthers, *Book of Memory*, pp. 84-5, 92 (Hugh of St Victor's 'principle of shortness').

[98] Carruthers, *Book of Memory*, p. 53 (using Avicenna and Aristotle, *De Anima* III.11.434a), and ch. 2, *passim*; David Summers, *The Judgment of Sense: Renaissance Naturalism and the Rise of Aesthetics* (Cambridge, 1987) chs 4-6 and 9.

[99] Vivian Salmon, *The Works of Francis Lodwick* (London, 1972) pp. 60-67, 110-116, 144-145: shorthand was almost exclusively developed in England until the late seventeenth century. John Willis, author of the 1618 *Mnemonica*, had previously (1602) written on stenography. The *O.E.D.* refers to 'the Shorthand of the Mind' which 'crowds a great deal into a little space' (J. Collier, *O.E.D.* s.v. 'shorthand'). Earlier Renaissance theorists including Trithemius (d.1516) and Agrippa had linked ciphers and spirits as media of information-processing and representation: Walker, *Spiritual and Demonic Magic* (Note 15) pp. 85-96.

[100] For this point in the modern context see Andy Clark, 'Connectionism, competence, and explanation', *British Journal for the Philosophy of Science* 41 (1990) 195-222; and more generally Hubert and Stuart Dreyfus, 'Making a mind versus modelling the brain', in S. Graubard (ed.), *The Artificial Intelligence Debate* (Cambridge, MA, 1988) pp. 15-41.

[101] Sound general work on metaphors of memory and mind includes John C. Marshall and David M. Fryar, 'Speak, Memory! An introduction to some historic studies of remembering and forgetting', in M.M. Gruneberg and R. Morris (eds), *Aspects of Memory* (London, 1978) pp. 1-25; H.L. Roediger, 'Memory metaphors in cognitive psychology', *Memory and Cognition* 8 (1980) 231-46; Robert F. Belli, 'Mechanist and organicist parallels between theories of memory and science', *Journal of Mind and Behavior* 7 (1986) 63-86; Michael S. Kearns, *Metaphors of Mind in Fiction and Psychology* (Lexington, 1987) chs 1-2; David Leary (ed.), *Metaphors in the History of Psychology* (Cambridge, 1990).

[102] Intellectual memory: Eckhard Kessler, 'The intellective soul', in Schmitt and Skinner (eds), *Cambridge History of Renaissance Philosophy* (Note 26) pp. 485-534, at pp.510-7; Descartes to Mersenne, 1.4.1640 and 6.8.1640, AT III.45 and 142, with the work on Descartes cited in Note 86; also Paul Landormy, 'La mémoire corporelle et la mémoire intellectuelle dans la philosophie de Descartes', *Bibliothèque* 4 (Paris, 1902) 259-98.

[103] This point was both motivation and support for belief in bodily resurrection, an important context for all these debates: see Stephen Davis, 'Traditional Christian Belief in the Resurrection of the Body', *New Scholasticism* 62 (1988) 72-97, and Caroline Walker Bynum, 'Material continuity, personal survival, and the resurrection of the body', in her *Fragmentation and Redemption* (New York, 1992) pp. 239-97.

[104] As Hume knew, the emergence of this kind of philosophy of self was historically and culturally specific (*Treatise on Human Nature* I.iv.6 on why it has 'become so great a question in philosophy, especially of late years in England...'). On accounts of modern 'origins' of 'the individual' see Sylvana Tomaselli, 'The first person: Descartes, Locke, and mind-body dualism', *History of Science* 22 (1984) 185-205.

[105] Compare Georges Canguilhem, *The Normal and the Pathological* (1966, reprinted New York 1991) pp. 33-4; and Jamie Kassler's account of Hobbes' physics of the self in this volume.

[106] On ecstasis in Renaissance philosophy see M.A. Screech, *Erasmus: Ecstasy and the Praise of Folly* (London, 1980). On the philosophical dimensions of Averroism in Italy see Skulsky, 'Paduan epistemology...' (Note 57); Martin L. Pine, *Pietro Pomponazzi: Radical Philosopher of the Italian Renaissance* (Padua, 1986) pp. 55-60 and 78-88.

[107] Pietro Pomponazzi, *De immortalitate animae* (1516), trans. W.H. Hay in E. Cassirer, P.O. Kristeller, J.H. Randall (eds) *The Renaissance Philosophy of Man* (Chicago, 1948) pp. 280-381, with Randall's introduction, 257-79; Pine, *Pietro Pomponazzi*, chs 1-2; Brian D. Copenhaver, 'Did science have a Renaissance?', *Isis* 83 (1992) 387-407, especially pp. 388-401 on Pomponazzian method and historiography. Accepting only 'relative' immortality for the intellect (as a shadow of fully immaterial intelligences) since the early 1500s, Pomponazzi's position grew increasingly less orthodox, despite the controversy, and in *De Nutritione* (1521) he argued that intellect is both extended and divisible.

[108] For example, Etienne Dolet, who celebrated in Latin verse the good fortune of a hanged criminal dissected in 1537 by Dolet's friend Rabelais, was a pupil and follower of Pomponazzi: see Mikhail Bakhtin, *Rabelais and His World*, trans. H. Iswolsky (Bloomington, 1984) pp. 355-62.

[109] 'Within are secrets': Middleton and Rowley, *The Changeling*, I.1.166, quoted and applied in this context by Whigham, 'Reading social conflict in the alimentary tract' (Note 41) 341.

[110] Charles Singer, *Vesalius on the Human Brain* (Oxford, 1952) pp. 1-6, 39-40, 49-50 (Book 7, chs 1, 6, 10).

[111] *Vesalius on the Human Brain*, 49: the word is 'fures', which as Singer notes (n.72, p. 81) can mean captives, thieves, or thefts.

[112] *Ibid.*, pp. 49-50. Vesalius' central argument against such a view is theological. Since the relevant anatomical structures (for the passage of animal spirits between the ventricles) are common to humans and beasts, either a) the localisation of cognitive function in these structures is true only for humans, leaving beasts, impossibly, with useless structures, or b) beasts too have rational capacities for control of the relation between memories and reason (49-50: for comparative brain anatomy compare 6-7, 24). Note the assumption here: if memory and reason are localised, with a physical 'passage' between them, it must, at least in principle, be reason which 'commands' and operates the passage, refusing or permitting entrance to individual memory representations (49-50). It is not that Vesalius believes his denial of the antecedent of this conditional to solve the theoretical problems: he simply rejects the idea that such problems can (or should) be solved.

[113] Vesalius' own view of the arts of memory, mentioned by Kemp (this volume), seems to be in line with that of humanists who encouraged the use of external images and aids to strengthen natural memory while remaining sceptical of traditional schemes for the artificial projection of such aids inside the mind (on Erasmus and moderate humanist hostility to place memory systems see Yates, *Art of Memory* pp. 137-9, 160-2, 169, and compare Montaigne, *Essays* II.17 ['On Presumption'] on how excessive prescriptive attention to remembering, as to the bodily organs, can actually cause the failure of the function in question). Neither psychology nor neurophysiology, for Vesalius, could give such insight into functional architecture and mental representation. This gives some counter to praise like Walker's (Note 15) for Vesalius and other sixteenth-century scientists who avoided hypothetical internal entities like animal spirits: the cost was inability even to acknowledge as explananda any complex cognitive functions.

[114] Marie-Christine Pouchelle, *The Body and Surgery in the Middle Ages* ([1983] New Brunswick, 1990) chs 6-11, especially pp.132-6. Also useful is Danielle Jacquart and Claude Thomasset, *Sexuality and Medicine in the Middle Ages* ([1985] Cambridge, 1988), especially ch. 2.

[115] Pouchelle, *The Body and Surgery*, Part Two *passim*, especially pp. 158-9 and the metaphorical tables at 207-217. Pouchelle's detailed study of Henri de Mondeville also extends to a range of other medical writers.

[116] This relation between theory and experience, both quite alien to us, is most strongly argued for by Gail Kern Paster, *The Body Embarrassed: Drama and the Disciplines of Shame in Early Modern England* (Ithaca, 1993) pp. 1-22 and passim. It is a natural consequence of the dependence in humoural physiology of bodily *krasis* (proper blending) on temperature, climate, diet and the other non-naturals, with treatment being in part aiding the body's limited capacities temporarily to resist the immediate environment: see for example Klibansky *et al., Saturn and Melancholy* (Note 74); Owsei Temkin, *Galenism: Rise and Decline of a Medical Philosophy* (Ithaca, 1973) ch. 4; Nicholas Steneck, *Science and Creation in the Middle Ages* (Notre Dame, 1976) ch. 8; Nancy Siraisi, *Medieval and Early Renaissance Medicine* (Chicago, 1990) ch. 4. Much more of this permeability survived into seventeenth-century physiology, even of the most mechanistic variety, than is usually recognized: see Descartes, *Treatise of Man* AT XI.166-9, T.S. Hall (trans.) (Cambridge, MA, 1972) pp. 72-5 on the enormous variety of factors, within the external and internal circulatory schemes, which influence psychophysiological function by way of the blood and subtle animal spirits.

[117] Thomas Laqueur, *Making Sex: Body and Gender from the Greeks to Freud* (Cambridge, MA, 1990) pp. 35-43 and 103-8.

Digby and Hooke made valiant attempts to anchor independent local memories in cells (Digby) or along the coils and spirals of memory (Hooke): but interconnectedness and lack of independence within brain substances was recognized by most. On later problems of neural localization see for example W.F. Bynum, 'The Anatomical Method, Natural Theology, and the Functions of the Brain', *Isis* 64 (1973) 445-68; Kenneth Dewhurst and Edwin Clarke, *An Illustrated History of Brain Function* (London, 1973); Mary A.B. Brazier, *A History of Neurophysiology in the Seventeenth and Eighteenth Centuries* (New York, 1984); Robert G. Frank, 'Thomas Willis and His Circle: Brain and mind in seventeenth-century medicine', in G.S. Rousseau (ed.), *The Languages of Psyche* (Berkeley, 1990) pp. 107-46.

[119] Malebranche, *Search After Truth* (Note 66) V.8, p.388.

[120] Jan C.C. Rupp, 'Matters of life and death: The social and cultural conditions of the rise of anatomical theatres', *History of Science* 28 (1990) 263-87; Rupp, 'Michel Foucault, body politics, and the rise and expansion of modern anatomy', *Journal of Historical Sociology* 5 (1992) 31-60.

[121] Francis Barker, *The Tremulous Private Body* (London, 1984). Barker argues that these violences disappeared, or were hidden, as a distinction between public and docile or private bodies was carved through the seventeenth century. Yet it was not only in texts which explicitly display the cruelty of modern reason, like Swift's *A Tale of A Tub*, that new docility was challenged: even in what traditional histories of philosophy cast as passively mechanistic conceptions of an inert body housing a ghostly soul, the body was in fact always urgently active, its microprocesses being obsessively theorised (not forgotten), with vigilant theorists producing, rather than neglecting, its interventions between will and world.

[122] Robert Burton, *The Anatomy of Melancholy*, A.R. Shilleto (ed.), 3 vols. (London, 1926-7) I.298.

[123] K.B. Roberts and J.D.W. Tomlinson, *The Fabric of the Body* (Oxford, 1992) ch. 7, with an uneasy discussion of the convention on pp. 255-9.

[124] Jonathan Sawday, 'The Fate of Marsyas: Dissecting the Renaissance body', in L. Gent and N. Llewellyn (eds), *Renaissance Bodies* (London, 1990) pp. 111-35.

[125] Useful historical and contemporary papers in D. Wegner and J. Pennebaker (eds), *Handbook of Mental Control* (Englewood Cliffs, 1993).

[126] Devotions (1623, Michigan 1965), Meditation 9, p. 56 (cf. p. 60); compare Lyly's (1578) Euphues going off 'to macerate my Selfe with melancholye' (quoted in Devon L. Hodges, *Renaissance Fictions of Anatomy* [Amherst, 1985], ch. 2, an invaluable study on beliefs about confusion in words, method, and matter), and Jonson, *Volpone* II.5.69-72.

[127] 'Quid me mihi detrahis?', Ovid, *Metamorphoses* 6.385; D.T. Starnes and E.W. Talbert, *Classical Myth and Legend in Renaissance Dictionaries* (Chapel Hill, 1955) pp. 194-210; Edgar Wind, *Pagan Mysteries in the Renaissance* (New Haven, 1958) ch. 11; Leonard Barkan, 'Diana and Actaeon: The myth as synthesis', *English Literary Renaissance* 10 (1980) 317-59, especially pp. 339-45 on cruel hounds as desires; Nancy Vickers, 'Diana described: Scattered woman and scattered rhyme', in E. Abel (ed.), *Writing and Sexual Difference* (Brighton, 1982) pp. 95-109; Sawday, 'The fate of Marsyas' (Note 124). See also Christopher Ricks, 'Sejanus and dismemberment', *Modern Language Notes* 76 (1961) 301-8.

[128] Phillip Stubbes, *A Perfect Pathway to Felicitie* (1610), quoted by Peter Stallybrass, 'Reading the body: *The Revenger's Tragedy* and the Jacobean theater of consumption', *Renaissance Drama* 18 (1987) 121-48, p. 124: compare Paster, *The Body Embarrassed* (Note 116).

[129] Lawrence Babb, *The Elizabethan Malady* (Michigan, 1951) pp. 17-20, quoting Vives and Wright. Babb's account (e.g. pp. 175-85) of the double movement within the mass of Renaissance thought on melancholy tallies closely with the argument of this paper: fear of the extremities and disorders of manic and mopish melancholy could coexist with fascination at the latent secrets available to melancholy genius within the Neoplatonic tradition. Compare Klibansky *et al., Saturn and Melancholy* (Note 74); Bridget G. Lyons, *Voices of Melancholy* (London, 1971); Michael Macdonald, *Mystical Bedlam* (Cambridge, 1981) ch. 5; Stanley W. Jackson, *Melancholia and Depression* (New Haven, 1986) chs 5-6.

[130] The Puritan Richard Sibbs, *The Soules Conflict with Itself and Victory over Itself by Faith* (London, 1635) p. 143, quoted by Jonathan Sawday, "Mysteriously divided': Civil war, madness, and the divided self', in Healy and Sawday (eds), *Literature and the English Civil War* (Note 4) 127-43, p.135, in a paper which gathers wonderful material on this topic from a variety of moral and religious traditions.

[131] Greenblatt, *Renaissance Self-Fashioning* (Note 6) p. 9 and *passim*.

[132] John Carey, *John Donne: Life, Mind, and Art* (2nd ed., London, 1990) p. 265: also ch. 7 in general on Donne's 'passion for fusion or interpenetration'. For related themes in Donne compare Leonard Barkan, *Nature's Work of Art: The Human Body as Image of the World* (New Haven, 1975) pp. 51-60.

[133] Jonson, *Cynthia's Revels*, V.2.52-3.

[134] Robert M. Durling, *Petrarch's Lyric Poems* (Cambridge, MA, 1976) Rime XXIII, line 157, with Barkan, 'Diana and Actaeon' (Note 127) pp. 335-8, and Vickers, 'Diana Described' (Note 127), especially pp. 103-5.

[135] There are outstanding discussions of the shifting lines of individuality in Jacobean drama in Dollimore, *Radical Tragedy* (Note 4) chs 1, 10, 15, and Catherine Belsey, *The Subject of Tragedy* (London, 1985) chs 2 and 4.

[136] Scarry, 'Donne' (Note 5); there is more rich material in D.C. Allen, 'John Donne's knowledge of Renaissance medicine', *Journal of English and Germanic Philology* 42 (1943) 322-42, especially sections 4-6 on dissection, physiology, and theories of reproduction. On the ambiguous Jacobean kiss Stallybrass, 'Reading the Body' (Note 128) pp. 133-4, and also pp. 139-42 on unmasking and dissolution.

[137] Descartes to Mersenne, Nov. 20, 1629, *The Philosophical Writings of Descartes* (Note 86) vol. 3 (1991), p. 12, AT I.80, in the context of a discussion of a project for a new language: see also Slaughter, *Universal Languages* (Note 4) pp. 127-9.

JAMIE C. KASSLER

ON THE STRETCH: HOBBES, MECHANICS AND THE SHAKING PALSY

... it is not when parts are more relaxed than usual that spasms and tetanus supervene, but when they are more on the stretch.
Hippocratic writings 'On Fractures'

1. PREFACE

In 1647 the much maligned English philosopher, Thomas Hobbes (1588–1679), had a protracted illness, which he afterwards described briefly in a letter and from which description we may identify four stages. First, there was a painful and continuing fever, during which period Hobbes was delirious. This lasted six weeks. Next, as the fever waned, abscesses ('apostemata') broke out, so that he was confined to bed for four more weeks. Then, after the abscesses healed, he suffered excruciating pain, which he attributed to sciatica ('ischidiaca'). Indeed, the attending physician recorded that the pain was such that Hobbes wished to kill himself. And finally, there was apparent recovery, for as the pain became milder, the will to live reasserted itself.[1]

Shortly after this illness Hobbes' hands began to tremble, thus manifesting the first symptom of a 'shaking Palsey'. According to Hobbes' friend and biographer, John Aubrey, the tremor 'began in France before the year 1650, and 'haz growne upon him by degrees, ever since, so that he haz not been able to write very legibly since 1665 or 1666, as I find by some letters he hath honoured me withall'.[2] By 1663, according to another friend, Robert Hooke, Hobbes' hands 'shook as fast one way as his head did the other'.[3] And Aubrey also reported that Hobbes was 'for severall yeares before he died so Paralyticall that he wase scarce able to write his name, and that in the abscence of his Amanuensis not being able to write anything, he made Scrawls on a piece of

G. Freeland and A. Corones (eds.), 1543 and All That, 151 –187

paper to remind him of the conceptions of his Mind he design'd to have committed to writing.'[4]

From this evidence, we may conclude that Hobbes' shaking palsy followed the pattern first described in detail by James Parkinson, after whom the syndrome now is named.[5] If we adapt Parkinson's description to the case in question, we may gain some insight into Hobbes' experience, which would begin with a slight sense of weakness and a proneness to tremor in one of his hands and arms. Although these symptoms would have gradually increased in the part first affected, they would have been felt in some other part, presumably the head, within less than a year. Not long afterwards Hobbes would have found difficulty in preserving an upright posture; and as this symptom increased, one and then the other of his legs would have been affected with tremor and loss of power.

Up to this point, Hobbes would have experienced little inconvenience because of the strong influence of habitual endurance. Soon, however, the performance of skilled tasks would require increased attention and greater effort. Being harassed by a tormenting round of tremor, Hobbes would seek relief in exertions such as walking or tennis, for these would divert his attention from the unpleasant feelings. But, eventually, even this temporary relief would be denied, because Hobbes would develop a propensity to lean forward, thus being forced to walk on tip-toe and, at the same time, irresistibly impelled to take much quicker and shorter steps, and thereby to adopt unwillingly a running pace. At last, the trunk of Hobbes' body would become almost permanently bowed, the influence of the will over the muscles would decay little by little, and tremor would become so vehement as to make sleep impossible. Since the disease rarely kills, release would come from other causes, which in Hobbes' case was a stroke.[6]

To understand the disorders of Parkinsonism, one needs to understand normal movement to see where the strategy has fallen down. Such an understanding, I believe, is to be found in Hobbes' neglected philosophical masterpiece, which was partially written before 1647 and was subsequently completed and published, first in Latin in 1655, and then in English in 1656, under the title, *De corpore, Concerning Body*.[7] In this work Hobbes sought to describe nature by recourse to the concept of continuity. In so doing, he rejected the species theory of the Peripatetics, the continuous atomism of Descartes, and the emission theory of the neo-Epicureans, for all these theories described nature in terms of recognisable individual elements, like the numbers 1, 2, 3,....[8] But individualised elements, like numbers, are not the appropriate image for the concept of continuity.[9] For this concept we require an image like a string stretched in three dimensions with resistance.

The Stoics had used this image of a stretched string;[10] but Hobbes reinterpreted the Stoic image, for he conceived functions immanent in organisation like bundles of strings of different lengths and tensions. He then developed this conceptual model by drawing on the new science of music that emerged in the early part of the seventeenth century from problems relating to musical strings, elastic substances that vibrate.[11] It is noteworthy, therefore, that during his third residence in France, which extended to eleven years, Hobbes paid daily visits to Marin Mersenne, who in 1636/37 established the chief laws of the stretched string: that the frequency of the string is directly proportional (a) to its length, (b) to the square root of the weight which stretches it, (c) to the square root of the weight of the string itself, (d) to the reciprocal of its diameter, and (e) to the reciprocal of its specific gravity.[12] And in 1644 Mersenne published a collection of tracts, in which he expounded some of Hobbes' ideas.[13]

De corpore, then, is a theory of mechanics; but it is a practical theory, for it presents a methodical procedure for investigating motion. This procedure is threefold, for first, 'we are to search out the ways of motion simply'; next, 'the ways of such generated motions as are manifest'; and lastly, 'the ways of internal and invisible motions'.[14] The initial study is that of geometry, which treats ideal motion or motion in the abstract. The second study is kinematics, which deals with the motion of particles and bodies. And the third study is that of dynamics, which is concerned with forces and how they affect the kinematics. According to Hobbes, only the last study is 'the enquiry of natural philosophers'.[15] And in the epistolary dedication to *De corpore*, he observed that, in England, 'the only true natural philosophers' are physicians, especially 'our most learned men of the College of Physicians in London'.[16]

You will note that Hobbes' method proceeds inward, like the great anatomy of Andreas Vesalius, who, in *De humani corporis fabrica libri septem* (1543), stripped the human creature layer by layer in an investigation that also was a method of discovery.[17] Moreover, this stripping away was depicted in a number of magnificent illustrations, some of which revealed bundles of strong cords stretched tight, like the tuned strings of Hobbes' own instrument, the bass viol

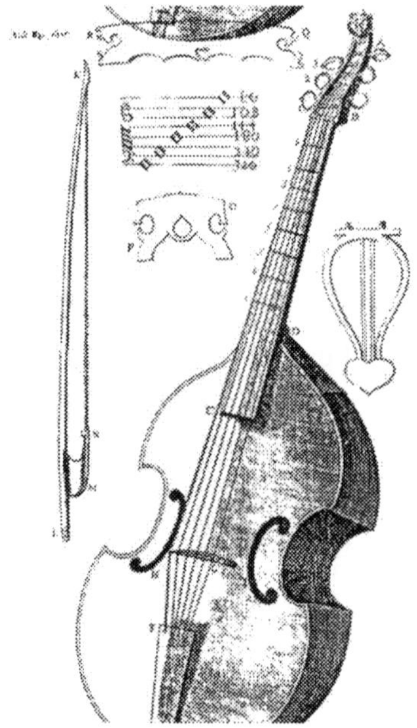

Figure 1. *Bass viol,* from M. Mersenne, *Harmonie universelle* (Paris, 1636/7).

(see fig. 1).[18] In the early 1640s, in Paris, Hobbes read Vesalius' book with his friend, William Petty,[19] and it is possible that during this period the idea took root that the cohesion of body—its holding strength and degrees of resistance—could be explained by recourse to stretched strings of different lengths and tensions.

That Hobbes was investigating problems of cohesion appears from two tracts written before and after his reading of Vesalius. In the first tract, *De mundo dialogi tres* (1642), he attributed the cohesiveness of body to the resistance of its parts to any exterior motion tending to destroy it and explained the motions of the earth as the effect partly of wave fronts created by the contraction and dilation of the sun and partly of the earth's turning like an animal before a fire, so that each section successively receives solar heat.[20] And in a second tract, now known as *Tractatus opticus* (1644), he argued that body has degrees of resistance which affect the kinematic propagation of physical rays from the contracting and dilating sun; and he conceived physical rays as infinitesimal portions of an expanding wave front.[21]

The term, resistance, utilised in both tracts, implies cohesive force and, hence, is an 'enquiry of natural philosophers'. It is noteworthy, therefore, that Hobbes' interest in such enquiries predates the 1640s, as appears from his friendship with the physician, William Harvey.[22] It is well known that the two men became close friends during the last years of Harvey's life, when Hobbes was completing *De corpore*. But their first meeting seems to have occurred between c.1621 and c.1626, when Hobbes was amanuensis to Francis Bacon, one of Harvey's patients. Harvey's researches on muscle and on animal movement were carried out between c.1616 and 1628, and there are striking similarities between some portions of Harvey's draft treatise on animal movement and Hobbes' chapter on the senses in *De corpore*.[23] From internal evidence, that chapter appears to have been written earlier than other portions of the treatise, some of which are developments from Harvey's last work, *De generatione animalium* (1651).[24]

Harvey began writing *De generatione animalium* in the early 1630s, when he also performed dissections on the King's deer.[25] During this period Hobbes witnessed Harvey's dissections, which were vivesections—in the tradition of Vesalian anatomy. Indeed, Hobbes recorded that:

> At the breaking up of a Deer I have seen it [pulsating motion] plainly in his Bowels as long as they were warm. And it is called the *Peristaltique* Motion, and in the Heart of a Beast newly taken out of his Body; and this Motion is called *Systole* and *Diastole*. But they are both of them this compounded [i.e., pulsating] Motion, whereof the former causeth the food to Winde up and down through the guts, and the latter makes the Circulation of the Blood.[26]

The gastrointestinal tract, as well as the heart, are cordlike substances that pulsate. Although the sensory 'image' of cords pulsating may suggest a regularity in nature, the image must be brought to certainty by testing through experience. For this, Hobbes could not rely on images from vivesection or from anatomical drawing. Instead, he would require experimental analogues such as lute or bow strings, for these may be stretched and set vibrating in different ways. But the lute and bow are important for another reason, for when a stretched string is plucked or otherwise acted upon, there is an interaction between *two* bodies: in the case of the lute, between string (or strings) and soundboard; in the case of the bow, between string and lath. Such analogues, therefore, might suggest an explanation of mind/body interaction as a conflict of forces. And the same analogues might suggest an explanation of tremor, which, physiologically, is a more or less regular rhythmic oscillation of part of a body around a fixed point or points involving alternate contraction of agonist and antagonist muscles.

Contrary to current opinion, Hobbes did utilise experimental analogues, as is evident from a number of his writings, including passages in *De corpore*.[27] Moreover, these analogues provided the model for his theory of motion and, especially, that part of it devoted to 'internal and invisible motions'. By way of introduction, therefore, I begin with a brief historical review of bio-mechanics, with special reference to forces that control body movement. But my main purpose is to elucidate Hobbes' chief dynamical construct, as it is found in *De corpore*, since this is the fundamental construct of his last philosophy.[28] Indeed, when properly understood, this construct explains Hobbes' physiology as well as his pathology. In *De corpore* Hobbes had little to say about pathology, which, for him, was a study of the passions. Although he presented several such studies, I draw on that which he presented in *Leviathan,* Hobbes' rhetorical masterpiece, which was written during the onset of Parkinsonism and published in 1651.[29] Finally, I treat the reception of Hobbes' theory by some of his contemporaries, with particular attention to Robert Boyle, who sought to undermine the theory, and Robert Hooke, who promoted and developed the theory.

2. INTRODUCTION: *TONOI, TONOS, TONIKE KINESIS*

From early times up to, and including, the seventeenth century, there was confusion about the cordlike substances in the body, which traditionally were called *neura* (from *neuein*: to bend) or *tenon* (from *tenein*: to stretch). In the Hippocratic corpus, however, the term *tonoi* is introduced and used interchangeably with *neura* to denote strong stretched cords, usually tendons or ligaments, that serve as a bond for the joints and give the body flexibility as well as the ability to bend and stretch.[30] Even though, in the same corpus, the heart is described as 'an exceedingly strong muscle—'muscle' in the sense not of 'tendon' but of a compressed mass of flesh', there is no concept of muscle as an organ with a capacity for contracting and thereby performing work.[31]

Nor does Aristotle adumbrate the concept of muscle as an organ with a special function, even though he recognised muscles, but not nerves, as a separate tissue group.[32] Instead, he explained the mechanical aspect of movement by the tendons attached to the bones, whereby the joints are either pushed or pulled.[33] But, for him, tendons and other bodily parts function in virtue of their connection to a motor centre, which Aristotle conceived as a fulcrum, the unmoving point from which opposite motions arise. Hence, functions immanent in organisation operate on the principle of the lever: Just as the extremities of the balance arm move in a greater arc than do points closer to the centre, so too a small change at the centre of a symmetrical body initiates motion which becomes ample in the extremes.[34]

In Aristotle's bio-mechanics, there are three classes of animal movement—voluntary, involuntary and non-voluntary, the last being partly in our power, as when we decide to hold our breath for a short period of time. But a phantasm is involved in all three classes, the main difference being that involuntary and non-voluntary movements are initiated by appetite—phantasms without reasoning, whereas voluntary movements are initiated by desire—phantasms with reasoning. Hence, self-motion is possible only insofar as a living body has appetite; and soul, a force that Aristotle located in the heart, is the source of every vital change in, or movement of, a living body, although every such change involves both body and soul. But the instrument of soul, as well as the instrument of instruments, is heat—connate *pneuma*, so that the brain functions as a counterbalance by providing coolness.[35]

In their criticism of Aristotle, the Stoics retained the notion of the heart as a motor centre and seat of the principal part of an extended soul or warm breath—*pneuma*.[36] But the Stoics replaced Aristotle's lever principle with that of stretched strings to illustrate their doctrine that *pneuma* pervades and totally mixes with unformed body so as to act as a cohesive and binding force—*tonos*.[37] In one important illustration, for example, Chrysippus compared the universe to a lyre that coheres and is harmonious, because it is bound together by *tonoi*.[38] Like lyre strings, *tonoi* function as a binding agent when they possess tension; and when they lack tension, the cosmic lyre, no longer bound up begins to dissolve. But *tonos* also causes cohesiveness in humans, for Chrysippus compared the *tonoi* of the body to the *tonoi* of the soul. The former are said to be well or ill tuned in respect to cordlike substances, whereas the latter may be spoken of as in or out of tune.[39]

The Stoics, therefore, developed a theory of the soul as an attunement. But their doctrine must be distinguished from the Platonic (Pythagorean) one, in which the soul is conceived as a harmony of abstract numbers and the relation of soul to body as an organisation by a pre -existing harmony of numbers that outlasts the body. Against this dualist theory, the Stoics conceived soul as a body—*pneuma*, thereby treating the relation of soul to body as an interaction or sympathy between two bodies, the cohesion of which is maintained by *tonos*. Hence, *pneuma* functions not only as a binding force but also as an instrument that generates physical states in body, the cordlike substances of which contract and expand with temperature changes. Such substances, therefore, require continual attunement, like strings of a lyre; and this attunement is affected by *tonike kinesis*—tonic motion.[40]

Doxographers do not agree on the meaning of this term,[41] despite the testimony of Galen, who conceived tonic motion as a local motion, when he identified four kinds of muscular motion: (1) the active motion of contraction; (2) the passive motion of relaxation due to the contraction of an antagonist muscle; (3) motion that involves neither contraction nor relaxation but which is the result of gravity—as when a raised arm falls; and (4) apparent immobility that still involves the active motion of contraction—as when the muscles of an extended arm under strain are in a state of excitation (which implies motion) and yet at rest as a whole.[42] It was this last kind of motion that Galen called '*tonike kinesis*' and that we now call 'active posture'.[43]

Galen thus identified contraction as the chief function of muscle; and he also recognised that muscles are innervated: nerves are resolved in the muscle, they assemble again into thicker branches and finally leave the muscle in the form of a tendon, to which is added ligaments and bands by which the muscle issues from and is attached to the bone. Nevertheless, he treated muscle as a kind of supporting tissue or packing between skin and bones, because he retained Aristotle's doctrine that tendons bend the joints. And Galen, too, believed that body is wholly in service of the soul, since muscle can never reach complete contraction without the cooperation of *tonos tes psyches*, a force that Galen seated in the brain and distributed to the various muscle groups via nerve channels.

Despite relocating the motor centre, Galen retained Aristotle's notion of the heart as a container by distinguishing between two types of fibre, voluntary and involuntary. The former, which run in the same lengthwise or crosswise direction, are muscle; and the latter, which run in several directions—as in the heart—are muscle-like. Nevertheless, he believed that an innate tension or *tonos symphytos* is present in all parts of the body, though variable in strength, since both types of fibre contract and, hence, perform work. But Galen referred the work of involuntary fibres to gross peristaltic motions of whole organs so as to explain the attraction of food, its assimilation and retention, and the discharge of surplus material, thereby restricting the work of voluntary fibres to the four muscular motions previously enumerated.

From the foregoing, we may identify three problems requiring solution. First, there is the nature of contractile substance: is it muscle or tendon? According to Galen, muscles consist of a fleshy part, serving as packing material, and a fibrous part that passes into and is continuous with the fibres of the tendon at each end, the tendon being the contractile substance. In the sixteenth century a small number of writers, including Vesalius, disputed the tendon theory by arguing that the fleshy part contracts.[44] But if muscle itself is the contractile substance, what is the contractile element: parenchyma or fibre? Galen's answer was: the ligamentous elements which spread into the muscle from the tendon. Yet, no definitive answer could be given to this question until the microscope revealed finer anatomical detail; and even then the notion of fibre as the element of form and function of the organism was slow to develop.[45]

Second, there is the characterisation of muscles as organs of voluntary movement, since, according to Aristotle and Galen, there are no mindless movements; and movement itself is not a primary expression of life, since it is a process emanating from the soul in which muscles have no share. Yet, there are many movements of which we are aware but cannot control, as the Stoics seem to have recognised when they classed tremor from fear as a voluntary reaction and tremor from sudden shock as an involuntary reaction.[46] And much later, in the notebooks of Leonardo da Vinci, we find the observation that paralytics or those who are shivering or benumbed by cold move their trembling limbs such as the head or the hands without permission of the soul; which soul with all its powers cannot prevent these limbs from trembling. The same happens in the case of epilepsy or with severed limbs such as the tails of lizards.[47]

For a solution to this second problem, two things were required. First, it was necessary to redefine muscles as organs of movement. This characterisation is implicit in the writings of Jean Fernel, who pointed out that even the strongest will is unable to cause movement in a paralysed limb and offered this as proof that movement has relation not to appetite or desire but to the capacity of a muscle.[48] Second and related to the first, it was necessary to treat some muscular movements as mindless, that is to say, as reflex. In 1632 René Descartes wrote a tract devoted chiefly to mindless movements; and in 1664 Thomas Willis utilised the term, reflex, when attempting to describe the neural processing of various reflex actions.[49]

Finally, there is a third problem—*tonike kinesis*, the motion underlying static equilibrium, as in active posture or tetanus. Galen supposed that static equilibrium was caused by equal but opposing forces inside, not outside, the body, although the rapid succession of motions in opposite directions occur at such speed that, to the observer, the body seems to be in a state of rest. Although Galen thus described *tonike kinesis* as an oscillatory motion, he believed that an explanation of such motion was to be sought from mechanics, not from muscle physiology. But mechanics from the time of Galen to the seventeenth century was Aristotelian, and the Aristotelian doctrine of contrary motions denied that oscillations are, or can be, continuous and regular. It was Galileo Galilei who, in 1638, refuted the Aristotelian doctrine by demonstrating, from his work on the pendulum, that opposite motions can be continuous and regular.[50]

In the 1620s, before Galileo published his refutation, William Harvey reviewed the classical, as well as some sixteenth -century sources in his draft treatise on animal movement.[51] Against the tendon theory of Aristotle and Galen, he maintained, with Vesalius, that contractile substance is muscle flesh and contractile element, fleshy fibre.[52] But, in this text, he failed to resolve the problem of the motor centre, as is evident from his explorations of the role of the heart and brain by means of various musical and other analogies.[53] Nevertheless, there is at least one hint that Harvey may have been dissatisfied with the doctrine that, for self-motion, body is wholly in service of the soul, for he queried what kind of movement scratching is.[54] Moreover, he did not repeat the classical definition of muscles as organs of voluntary movement; instead, he emphasised three things: all movements begin from contraction;[55] muscles 'pulsate' when working;[56] and coordinated movement relies on many muscles working together in 'harmony and rhythm'.[57]

In considering the problem of tonic motion, Harvey provided a number of instances, mostly from Galen—for example, tetanus, an arm outstretched, a bird hovering in air, a fish swimming upstream. But he also included leaning and standing when awake, as well as sitting, standing, walking, retaining excrements, keeping the mouth closed when asleep.[58] In quoting from, or paraphrasing part of, Galen's *De motu musculorum*, Harvey wrote that muscles 'act and are not moved in tonic motion'.[59] But he also recorded, that tonic motion, when it waxes and wanes (*vacillat*), creates tremor.[60] And in a separate section on tonic motion, Harvey commonplaced Galen's definition that 'tonic motion occurs when both muscles act against each other and do not yield'.[61] But he then added: 'this is only one kind of tonic motion. The other kind is seen when a muscle is kept contracted as happens in an extension made by a single or coincident action'.[62]

Of the greatest importance, however, are three other brief additions to the Galenic doctrine of tonic motion. First, when paraphrasing Galen that tonic motion of the parts occurs 'when two muscles act in opposition to each other whose weakness gives tremor', Harvey added: 'balance needle'. Second, when surveying his teacher's version of the Galenic doctrine that holding is effected 'by having one part without movement or partly without movement as in tonic motion' and weakness is revealed as tremor, Harvey added: 'balance, compass'. And third, in considering perhaps his own notion of the pulsation of muscles 'according to rhythm', Harvey queried: 'Is alternation like iron in equilibrium? Or like the compass to the lodestone?'.[63] All these additions suggest that Harvey conceived the motion that gives tone by analogy to the minute jiggle of the needle of a compass when excited by magnetism.[64] But there is an added implication that tremor appears when tonic motion lacks harmony and rhythm.

With this background in mind, we may now turn to Hobbes' own movement disorder, for the most disabling symptom of Parkinsonism is an impairment of the voluntary control of movement. It is this impairment that the patient experiences as an inner conflict between the forces of impulse and resistance or to use psychological language, one's body won't do what one tells it to.[65] Sometimes the patient may experience the conflict as an obstructive will, when the performance of normal actions is rendered difficult or impossible; other times, as an explosive will, when dominant, abnormal actions are irresistible. In both types of experience, patients find 'that as soon as they "will" or intend or attempt a movement, a "counter-will" ... rises up to meet them'.[66] From the patient's point of view, therefore, Parkinsonism is a conative disorder, that is, a disorder of effort or endeavour.[67]

3. ENDEAVOUR AS TENSION

3.1. Organised body: Hobbes' dynamic physiology

The term, endeavour (*conatus*), is Hobbes' fundamental dynamical construct; and in *De corpore* he provided its only complete exposition. According to the developing argument presented there, endeavour is a process entailing deformation ('action') and restitution ('reaction') through continuous intervals to and through the infinitesimal vibratory motion that constitutes rest. To explicate this process, Hobbes had recourse to a number of different kinds of stretched strings, including the archer's monochord, as follows:'

> ... when the lath of a cross-bow bent doth, as soon as it is at liberty, restore itself, though to him that judges by sense, both it and all its parts seem to be at rest; yet he, that judging by reason doth not account the taking away of impediment for an efficient cause, nor conceives that without an efficient cause any thing can pass from rest to motion, will conclude that the parts were already in motion before they began to restore themselves.[68]

All bodies, according to Hobbes, have 'the beginning of their restitution within themselves, namely, a certain motion in their internal parts, which was there, when, before the taking away of the force, they were compressed, or extended'.[69] Hobbes, therefore, conceived rest not as the opposite of motion but as thoroughly kinetic—like the indiscernible motion or *z* force that maintains tone.

In this neat example we have a demonstration of elasticity, as well as conservation. Elasticity relates to the possibility of getting mechanical work back from a system after it has been deformed by mechanical work done on it, whereas conservation relates to the possibility that upon deformation, energy is not lost but is transformed into the internal motion of the body's constituent particles. In practical cases, some of the mechanical energy put in is dissipated in overcoming internal frictional forces during the deformation. And some of the stored energy is dissipated against internal friction during the relaxation phase. Consequently, we cannot get back quite as much mechanical work as we put in, and the work 'lost' is dissipated as heat.[70] In ideal cases, the whole of the work done on the body during deformation can be recovered during relaxation. This implies not only that the body returns precisely to its original shape and dimensions when the deforming force is removed, but also that at each value of the deforming force the magnitude of the deformation is the same no matter whether the force is increasing or decreasing.

The study of deformable media is called 'continuum mechanics', the aim of which is to give theoretical insight into deformation and flow. In this kind of study, the kinematics deals mainly with motions that are deformations, whereas the dynamics deals mainly with stress, that is, with the forces exerted on one part of a body by a neighbouring part. The detailed discrete nature of the atomic and molecular structure is ignored, because it is assumed that the net effect of the atomic and molecular forces and motions can be adequately represented by the stress, the strain or rate of strain, and an appropriate constitutive relation. Although Hobbes introduced continuum mechanics to English readers, it was his younger contemporary, Robert Hooke, who provided the first constitutive relation known as Hooke's Law: *ut tensio, sic vis*, as the stretching, so the force.[71]

Nevertheless, Hobbes' example of the crossbow demonstrates the relation between stress and strain, for until the bow is strung, the string lies slack. As the bow is bent, its 'spring' or elastic force is felt. When the string is fitted to the bow, it then bears the elasic force and acquires tension. When the string is pulled, it resists the pull; and when the string is released, it tends to spring back to its equilibrium position. The same is true of the lute, with its bundles of strings. And, according to Hobbes, it is true also of particles of matter, for when urged against its neighbour by some exciting force, a particle may, by a sharp effort and sudden recoil, deliver up its motion in virtue of the elastic force exerted between the two particles.[72] The greater the elastic force, the more rapid the delivery and recoil, and vice versa. Elastic force, therefore, is like a sharp muscular effort and sudden recoil, for it is action and reaction 'in the same instant'.

The phrase, 'in the same instant', occurs frequently in *De corpore* and, like Hobbes' other key terms, has a number of different applications. When understood in the context of processes that are perfectly inverse, the term, regardless of its application, signifies that point when the conclusion of one movement is identical with the beginning of the opposite and conversely.[73] For example, in continuous processes, such as oscillatory systems, 'in the same instant' is that point when kinetic motion ('action') and dynamic tension ('reaction') conjoin, thereby producing the 'entire cause' or turning point, in which action is reversed. And in continuous processes, such as the flow of impulses that constitute our sensory life, the turning point is that instant when we decide to act or not to act. Hobbes called this point 'the beginning of voluntary motion', and it is like the Stoic apprehension or grasping, that moment when the hand contracts into a fist.[74]

If we apply Hobbes' endeavour concept to the psychomotor domain, there are three aspects to note. First, rest is only an appearance, because body is maintained in a permanent state of tension by a real but infinitesimal vibratory motion, the *z* force necessary for the performance of all action. This is like muscular tonus, the slight resistance that normal relaxed muscle offers to passive movement. Second, when a body is set vibrating by an impact, blow or friction, the reaction is equal to the action. This is like muscular contraction, the active resistance of a muscle to an extending or compressing force. And third, the two processes, taken together, constitute Hobbes' 'entire cause', that which adjusts the motions of deformation, thereby enabling the restoration of equilibrium. This is like the muscle sense, the feeling of movement, weight, resistance and position which gives us not only an image of our body and its relation to surrounding objects in space but also an awareness of ourselves.[75]

Endeavour, therefore, is a condition of physiological equilibrium which, for Hobbes, is always associated with the presence of life. But equilibrium is not static, for internal tension is a constant infinitesimal vibratory motion that changes with time. Thus, any disturbance of this constancy, resulting in an excess of tension and leading to the attainment of goal and quiescence, can be said to exhibit the operation of a dynamic equilibrium. A recurrence of the same disturbance leads to the perception of cues, to increased tension and to appropriate action. In other words, repeated disequilibrium leads to learning by experience, which Hobbes exemplified by an aspiring musician learning to finger a musical instrument.[76] The mechanisms of equilibrium ensure that the sentient, so long as it remains alive, will tend toward optimal functioning; when optima are attained, so too is equilibrium. Indeed, for Hobbes, the underlying goal of living systems is self-optimisation.

But how is this goal to be achieved? In *De corpore* Hobbes provided a hint, when, in the concluding chapter on gravity, he made an analogy between the force of attraction and the sympathetic vibration of lute strings.[77] When stretched, a musical string becomes an extended object that can store energy in a series of discrete modes like musical pitches.[78] And if we pluck a second musical string that is in tune with the first, the latter will begin to sound in sympathy. It is possible in this way to transfer sound energy from one musical string to another, and pitches that take up energy in this way are said to resonate. Resonance frequency requires minimum force to maintain the vibrations, thereby overcoming gravitational, damping and stiffness forces that tend to diminish them, so that resonance is a periodic forcing function which not only supplies energy but also supplies it rhythmically with temporal consistency. But to do this, the strings must be tuned to the same tension, for if the strings are out of tune, a transfer of energy cannot take place.[79]

If we apply Hobbes' mechanics to a living organism, then the basic criteria for self-preservation and self-restoration recur as types of vibratory motion. Since the organism, like a lute when played upon, has continually changing degrees of tension, resistance (cohesive force) is important in Hobbes' conception of function. Indeed, the development of a living organism takes place because of the resisting power of the central systems (vascular and nervous), which are in a state of static tension. Like a lute with its bundles of stretched strings, each part is in relation to every other, so that any change in static tension will be due to elastic stress or strain, and the restoring force will be instrumental in all autonomous activities, including voluntary movement.

3.2. Disorganised body: Hobbes' dynamic pathology

From the foregoing, we may conclude that Hobbes did not rely on a simple mechanism that produces tension in response to extension. Rather, he conceived body as a system of units—bundles of strings of different lengths and tensions. He then defined work as a change in the configuration of such a system in opposition to the forces resisting it, and he treated energy as the capacity to do work. In passing through any cycle of changes of configuration, a physical system ('body') does the same quantity of external work which is done upon it, so that the energy derived from systems without is compensated for by an equal amount of energy communicated to external systems. When forces are in disequilibrium, action continues until equilibrium is reached. In this way physical systems are conserved. But rest, conceived as a dynamic equilibrium, presupposes a certain degree of cohesiveness of the system in question, although this cohesiveness may be studied without making any special assertions about the material of the system concerned.[80] Instead, the fundamental premisses will derive from Hobbes' endeavour construct.

The dynamic aspect of this construct is the elastic force we call 'tension'. In physiological terms, it is the healthy tautness that our muscles keep, even when not engaged in maintaining posture or executing movements. Since antiquity this tautness has frequently been likened to the tension of a tuned lyre or bow string and spoken of as tonus. But pathologists did not, and still do not, limit the term in this way. Rather, they employ tonus in connection with all conceivable parts, thus giving rise to a large number of related terms, as one may discover from consulting any medical dictionary.[81] If we disregard atonia, which, for Hobbes, would mean death—even a living death,[82] we may identify two main pathological states, of which there are many different degrees. One state is diminished tonus or hypotonia, which means not merely a weakening of contractile parts but also a loss of elasticity or a decrease of cohesiveness in general. The other state is increased tonus or hypertonia, which denotes a strengthening of contractile parts, a gain in elasticity and an increase of cohesiveness. The one state is a deficit; the other, an excess of endeavour.

In *De corpore* Hobbes pointed out that lack of cohesion may deprive us of intelligence, since 'we observe our own body, and find that by the indisposition of the eyes, the brain, the nerves, and the heart, that is, by obstructions, stupidity, and debility, we are deprived of light, so that a fitting disposition of the organs to receive impressions from without is ... a necessary part of the cause of light'.[83] But in *Leviathan* he supposed that defects of constitution, as well as of character, arose chiefly from three types of passions that are contrary to nature, namely, weak, indifferent and excessive passions. For 'to have weak Passions, is Dulnesse; and to have Passions indifferently for every thing, GIDDINESSE, and *Distraction*; and to have stronger, and more vehement Passions for any thing, than is ordinarily seen in others, is that which men call MADNESSE'.[84]

If we recur to Hobbes' stretched-string analogues, we may understand weak and excessive passions as too little or too much tone. In the former state, the self will be slack and motion, sluggish, so that reactivity will produce too little impulse. That this is Hobbes' meaning is clear from his definition of 'DULNESSE, *Stupidity*' and the like, the names of which 'signifie slownesse of motion, or difficulty to be moved'.[85] But in the latter state, too much tone, the self will be so highly strung that reactivity will produce extravagant impulse, and the result will be 'strange and unusuall behaviour'.[86] Both types of passions exhibit lack of control, since weak passions may result in '*self-conceipt*', which is 'great *Dejection* of mind', whereas excessive passions may result in '*vaine-Glory*; which is commonly called Pride'. The 'violence, or continuance' of both 'maketh Madness', because in one state we are enslaved by causeless fears and anxieties and in the other state, held in bondage to anger and its excesses, rage and fury. And between these extremes, Hobbes wrote, there is a legion of other passions, so that 'if the Excesses be madness, there is no doubt but the Passions themselves, when they tend to Evill, are degrees of the same'.[87]

But what about Hobbes' indifferent passions? How are these to be understood? Since Hobbes defined them as distraction—a divided self—we may conjecture that his analogue was a false string, which is thicker in one part of its length than in another.[88] When plucked, the thicker part vibrates more slowly and, hence, sounds lower, whereas the more slender part vibrates more swiftly and, hence, sounds more acute. The unequal degrees of motion and rest in such a string would interfere with the motion of other strings to which it might have a relation, thus affecting resonance frequency or preventing sympathetic vibration altogether. In pathological terms, there would be a delay, as well as an exaggeration of reactions, so that many voluntary motions would be accompanied by tremor. And if the movement is complex, there would be defects of adjustment, thereby producing incoordination.

Like weak and excessive passions, tremor and incoordination are signs of an impairment in the control of voluntary movements. It is important to note, therefore, that a taut string is a rudimentary servomechanism: the output of the string is made to control its operation with a view to not allowing the output at any time to exceed or to be less than a certain value. Hence, there is a physical limit, for if the tension of the string exceeds a certain value, it is likely to break, whereas if it is less than a certain value, there will be no reactivity.[89] Just as there is a limited scale of tones in music, so too there is a limited scale of tensions in health. In both cases, music and health, the scale will consist of different frequencies of periodic vibrations, and the limits of the scale will be determined by the nature of the instruments, some of which will be more, others less resilient.

If the scale is exceeded on the minus side, the level of tonus will be too low for control to be maintained. When this happens, there will be a tendency to flaccidity, diminished strength, an obstructed will. As the degrees of tension rise, so too does power, considered as rates of energy expenditure. As the scale continues beyond measure, our complex internal tensions will consist of a mixture of a very large number of frequencies, so that if the scale is exceeded on the plus side, the level of tonus will be too high for control to be maintained. When this happens, there will be a tendency to rigidity, increased strength, an explosive will. Hence, a person may become a fast-moving projectile, for example, by shooting out the arm in anger or by flinging about the entire body in rage.

For Hobbes, therefore, passions are tendencies conceived as different degrees of tensions; and these, in turn, may incite affective states, the impulses or feeling tones by which we become aware of ourselves and the external world. But this kind of awareness is pre-rational (*pathos*), not rational (*ethos*), because passions are transient not coherent tensions. To become coherent, or in sympathy with nature, the passions need tuning by 'Method, Culture, or Instruction', for by these we 'bend' our minds to the pursuit of truth.[90] In so doing, we become aware of ourselves reflexively as the subject *and* object of impulse, for we learn to test the generalisations we make from experience. Thus, it is in our power to develop conscience or right reason, the cohesive force that enables us to live a virtuous life. But virtue is not a static state; rather, it is a continual effort to fine tune the passions, since, according to Hobbes,

> :... there can be no such thing as perpetuall Tranquillity of mind, while we live here; because Life it selfe is but Motion, and can never be without Desire, nor without Feare, no more than without Sense. What kind of Felicity God hath ordained to them that devoutly honour him, a man shall no sooner know, than enjoy; being joyes, that now are ... incomprehensible ....[91]

Many commentators have puzzled over Hobbes' remarks about God; and many, even those of his own day, have suspected him of atheism. Yet, Hobbes' remarks are neither puzzling nor impious, for we find their parallel in the book of Job, the key biblical text for those in crisis and from which Hobbes took the title, *Leviathan*.[92] In the biblical text God and Satan join in a wager to shake

Job's faith and unseat his undoubted piety. God allows Satan to scourge Job but to spare his life. In the ordeal that follows Job never wavers from his belief *that God is*, though he fails to understand God's silence and indifference. Hence, Job seeks, indeed demands, accountability for his suffering; but God, who demands unconditional worship, does not render accounts. By accepting that God is beyond causality, Job is restored to his previous condition. But he now understands that good and evil rest solely on conscience, that human suffering is not a punishment of guilt, that 'Perfect love casteth out fear', even though 'No tremendous emotion can be had without fear'. Thus does Job expand his conception of the world which is sensed, for his scale of values includes a recognition of a moral order. Although people cannot explain the source of this order, an affirmation of it is a basis for courage.

You may remember that there are two stages to Job's ordeal. In the first, Satan destroys Job's sons; but this does not undermine Job's faith, and God offers this as a proof of Job's perfect understanding. In response Satan remarks that a man will suffer anything as long as he himself is left intact. Thus follows the second stage of Job's ordeal, for his body is covered 'with sore boils from the sole of his foot unto his crown'. Job then curses the day of his birth and longs for death, but 'it cometh not'. In Hobbes' life there also were two main crises—the Civil War in which the sons of England were killed, and the fever, abscesses and pain during which Hobbes wished to end his life. But there was this difference, for, unlike Job, Hobbes would never be restored to a former self. Indeed, for thirty-two years, he would have to endure a chronic and progressively disordering syndrome, one that would steadily diminish his ability to act in the world.

4. WHICH ENDEAVOUR: PRESSURE OR TENSION?

4.1. Boyle contra Hobbes

From the foregoing we are now in a position to summarise the chief features of Hobbes' mechanics as a theory of vibration. According to that theory, action takes place in a strictly continuous medium, in which the propagation of action takes time, and its velocity depends on the mechanical properties of the medium. Since Hobbes conceived bodies as highly deformable and mutually superpositional ('penetrable'), elasticity and compressibility are properties of every portion of the medium, however small, so that even if the medium is divided indefinitely, it still will be strictly continuous. Hence, force is all-pervading but continuously differentiated, like bundles of stretched strings of different lengths and tensions. On this approach, Hobbes could provide a mechanism for conservation by conceiving rebound as an elastic process equalising action and reaction.

Hobbes began to develop this theory from about 1636 as a reinterpretation of Stoic naturalism. But Hobbes' older contemporaries, René Descartes and Pierre Gassendi, sought to protect a supernaturalistic ontology against the naturalism of the Renaissance.[93] To accomplish their aims, the supernaturalists and their followers restricted the kinds of explanatory analogy admitted into natural philosophy by arguing two things: first, that only those actions which produce, or tend to produce, locomotion were to be accepted as fundamental; second, that change of motion in bodies was to be explained only by communication from outside and not by any innate power or striving within bodies themselves.[94]

In conceiving how mechanical action might be transmitted from one point to another, the supernaturalists chose an explanation of action by impact and pressure, in which impenetrability was taken to be an ultimate property of real but indiscernibly small, individualised bodies. Hence, force manifests itself as a series of contiguous impacts, motion proceeds by jerks from point to point in a line or trajectory, and mechanics is defined as a theory of percussion. By excluding the possibility of striving in bodies, the supernaturalists were unable to provide a mechanism for the conservation of motion. Hence, they proceeded on the assumption that, upon impact, motion will be lost, because impenetrable bodies will stop, not rebound.[95]

Here, then, we have two competing mechanics—that of the continuum (superposition, infinite divisibility) and that of the discrete (impenetrability, indivisibility). But in criticising Hobbes, the supernaturalists focused chiefly on two aspects of his mechanics. First, on his premiss of more or less elasticity, organic as well as inorganic bodies will have a power of recovery, so that Hobbes' theory of restitution seemed to reinstate innate powers. Second, on his premiss that rebound is an elastic process equalising action and reaction, Hobbes' theory seemed to lend itself to a possibility observed in antiquity, when Cicero's Stoic mouthpiece, Balbus, observed that 'nature persists and coheres by its own power without any help from the gods. There is indeed inherent ... a kind of harmony or "sympathy" as the Greeks call it. But the greater it is in its own right, the less need it be regarded as the work of some divine power'.[96]

Hobbes' theory, therefore, seemed to identify the creating principles of the universe with God and, hence, to accept that the universe is governed by necessity and not by a God who transcends creation. It is not surprising, therefore, that a number of English supernaturalists devoted considerable energy to promoting and developing percussion mechanics, so as to ensure that Hobbes' mechanics did not succeed. But the history of science teaches us that there was only one mechanics before Newton;[97] and this teaching is tacitly adopted by Steven Shapin and Simon Schaffer, in a widely acclaimed book, published in 1985 under the title, *Leviathan and the Air-Pump*.[98] These authors proceed on the assumption that the hostility between Hobbes and one of his fiercest critics, Robert Boyle, arose because of different approaches to scientific method. From what has been said, I do not think the case is so simple, as will appear from a closer examination of Hobbes' theory.

According to that theory, nature ('body') is conceived as an indefinitely continuous scale of degrees of tensional forces filling space. Hence, Hobbes could argue that empty space is not empty; rather, it is filled with kinetic energy, an energy that may be likened to the zero-point jiggling of a stretched string caused by the mote of energy remaining in the system when nothing is left. From this potential energy or *z* force, nature creates something from apparent nothing. Hobbes thus laid down a challenge to religious orthodoxy by presenting a solution to the paradox, *ex nihilo nihil fit*.[99] Indeed, Boyle singled out this very solution, when he asserted that Hobbes had built 'several things in his philosophy upon the creation of the world, and an infinite power: and how a thing material can create matter, and have an infinite power, I confess I do not understand'.[100]

Boyle's comment appears in an appendix, issued in 1662 with the second edition of his tract, *Spring of the Air*, in which he provided the first complete statement of what now is commonly but not very justifiably referred to as 'Boyle's Law': that the pressures and expansions of the 'air' are in reciprocal proportion.[101] By the term 'pressure' Boyle denoted an 'endeavour outwards', and he conceived this state as a tendency to and not as an actual motion of particles constituting a fluid. A spring, not a crossbow, is the appropriate analogue for Boyle's construct, since, if squeezed, a spring tries to recover itself and exerts a certain force. In this instance force is considered as a pressure; and you can measure the force by seeing how much it will press things. If you take a spring, and press it an inch, it takes perhaps a force of 1 lb.; and it will take a force of 2 lbs. to press it in another inch. Or again, if you pull it out an inch, it takes a force of 1 lb.; and if you pull it out another inch, it takes a force of 2 lbs. This is Boyle's endeavour construct, for he conceived force as producing pressure and being measured by pressure.

Such a conception is suitable for observations on bodies at rest, but it is not suitable for observations on bodies that are free to move. In the latter case we equire a different conception of force, according to which every change of motion, either in direction or speed, must be the result of force and must be proportional to that force. This was the definition of Galileo, who, in 1638, considered forces as proportional to the motions, and the motions as proportional to the forces. When we wish to measure a force in this way, we do not press it against springs to see how much it will press them in. What we do is to cause it to act on bodies that are free to move and see what motions it will produce in them.

Galileo was concerned chiefly with the action of a body (e.g., a pendulum) under the accelerating or speed-quickening force due to gravity, the attractive force of which on any body is always proportional to the mass of that body. But he also investigated what happens when a heavy weight is hung at the end of a horizontal beam, thereby introducing the new science called 'strength of materials'. In the discourse relating to this new science, Galileo did not consider either the bending or the compression of fibres that takes place on the under side of the beam.[102] But Hobbes did just this, when he considered 'all flexion', that is, what happens in both convex and concave sides of a stressed structure.[103] And he developed Galileo's strength of materials in another respect, for he considered the action of a body (e.g., a crossbow) under the force due to elasticity.

But Hobbes also used elastic bodies for a different purpose, when he proposed that the sympathetic resonance of lute strings might explain the action of a magnetic body under the force due to attraction. In 1662 Boyle rejected Hobbes' continuum theory in favour of an emission theory, when he ascribed attraction to the impact of specially shaped particles of subtle matter, whose stream lines (magnetic 'effluvia') issue from the lodestone and pass through the pores of nearby bodies. Nevertheless, he admitted that 'Mr. *Hobbes* has another hypothesis ... but I know, that divers learned writers have absolutely rejected it, and not one such, that I have heard of, has approved it'.[104] Amongst such learned writers were Joseph Glanvil and Henry More, who explained all action at a distance by recourse to a matter-governing *anima mundi* or subtle matter. According to this explanation, the attraction between lodestones, as well as the sympathy between strings and other physical objects, takes place because a subtle matter conveys information by pushing impenetrable particles in a straight line.[105]

Glanvil and More, therefore, advocated a translation theory, according to which energy is transferred directly into the work of running the machine ('body'). The resonance theory of Hobbes is different, for it holds that actions not evidently accomplished by direct contact of objects are transmitted over long distances by the propagation of disturbances (oscillations, vibrations, undulations) through a continuous medium that is more or less elastic. This is a 'thermodynamic', not a machine theory, because energy passes through an intermediate stage of heat, the *z* force that Hobbes conceived as an infinitesimal vibratory motion. This kind of motion implies reversible change, in which each state differs infinitesimally from that preceding it. But the energy called 'heat' does not run the machine; rather, it integrates the activities of the organism ('body').

Let us now recall Hobbes' theory that bodies seemingly at rest still have *z* force, because their internal parts jiggle. According to this theory, 'body' is governed by the strict continuity of the *z* force, which is Hobbes' version of the Stoic *tonos,* the heat or active force that binds the universe into a dynamic whole. Hobbes attributed changes in degrees of elastic force to temperature changes, which, in turn, are due to the infinitesimal vibratory motion that produces more or less heat. Depending on the degree of its 'fervour', therefore, *z* force causes the mercury in the thermometer to rise and fall, just as more or less heat makes a taut string swell or shrink.

The implications of this approach are spelled out by Samuel Sambursky, who states that the Stoics had

> ... hit upon an important physical law which applies to closed systems that are not subject to any interference. All the forces acting in such a system are inner forces and their sum total vanishes. Any partial system, being open to

> influences from other parts, is imperfect in the sense that no such conservation law can be formulated for it and that changes occurring in it are the result of forces which under the prevailing circumstances are external and imposed from the outside.[106]

In Hobbes' philosophy, all functions are immanent in organisation (hence, his emphasis, in *De corpore,* on internal freedom from contradiction). But in supernaturalism only some functions are immanent, so that a transcendent God must intervene to keep the universe from running down. It seemed, therefore, that Hobbes had 'struck at the root of morality', a phrase utilised by Isaac Newton in a letter of apology to John Locke, when, during an illness, he (Newton) had taken Locke for a 'Hobbist'.[107]

4.2. Hooke pro Hobbes

Historians of seventeenth-century science have devoted considerable attention to Boyle's conception of elasticity as pressure but have paid little attention to the history of elasticity as tension. Indeed, the forces called 'tension' and 'pressure' are often confused, and the problem is compounded by inexact usage.[108] In physics, elastic force is defined as a constrained condition of the particles of a body when subjected to forces acting in opposite directions away from each other (usually along the body's greatest length), thus tending to draw them apart, balanced by forces of cohesion holding them together. This was Hobbes' definition. In common parlance, we refer to this force, or combination of forces, as tension. The same term is used inexactly to denote the expansive force of gas or vapour, properly called 'pressure'. This was Boyle's definition. To make the distinction precise, therefore, it was necessary to establish a constitutive relation for elasticity as tension. In 1674 Hobbes' friend, Petty, suggested duplicate proportion, that (in his terms) the ratio of the 'depressing powers' equals the ratio of the squares of the 'spaces of depression'.

This suggestion occurs in a tract that includes practical applications of 'measure, number and weight' to sound, pendulums, strings and bells, along with an hypothesis for magnetism, in which Petty conceived atoms as magnets.[109] But it was Petty's mathematical, not his discrete, approach that Hobbes praised in a letter to Aubrey, written in February of the same year, where he stated:

> ... if I had seen his [Petty's] book before it went to the press I would not (as he thinks) have hindered it, but done as the [Royal] Society did, that is, urged him to print it. For the doctrine is easy to be demonstrated. The last Chapter which is of Elasticity is different from the Principle which I have taken for natural Philosophy; but I am of opinion that his supposition [about duplicate proportion] is very true, and will go a great way.[110]

Petty's mathematical formulation, which is not very clear, seems to relate to pressure ('depressing powers') and, thus, to bodies at rest. In 1677 Hooke offered a different version of duplicate proportion, one more appropriate to bodies in motion: 'that the proportion of the strength or power of moving any Body is always in a duplicate proportion of the Velocity it receives from it'. This relation, he stated, is a 'General Rule of Mechanicks' and is

> ... most certainly true in the motion of Bullets shot out of Cannons, Muskets, Pistols, Wind-guns, Cross-bows, Spitting-Trunks, and the like; as likewise in the motion of Arrows shot with Bows or Ballistaæ; of Stones thrown by the hand, or with Slings; of Pendulums moved by Gravity or Weights; of Musical Strings; of Springs, and all other vibrating Bodies; of the motion of Wheels, Flies, *&c.* drawn and turned by Weights or Springs; of the motion of Perpendicularly or Obliquely falling Bodies; and in a word, of all other Mechanical and local motions, allowance only being made for the impediment of the Air or other Fluid Medium, through which the Body is moved.[111]

Then, in 1678, Hooke announced the elastic law that bears his name.[112]

It is well known that Hooke formulated this law much earlier than 1678. But it is not well known that his musical experiments demonstrated that Hobbes' mechanics was correct: bodies are more or less elastic and consist of parts in continual vibration; there are continuous forces of tension in body; and action at a distance is due to these forces, as Hobbes had hinted when he compared the force of attraction to the sympathetic resonance of lute strings.[113] In developing this hint, Hooke laid down three suppositions about magnetism: first, that 'all magnetical Bodies have the constituent Parts of them of equal magnitude and equal Tone'; second, that 'the Motion or Tone of one magnetical Body is convey'd to that of another by means of a Dense Medium'; and third, that 'the motion of the Dense Medium is Circular and Vibrating'. From these three suppositions, he wrote, 'all the *Phœnomena* of Magneticks will be most evidently and clearly, even *a priori* deduced'.[114] He then conjectured that the planets are kept in their orbits by a force analogous to that of magnetism.[115]

Like Hobbes, Hooke modelled his representation of nature by recourse to different elastic bodies, including musical strings.[116] Once stretched to its tuned tension, it is impossible to make a musical string alter the pitch of a note without altering its elastic force. This is because the string, being elastic, obeys Hooke's Law, that the time of back spring is invariable, so that the pitch of the note produced remains invariable, whatever the amplitude of vibration may be. Hooke recognised that upon this law depends the correct going of clocks and watches; and he speculated that upon this law also depends the elastic vibrations giving rise to heat, light, sound, magnetism and gravitation. It was Hooke, therefore, who provided the whole basis for a theory of elasticity and for the science that treats wave motion in deformable material media; but it was Hobbes who opened the way for Hooke's achievement.

In carrying forward Hobbes' research programme, Hooke developed a more profound understanding not only of elasticity but also of resonance.[117] Indeed, in a remarkable series of experiments on bells, he revealed that these instruments 'echo' their own tone. This insight, in turn, led him to conceive 'body' as an incessantly-ringing bell, so that the whole universe 'chimes', that is, resonates.[118] It is noteworthy, therefore, that at least one supernaturalist, Henry More, wrote against Hooke's theory, whereby Hooke responded by pointing out that More's hypothesis of an *anima mundi* ('Hylarchick Spirit')

> ... tends to nothing but the discouraging industry from searching into, and finding out the true causes of the Phenomena of Nature: And incourages Ignorance and Superstition by perswading nothing more can be known, and that the Spirit will do what it pleases. For if all things be done by an Hylarchick Spirit, that is, I know not what, and to be found I know not when or where, and acts all things I know not how, what should ... I trouble my self to enquire into that which is never to be understood, and is beyond the reach of my Faculties to comprehend?[119]

Hooke's more immediate audience, however, were Fellows of the Royal Society, where he performed many experiments to demonstrate elasticity, modes of vibration and resonance. But his ideas seemed to fall on deaf ears.[120] Or did they? That such might not have been the case appears from a tract with the title, *An Essay of the Great Effects of Even Languid and Unheeded Motion.*

Published anonymously in 1685, the tract was reissued in the same year with Boyle's name and republished in 1690.[121] Ostensibly intended as part of Boyle's history of air, the *Essay* includes reports of experiments on sound generators and on the transmission of many different kinds of sounds—for example, the human voice, the tones of stretched strings, the ringing of bells, the report of ordnance, the noise of earthquakes and the clap of thunder, all of which were Hooke's special studies. According to an advertisement at the front of the tract, the portion relating to acoustics had been printed seven years earlier in 1679, the year of Hobbes' death.

It is highly probable, therefore, that the *Essay* constitutes a covert attack on the mechanics of Hobbes and Hooke; and this probability increases if we consider the following five points: First, in the two chapters relating to effluvia, Boyle reiterated an emission theory of magnetism.[122] Second, in the remaining chapters on acoustics, he included many of Hooke's experiments, though he failed to mention Hooke or to describe the experiments accurately (as will appear from the next three points).[123] Third, although Boyle allowed that solids might be more or less elastic (a real concession, since this would contradict the metaphysic of impenetrability), he suppressed the fact that different kinds of solids vibrate in different ways, as Hooke repeatedly demonstrated from 1665 onwards.[124] Fourth, in treating sound generation and transmission, Boyle conceived vibration as a random 'agitation' of particles and not as simple harmonic motion, the importance of which Hooke understood so well, even though his 1678 analysis—one of the earliest in an elastic context, was flawed (because based on his 'General Rule of Mechanics').[125] Fifth and finally, although Boyle agreed that many effects at a distance are due to 'sympathy', he retained a translation theory in opposition to the resonance theory of Hobbes and Hooke.[126]

Although, in the *Essay*, Boyle focused on 'sympathy' and, hence, the problem of action at a distance, his earlier 1662 attack was concerned chiefly with Hobbes' conception of a *z* force, according to which rest is a real but infinitesimal vibratory motion. Since this motion maintains body in a permanent state of tension, Hobbes propounded a dynamic conception of 'body', just as Harvey did, when he hinted that the indiscernible motion that gives tone is like the jiggle of the needle of a compass when excited by magnetism. It is tempting to believe that Hobbes recalled Harvey's insight when developing his conception of *z* force from experiments with stretched strings. Whether or not this was the case (for there is little evidence to prove it), Hobbes' experiments would have demonstrated that regular (i.e., periodic) vibratory motion produces 'tone' (pitch), but when that motion waxes and wanes, 'tremor' (beating) results.[127]

5. CONCLUDING POSTSCRIPT

Although a number of physicians may be numbered amongst Hobbes' friends, biographers have yet to discover the name of the physician who treated Hobbes after his return to England from France. Hence, Aubrey remains our best source for Hobbes' regimen, which, for thirty-two years, emphasised 'moderation and regularity' thus: After waking at seven, Hobbes would breakfast, then walk and meditate till ten, when he would return to his rooms to 'putt downe the minutes of his thoughts'. At eleven, he would eat the main meal of the day, finishing off with a pipe of tobacco and a thirty minute nap. In the afternoon he would write out the thoughts minuted in the morning; and before retiring, he would close his doors fast, so that no one could hear him. Then, when abed, he would pick up his 'bookes of prick-song lyeing on his table' and would sing 'aloud (not that he had a very good voice) but for his health's Sake: [for] he did beleeve it did his Lunges good, and conduced much to prolong his life'.[128]

NOTES

[1] For the letter, written in Latin to Samuel Sorbière and dated 27 November 1647, see G. Croom Robertson, 'Some newly discovered letters of Hobbes', *Mind* 15 (1890) 446-7. See also G. Patin, *Lettres ...*, ed. J.H. Reveillé-Perise (new edn., 2 vols., Paris, 1846) vol.2, pp. 593-4. Apparently, Patin treated Hobbes for several illnesses, according to F.R. Packard, *Guy Patin and the Medical Profession in Paris in the XVIIth Century* (New York, 1970) p. 89.

[2] J. Aubrey, *Brief Lives*, ed. O.L. Dick (Harmondsworth, 1978) p. 315.

[3] (Letter to) R. Boyle, *The Works*, ed. T. Birch (6 vols., London, 1965-66), vol. 6, p. 486. For this letter and evidence of Hooke's friendship with Hobbes, see J.C. Kassler, *Inner Music: Hobbes, Hooke and North on Internal Character* (London, 1995), p. 128.

[4] Aubrey, *Brief Lives*, p. 316.

[5] J. Parkinson, *An Essay on the Shaking Palsy* (London, 1817) pp. 3-9, who did not mention such symptoms as immobile facial expression, eye movement abnormalities and loss of strength in the modulation of the voice. But he did attempt to distinguish between the tremor of Parkinsonism and other kinds of tremor. For modern attempts, see L.J. Findley and R. Capildeo (eds.), *Movement Disorders: Tremor* (London, 1984), who identify three cardinal features of Parkinsonism: tremor, rigidity and bradykinesia.

[6] According to S. Mintz, 'Thomas Hobbes', in C.C. Gillispie (ed.), *Dictionary of Scientific Biography* (16 vols., New York, 1970-80) vol.6, pp. 444-51.

[7] T. Hobbes, 'De corpore', in W. Molesworth (ed.), *Thomae Hobbes Malmesburiensis Opera philosophica quœ Latine scripsit* (5 vols., London, 1839-45) vol.1; T. Hobbes, 'Concerning Body', in W. Molesworth (ed.), *The English Works of Thomas Hobbes of Malmsbury* (11 vols., Scientia Aalen, 1962) vol.1. Hobbes did not prepare the English version; nevertheless, I rely on it here for quotations.

[8] For details, see Kassler, *Inner Music*.

[9] E.T. Bell, *Men of Mathematics* (New York, 1961) p. 13, points out that:

> *All* the points on a segment of a straight line ... have no ... clear-cut individualities as have the numbers of the sequence 1, 2, 3, ..., *where the step from one member of the sequence to the next is the same* (namely 1: 1 + 2 = 3, 1 + 3 = 4, and so on); for *between* any two points on the line segment, no matter how close together the points may be, we can always *find*, or at least *imagine, another* point: *there is no 'shortest' step from one point to the 'next'*: in fact there is no *next* point at all.

See also E.T. Bell, *Mathematics: Queen and Servant of Science* (New York, 1951) pp. 297-8, 301-2.

[10] D.E. Hahm, *The Origins of Stoic Cosmology* (Ohio State University Press, 1977) pp. 3-10.

[11] See Kassler, *Inner Music*, Chapter 2.

[12] M. Mersenne, *Harmonie universelle* ... (Paris, 1636/37), whose experiments are described by F.V. Hunt, *Origins in Acoustics: The Science of Sound from Antiquity to the Age of Newton* (New Haven, 1978) pp. 89-94. See also S. Dostrovsky, 'Early vibration theory: Physics and music in the seventeenth century', *Archive for the History of Exact Sciences* 14 (1974-75) 169-218.

[13] M. Mersenne, 'Ballistica' and 'Hydraulica', *Cogitata physico mathematica* ... (Paris, 1644), contains expositions of Hobbes' ideas as follows: 'Ballistica, Præfatio...ad Lectorem' (unpaginated on faculties of the soul); 'Ballistica, Corollarium post Prop. XI' (p. 29-30 on dynamics and 'reditus arcus'); 'Ballistica, Proposition XXIV et Monita I et II' (pp. 74-82 on theory of light and planetary motion); 'Hydraulica, Corollarium II post Propos. XXV' (pp. 129-31 on 'De parabola helici Archimedeæ æquali'). The first treatise was described by J. Wilkins, 'Mathematical magic (1648)', *The Works* ... (2 vols., London 1802) vol.2, p. 162, who wrote:

> He that would be informed in the nature of bows, let him consult *Mersennus de Ballistica et Acontismologia*, where there are diverse subtile enquiries and demonstrations, concerning the strength required to the bending of them to any distance, the force they have in the discharge, according to several hints, the strength required to be in the string of them, the several proportions of swiftness and distance in an arrow shot vertically, or horizontally, or transversally.

For Hobbes' own contribution to Mersenne's tracts, see below (Note 21).

[14] Hobbes, 'Concerning Body', p. 73.

[15] *Ibid.*

[16] *Ibid.*, p. ix.

[17] According to R. Colie, *Paradoxia epidemica: The Renaissance Tradition of Paradox* (Princeton, 1966) p. 431, the subject of Vesalius' book provided a metaphor for all sorts of 'uncoverings', in which the investigators sought to articulate, as anatomy does, the disparate parts of a system or systems into a fitting whole. For examples of such 'uncoverings' in Vesalius' book, see the muscle illustrations in the frontispiece to this volume, and fig. 8 in Martin Kemp's paper (also this volume).

[18] See Mintz, 'Thomas Hobbes'; but from the musical inventory, he also seems to have played the lute. See L. Hulse, 'Hardwick MS 29: A new source for Jacobean lutenists', *The Lute* 26 (1986) 63-72.

[19] Aubrey, *Brief Lives*, p. 399: 'At Paris he [Petty] studyed Anatomie, and read Vesalius with Mr. Thomas Hobbes, who loved his company. Mr. H then wrote his *Optiques* [see below Note 21]; Sir W.P. then had a fine hand in drawing and limning, and drew Mr Hobbes Opticall schemes for him, which he was pleased to like'.

[20] T. Hobbes, *Critique de 'De mundo' de Thomas White,* ed. by J. Jacquot and H.W. Jones (Paris, 1973), which includes, besides the edition of Hobbes' manuscript, his Latin poem on the motion of the earth and his English notes on an early draft of some chapter of *De corpore*; see also T. Hobes, *Thomas White De mundo examined* ..., trans. H.W. Jones (London, 1976).

[21] T. Hobbes, 'Opticorum libri septem Præfatio ad Lectorem IV', M. Mersenne, *Universæ geometriæ, mixtaque, synopsis et bini refractionum demonstratum tractatus* (Paris, 1644), pp. 472-5. For a detailed treatment of Hobbes' treatise and its reception, see A.E. Shapiro, 'Kinematic optics: A study of the wave theory of light in the seventeenth', *Archive for History of Exact Sciences* 11 (1973) 143-72, who is the first to point out that Hobbes began the kinematic tradition in continuum theories of light.

[22] Aubrey, *Brief Lives*, pp. 292, 318.

[23] G. Keynes, *The Life of William Harvey* (Oxford, 1966). For Harvey's treatise, see below (Note 51).

[24] Kassler, *Inner Music*, and 'The paradox of power: Hobbes and Stoic naturalism', in S. Gaukroger (ed.), *The Uses of Antiquity: The Scientific Revolution and the Classical Tradition* (Dordrecht, 1991) pp. 53-78.

[25] R.G. Frank, *Harvey and the Oxford Physiologists: Scientific Ideas and Social Interaction* (Berkeley, 1980) pp 38-42.

[26] Quoted in Keynes, *The Life*, p. 388n. Today, peristalsis is understood as a progressive wave of contraction seen in tubes provided with longitudinal and transverse muscular fibres (e.g. the gastrointestinal tract). It consists in a narrowing and shortening of a portion of the tube, which then relaxes, while a distal portion becomes shortened and narrowed. By means of this motion the contents of the tube are forced toward the opening.

[27] E.g., R. Tuck, *Hobbes* (Oxford 1989) pp 49-50, who repeats, uncritically, the interpretation of S. Shapin and S. Schaffer, *Leviathan and the Air-Pump: Hobbes, Boyle, and the Experimental Life* (Princeton, 1985). The latter authors focus on Hobbes' criticisms of Boyle's vacuum experiments, so as to depict Hobbes as an anti-experimentalist. But see Kassler, *Inner Music*, Chapter 2.

[28] I.e., works written after c.1636.

[29] T. Hobbes, *Leviathan,* ed. C.B. Macpherson (Harmondsworth, 1986). The terms, physiology and pathology, were coined in the sixteenth century by the French physician, Jean Fernel.

[30] For the nomenclature, see Hahm, *The Origins of Stoic Cosmology*, pp. 154-5, 168, 170-2, 179 n.45, and E. Bastholm, *The History of Muscle Physiology* ... (Copenhagen, 1950) p. 21 *et passim*.

[31] G.E.R. Lloyd (ed.), 'The heart', trans. by I.M. Lonie, *Hippocratic Writings* (Harmondsworth, 1986) 347-51, p. 348.

[32] In Aristotle's technical vocabulary, *neura* denotes sinews, ligaments and muscles but not nerves. See F. Solmsen, 'Greek philosophy and the discovery of the nerves', *Museum Helveticum* 18 (1961) 150-67, 169-97 and 'Tissues and the soul: Philosophical contributions to physiology', *Philosophical Review* 59 (1950) 435-68.

[33] Aristotle, *De motu animalium*, trans. by M.C. Nussbaum (Princeton, 1978). See also Bastholm, *The History of Muscle Physiology*, pp. 41-52.

[34] For the importance of the lever principle in Aristotle's philosophy, see T.J. Tracy, *Physiological Theory and the Doctrine of the Mean in Plato and Aristotle* (Chicago, 1969).

[35] Aristotle, *De motu animalium,* seems to introduce connate *pneuma* as a development of his earlier doctrine of innate heat. See E. Mendelsohn, *Heat and Life: The Development of the Theory of Animal Heat* (Cambridge, Massachusetts, 1964) pp. 11-4, who remarks that vital heat, as defined by Aristotle, was utilised in explanations by the foremost physicians and physiologists well into the seventeenth century.

[36] See B. Inwood, *Ethics and Human Action in Early Stoicism* (Oxford, 1985).

[37] For illustrations relating to bow and lyre strings, see Hahm, *The Origin of Stoic Cosmology*, pp. 153-4; for illustrations using the taut threads of a spider's web, see S. Sambursky, *Physics of the Stoics* (London, 1987) pp. 24-5, 123-4. The pre-Socratic philosopher, Heraclitus, used all three analogues to illustrate 'back-stretched connection' (i.e., tension).

[38] J. Rist, 'On Greek biology, Greek cosmology and some sources of theological pneuma', *Prudentia* supplementary number (1985) 27-47, p. 45. Chrysippus did not give up the imagery of a lyre, as Hahm, *The Origin of Stoic Cosmology*, p. 169, claims; rather, he changed the emphasis of the lyre imagery from that of a plucked lyre string (Cleanthes) to that of the whole lyre bound together by *tonoi* (Chrysippus).

[39] See Hahm, *The Origin of Stoic Cosmology,* pp. 169, 171; J.B. Gould, *The Philosophy of Chrysippus* (Leiden, 1970) p. 195.

[40] For the Stoic definition of 'primary impulse' as the power—*tonike dunamis*—to set oneself in motion in pursuit of goals, see Inwood, *Ethics and Action*, pp. 27-8. Hence, the precondition for primary impulse is tautness—*syntonia*.

[41] See Hahm, *The Origin of Stoic Cosmology*, pp. 167, 182-3, n.77.

[42] Bastholme, *The History of Muscle Physiology,* pp. 74-96; Sambursky, *Physics of the Stoics*, pp. 32-4.

[43] C. Sherrington, 'A note on the history of the word "tonus" as a physiological term', in C.L. Dana (ed.), *Contributions to Medical and Biological Research dedicated to Sir William Osler* ... (2 vols., New York, 1919) vol.1, pp. 261-8; F. Fearing, 'The reflex maintenance of posture', *Reflex Action: A Study in the History of Physiological Psychology* (New York, 1964) pp. 218-31.

[44] O. Temkin, 'Vesalius on an immanent biological motor force', *Bulletin of the History of Medicine* 39 (1965) 277-80.

[45] In 1664 Niels Stensen argued that muscles do not consist of parenchyma and fibres; rather, they are collections of motor fibres which may be divided into minor fibrils bounded together by continuous transverse fibres of the membrane of the muscle. See E. Bastholm, 'Niels Stensen's

myology', in Gustav Scherz (ed.), *Steno and Brain Research in the 17th Century* (Oxford, 1968) pp. 147-53. Long after the seventeenth century nerves continued to be conceived not as bundles of fibres but as tubular prolongations of the substance of the central nervous system, although there were some early sceptics, including Vesalius and his pupil, Gabrielo Falloppio. See, e.g., E. Clarke, 'The doctrine of the hollow nerve in the seventeenth and eighteenth centuries', in L.G. Stevenson and R.P. Multhauf (eds.), *Medicine, Science and Culture: Historical Essays in Honor of Owsei Temkin* (Baltimore, 1968) 123-41.

[46] Inwood, *Ethics and Human Action*, p. 176. According to Aristotle, *De motu animalium*, p. 42, 'sense-perceptions are at once a kind of alteration[,] and *phantasiai* and thinking have the power of the actual things. For it turns out that the form conceived of the ... pleasant or fearful is like the actual thing itself. That is why we shudder and are frightened just thinking of something'.

[47] L. da Vinci, *The Notebooks*, trans. E. MacCurdy (2 vols., London, 1956) vol.1, p. 107.

[48] Bastholm, *The History of Muscle Physiology*, p. 117.

[49] See R. Descartes, *Treatise on Man,* trans. T.S. Hall (Cambridge, Massachusetts, 1972), which was written in 1632 but not published until 1662; and T. Willis, *Cerebri anatome: Cui accessit nervorum descriptio et usus* (London, 1664).

[50] See P. Ariotti, 'Aspects of the conception and development of the pendulum in the 17th century', *Archives for the History of Exact Sciences* 8 (1971-72) 329-410.

[51] W. Harvey, *De motu locali animalium 1627*, trans. G. Whitteridge (Cambridge, 1959). The draft is more like a common place book than a treatise, as is clear from the way Harvey recorded the etymology of *neura* and *tenon, ibid.,* pp. 68 n.1, 69 and n.4, 71 and n.7, 83, 97, 115, 117; repeated Aristotle's conception of fear (see above Note 46), *ibid.*, pp. 37, 99; reviewed ideas about tremor, *ibid.*, pp. 57, 83, 97, 99, 103, 121, 125, 141, 149; etc.

[52] *Ibid.*, pp. 5, 89 and 117, in contradistinction to his anatomy teacher in Padua, Girolamo Fabrici (Hieronymus Fabricius ab Aquapendente), who retained the tendon theory.

[53] *Ibid.*, pp. 111, 150.

[54] *Ibid.*, p. 122 (*quod scalpere*).

[55] *Ibid.*, pp. 45, 103.

[56] *Ibid.*, p. 111.

[57] *Ibid.*, pp. 67, 111, 143, 145, 147, 149, 153.

[58] *Ibid.*, pp. 27, 29, 61, 65, 83, 119, 121, which includes an instance of tonic motion 'in the French execution horse'. The translator glosses this as 'the instrument of torture known as a chevalet'. But the context suggests one of the 'standing' positions taken by a horse in dressage. The terms for these positions were explained by Hobbes' friend, W. Cavendish, Duke of Newcastle, *La méthode et invention nouvelle de dresser les chevaux* (Antwerp, 1658), for whom Hobbes wrote a paper entitled 'Considerations touching the Facility or Difficulty of the Motions of a Horse on straight lines and circular'. See C.H. Firth (ed.), *The Life of William Cavendish, Duke of Newcastle ... by Margaret, Duchess of Newcastle* (2d edn., London, n.d.) pp. xiv-xvi, xx-xxii, 62-3 nn.1-2, 106 and n.1, 206.

[59] Harvey, *De motu locali animalium*, p. 57.

[60] *Ibid.*, p. 57; see also pp. 83, 121.

[61] *Ibid.*, p. 119.

[62] *Ibid.*

[63] *Ibid.*, pp. 121, 141, 153.

[64] Perhaps from W. Gilbert, *De magnete* trans. P. F. Mottelay (New York, 1958) pp. 192, 353. For Harvey's ideas about action at a distance, see Kassler, *Inner Music*, and 'The paradox of power'.

[65] For an extensive treatment of Parkinsonism from the physician's, as well as from the patient's point of view, see O. Sacks, *Awakenings* (London, 1982).

[66] *Ibid.*, pp. 10, 38; cf. W. James, *Textbook of Psychology* (London, 1892) p. 454. The distress that results may call forth a variety of other symptoms, so there is need to alleviate distress by helping to make movement easy. One of the most effective ways of doing this, according to Sacks, *ibid.*, pp. 56, 99, 115, 148, 283, 317, is through the power of music. Since the disorders of movement and force result in a loss of naturalness in posture and action, music supplies externally what the Parkinsonian patient lacks internally, for it functions as an artificial pace-maker, thus restoring, temporarily, each patient's natural 'harmony and rhythm'.

[67] Sacks, *ibid.*, p. 11, asserts that Parkinsonism 'exhibits a formal analogy of conative structure'.

[68] Hobbes, 'Concerning Body', pp. 347-8. For details, see Kassler, *Inner Music*.

[69] *Ibid.*, p. 344.

[70] *Ibid.*, pp. 324-5 *et passim*, for Hobbes' kinetic theory of heat.

[71] *Tensio* (from *tendo*): to stretch, strain; to aim, shoot (an arrow); to bend (a bow); to tune (an instrument); to endeavour, exert onself. Hooke related stress and strain to a stretched elastic spring, but his constitutive relation applies also to stretched strings and other linear systems. Although Hooke was aware of this relation as early as 1658, he announced it in an anagram in 1676 and provided details in 1678. In contemporary bio-mechanics Hooke's Law appears in models of postural mechanisms and, especially, that mechanism which determines the tone of a muscle—the stretch reflex.

[72] For some historical background, see J.L. Russell, 'Action and reaction before Newton', *The British Journal for the History of Science* 9 (1976) 25-38, who states that Newton 'did not create a law of action and reaction out of nothing. It was there already, although associated to a large extent with an obsolete Aristotelian theory of motion'. But see Kassler, *Inner Music*, for Hobbes' rejection of the Aristotelian theory of motion.

[73] For the perfectly inverse process called 'analysis and synthesis', see W. Sacksteder, 'Hobbes' *Logistica*: Definition and commentary', *Philosophy Research Archives* 8 (1982) 55-94.

[74] The Stoics identified three states in the cognitive process—presentation, assent and apprehension, the first of which is like an open hand with the fingers outstretched, the second, like the fingers contracted a little, and the third, like the hand closed entirely. In Hobbes' version, 'Concerning Body', p. 392, the incoming sensory impulses reach 'the last yielding part [the brain]; which by reaction [i.e., resistance], in the same instant, if the reaction be strong enough, makes a phantasm; and a phantasm being made, perception is made together with it'. For Hobbes, therefore, the stages are sense, phantasm and impulse, the last of which involves a contraction of the pia mater as the end link in a chain of events between the incoming stimulus and the outgoing final reaction (e.g., to act or not to act).

[75] The old term for muscle sense, kinæsthesis, was coined in the nineteenth century by H. Charlton Bastian; the new term, proprioception, was coined early in this century by Charles Sherrington.

[76] Hobbes, 'Concerning Body', pp. 349-50.

[77] *Ibid.*, pp. 527-8; see also pp. 499-500 on the causes of the 'concent of sounds'. Today, we would say that the magnetic field vibrates transversely about the direction of propagation, just like a stretched string, with the magnetic field corresponding to the displacement.

[78] If a musician sounds a pitch on a musical string, which is a linear system, the result is a superposition of overtones on the fundamental tone. For Hobbes' treatment of superposition, see Kassler, *Inner Music*.

[79] I.e., interference will take place, and this can be heard as beats, which are periodic fluctuations of loudness produced by the superposition of tones of close, but not identical, frequencies. In some cases, however, sound will be interdicted altogether. Although beats have always been known to practical musicians, Mersenne seems to have been the first theorist to describe this phenomenon. See H.F. Cohen, *Quantifying Music: The Science of Music at the First Stage of the Scientific Revolution, 1580-1650* (Dordrecht, 1984).

[80] For Hobbes, as for the Stoics, coherence is the criterion. In the case of bundles of vibrating strings, the initial conditions are each string's modulus of elasticity and density.

[81] E.g., atonic, atonic bladder, atonic ulcer, catatonic, cerebrotonic, dystonic, epitonic, hypertonic, hypotonic, hypotonic diplegia, isotonic, myotonic, myotonic acquisita, myotonic congenita, myotonic paradoxica, myotonic dystrophy, somatotonic, sympathicotonic, syntonic, tonic-clonic, tonic postural epilepsy, tonic seizure, tonic spasm, viscerotonic.

[82] For Hobbes, as for the Stoics, activity and existence are one, so that tonus is the condition for life, and death is a slackening of tension, where living death would be a profound state of stupor, unconsciousness or arrested activity, as in a deep narcosis.

[83] Hobbes, 'Concerning Body', p. 78; see also pp. 392, 393, 397, 400.

[84] Hobbes, *Leviathan,* p. 139.

[85] *Ibid.,* p. 135.

[86] *Ibid.,* p. 142.

[87] *Ibid.,* p. 140.

[88] As a viol and, possibly, a lute player, Hobbes would have had direct experience of false strings, although he also could have learned about them from two other sources: music treatises, including Mersenne, *Harmonie universelle*; and musicians employed by his patrons—the Cavendish family, their relations and neighbours. For some of the musicians, see, e.g., Hulse, 'Hardwick MS 29'; D.C. Price, *Patrons and Musicians of the English Renaissance* (Cambridge, 1981); and G.A. Philipps, 'John Wilbye's other patrons: The Cavendishes and their place in English musical life during the Renaissance', *Music Review* 38 (1977) 81-93.

[89] Hobbes, 'Concerning Body', pp. 475-6, where rupture means loss of continuity. This may happen in one of two ways, for either the separation may begin in the outermost superficies and proceed successively to the innermost parts thereof; or it may proceed in the convex superficies of the bowed part of a body and proceed to the concave superficies. Both types of rupture may be demonstrated by a lute or viol string which is excessively stretched beyond its tuned tension or excessively pulled or pushed from its equilibrium position.

[90] Hobbes, *Leviathan,* p. 134; see also pp. 91, 95, where Hobbes attributed night and day dreams to the 'distemper'—lack of attunement—'of some of the inward parts of the Body'. For attunement in the poetry of Hobbes' contemporaries, see J. Hollander, *The Untuning of the Sky: Ideas of Music in English Poetry 1500-1700* (New York, 1970).

[91] *Ibid.*, pp. 129-30; see also, p. 430, where Hobbes asserted: 'the nature of God is incomprehensible; that is to say, we understand nothing of *what he is*, but *only that he is*; and therefore the Attributes we give him, are ... our desire to honor him with such names as we conceive most honorable amongst our selves'.

[92] See Kassler, *Inner Music*. The same biblical text supplies Hobbes with the title for his book on the Civil War, *Behemoth.*

[93] K. Hutchison, 'Supernaturalism and the mechanical philosophy', *History of Science* 21 (1983) 297-333.

[94] M. Hesse, *Forces and Fields: The Concept of Action at a Distance in the History of Physics* (New York, 1962).

[95] According to W.L. Scott, *The Conflict between Atomism and Conservation Theory 1644-1860* (London, 1970), Descartes assumed that hard bodies rebound; but the criticism of his laws of impact led to a different conclusion, and this was maintained by most of the supernaturalists up to and including Newton. Hobbes, 'Concerning Body', pp. 334-5 *et passim*, rejected the metaphysic of impenetrability.

[96] Cicero, *De natura deorum* (2: 27-30). Hobbes, *Leviathan,* pp. 473, 414, 471-8, defined miracle as 'a work of God, (besides his operation by the way of Nature, ordained in the Creation,) done for the making manifest to his elect, the mission of an extraordinary Minister for their salvation'; but since 'Miracles now cease', he cautioned against the 'Imposture of Miracles' by false prophets.

[97] E.g., E.J. Dijksterhuis, *The Mechanisation of the World Picture*, trans. C. Dikshoorn (London, 1961).

[98] Shapin and Schaffer, *Leviathan and the Air-Pump* (Note 27 above).

[99] According to Colie, *Paradoxia epidemica,* p. 223, *nihil* paradoxes 'were engaged in an operation at once imitative and blasphemous, at once sacred and profane, since the formal paradox, conventionally regarded as low, parodied at the same time as it imitated the divine act of Creation'.

[100] Boyle, *The Works*, vol.1, p. 187; see also p. 194, where Boyle again expressed his dissatisfaction with 'that principle of Mr. Hobbes, though it be the fundamental one of his philosophy'. According to the classification offered by J.R. Jacob, 'Boyle's atomism and the Restoration assault on pagan naturalism', *Social Studies of Science* 8 (1978) 211-33, Hobbes would belong to the second group of naturalists attacked by Boyle. The first group, consisting of Roman Catholic scholastic philosophers, adumbrated what Jacob calls an 'heretical naturalism' that 'may be and often is materialistic, but it is never mechanistic. It is animistic and asserts that the world is governed by non-mechanical vital forces or spiritual agencies of divinity'. The second group, believed in God and a divine order; but 'their God does not exist outside of and above nature, and this divine order is immanent in nature'. Nevertheless, they were regarded as atheists 'not because they deny God's existence but simply because they are not theists, though because they worship a God immanent in nature they might also be called deists of sorts'.

[101] C. Webster, 'The discovery of Boyle's Law, and the concept of the elasticity of the air in the seventeenth century', *Archives for the History of Exact Sciences* 2 (1962-66) 441-502.

[102] G. Galilei, *Dialogues concerning Two New Sciences*, trans. by H. Crew and A. de Salvio (New York, 1954), pp. 115-6. According to the translators: 'The one fundamental error which is implicitly introduced into this proposition [Proposition I] and which is carried through the entire discussion of the Second Day consists in a failure to see that, in such a beam, there must be equilibrium between the forces of tension and compression over any cross-section'.

[103] Hobbes, 'Concerning Body', p. 343, did not repeat Galileo's mistake of supposing that in a stressed structure fibres are inextensible.

[104] Boyle, *The Works*, vol.1, pp. 223-4.

[105] See, e.g., J. Glanvil, *Scepsis scientifica: Or, Confest Ignorance, the Way to Science; in an Essay of the Vanity of Dogmatizing, and Confident Opinion*, ed. J. Owen (London, 1885), and H. More, 'An antidote against atheism', *Philosophical Writings*, ed. F.I. Mackinnon (New York, 1969), both of which contain an attack on Hobbes. Although Glanvil and More may be described as emanation theorists and Boyle as an emission theorist, the two types of theorists shared a common approach, as is pointed out by E.A. Burtt, *The Metaphycial Foundations of Modern Physical Science* (Garden City, 1954) p. 193.

[106] Sambursky, *Physics of the Stoics*, p. 114.

[107] Quoted in K. Dewhurst, *John Locke (1632-1704) Physician and Philosopher: A Medical Biography* (London, 1963) p. 285.

[108] E.g., H. von Foerster, M. Mead, H.L. Teuber (eds.), *Cybernetics: Circular Causal and Feedback Mechanisms in Biological and Social Systems: Transactions of the Ninth Conference, March 20-21, 1952* (New York, 1953) pp. 6-47 and pp. xvi, 3, 51-2, 68, 70, 131, where the participants fail to recognise that bio-medical and social sciences track scientific conceptual changes and technological concepts. Hence, tension, when applied figuratively (say, to physiological psychology), may derive either from physical tension, as defined in this paper, or from electrical tension, defined before c. 1882 as potential, electromotive or mechanical force exerted by electricity and after c.1882 as the stress along lines of a force in a dialectric.

[109] W. Petty, *A Discourse made before the Royal Society ... Concerning ... Duplicate Proportion* (London, 1674).

[110] Quoted in Q. Skinner, 'Thomas Hobbes and his disciples in France and England', *Comparative Studies in History and Society* 8 (1966), p. 165. For Nicholas Mercator's undated explication of Petty's duplicate proportion, see J. Aubrey, *On Education*, ed. J.E. Stephens (London, 1972) pp. 106-8. Mercator also wrote treatises on other aspects of applied mathematics, including one on music, about which see R. Hooke, *The Diary*, ed. H.W. Robinson and W. Adams (London, 1935) pp. 242, 254, and W. Holder, *Treatise of the Natural Grounds and Principles of Harmony* (New York, 1967) pp. 104-6 (whose 'Friend' is Hooke).

[111] For Hooke's 'General Rule of Mechanics', see R.T. Gunther, *Early Science in Oxford* (15 vols., Oxford, 1923-67) vol.8, pp. 186-7. For an earlier statement (1669) resulting from Hooke's efforts to demonstrate elastic rebound, see Gunther, *ibid.*, vol.6, pp. 346-8.

[112] For Hooke's Law, see Gunther, *ibid.*, vol.8, pp. 331-56. M. Hesse, 'Hooke's vibration theory and the isochrony of springs', *Isis* 57 (1966) 433-41 outlines the development of Hooke's theory without spotting the influence of Hobbes, although she states clearly what other historians tend to gloss over, namely, that Boyle appeared 'to feel little obligation to ascribe his assistant's ideas to their true author until forced to do so'. For the influence of Hobbes, see Kassler, *Inner Music*, Chapters 3 and 4.

[113] In a series of lectures delivered between 1828 and 1830, Michael Faraday reported that Hooke, in his *Micrographia* (1665), made the 'first correct observations' about the transmission of sound through solid bodies (i.e., that sound was propagated swifter through wire than air) and that his results were confirmed in Berlin in 1788. Faraday's source for this statement is C. Wheatstone, 'On the transmission of musical sounds through solid linear conductors, and on their subsequent reciprocation' (1830), *The Scientific Papers*... (London, 1879) 47-63. See also J.C. Kassler, *The*

Science of Music in Britain, 1714-1830: A Catalogue of Writings, Lectures and Inventions (2 vols., New York, 1979).

[114] R. Hooke, *The Posthumous Works* (New York, 1969) p. 364.

[115] *Ibid.*, pp. 183-5.

[116] For Hooke's musical analogues, see Kassler, *Inner Music*, pp. 124-59.

[117] L. Brodsley, C. Frank and J.W. Steeds, 'Prince Rupert's drops', *Notes and Records of the Royal Society* 41 (1986) 10-26, point out that Hooke also was the first to understand the origin of thermal stress.

[118] Kassler, *Inner Music*, details the experiments with bells, as well as the development of Hooke's thought concerning body as a resonating system (i.e., a combination of acoustic vibrators and resonators). Perhaps following Hooke, Leonard Euler conceived the sun as an incessantly ringing bell. See A.J.L. James, 'Thermodynamics and sources of solar heat, 1846-1862', *The British Journal for the History of Science* 15 (1982) 155-81, p. 156, who claims, p. 158, that J.R. Mayer rejected the bell analogy. But see J.R. Mayer, 'Celestial dynamics', in R.B. Lindsay (ed.), *Applications of Energy: Nineteenth Century* (Stroudsburg, 1976) 151-286, who, after comparing sound and light, observes, p. 152, that the 'sun has often and appropriately been compared to an incessantly sounding bell'. The analogy explained the radiation of heat but not its source, as Hooke grasped before Mayer when studying the source of earth tremors from volcanic eruptions. The former, but not the latter, can be explained by the bell analogy, just as, today, we describe the earth, during tremors, as 'ringing' like a bell.

[119] For Hooke's answer to More, see Gunther, *Early Science in Oxford*, vol.8, p. 188.

[120] The case was similar with Hooke's combustion theory, for which, see D. McKie, 'Fire and the flamma vitalis: Boyle, Hooke and Mayow', in E.A. Underwood (ed.), *Science, Medicine and History ...* (2 vols., London, 1953), vol.1, pp. 469-88.

[121] Boyle, *The Works*, vol.5, pp. 1-37.

[122] *Ibid.*, p. 22.

[123] Boyle, *ibid.*, p. 4, mentions Hooke only once, when reporting the theory of an 'ingenious person' on the cause of a flint's sparking (a marginal note identifies the person as Hooke).

[124] E.g., in describing Hooke's experiments with glass bells, Boyle, *ibid.*, pp. 30-1, ignored the patterns of vibration of the contained substance, reporting only that the substance is agitated. By contrast Hooke's glass bell experiments provided a visible demonstration of patterns of vibration, a type of experiment Hooke devised as early as 1665 with sand on a stretched drum membrane. See Kassler, *Inner Music*, pp. 157-58.

[125] See, e.g. R. Westfall, *Force in Newton's Physics: The Science of Dynamics in the Seventeenth Century* (London, 1971) pp. 211-12, 260-61.

[126] Objections to the resonance theory of sympathy seem to have been made on the grounds that the motion of the untouched string had been caused not by the 'impulse' of the air but by the propagated motion of the instrument itself to which the touched string was also fastened. Boyle focused on this very problem, which I hope to treat in detail elsewhere.

[127] A periodic vibration is heard as a tone; for beats, see above (Note 79).

[128] Aubrey, *Brief Lives*, p. 315; for historical background concerning singing for health, see G. Finney, 'Medical theories of vocal exercise and health', *Bulletin of the History of Medicine* 40 (1968) 422-49.

GUY FREELAND

THE LAMP IN THE TEMPLE: COPERNICUS AND THE DEMISE OF A MEDIEVAL ECCLESIASTICAL COSMOLOGY*

The best known passage in Copernicus' *De revolutionibus orbium coelestium* (1543) is undoubtedly that in which Copernicus likens the sun at the centre of the universe to a lamp hanging in the midst of a temple. The significance of this arresting pericope is enhanced by the fact that it appears in juxtaposition with *De revolutionibus'* celebrated diagram of the heliocentric universe. So well known, and so frequently analysed, is the passage that the historian cannot but feel some embarrassment at referring to it, much less quoting it. Yet quote it we must, for it alone can serve as the text for the present paper:

> In the middle of all is the seat of the Sun. For who in this most beautiful of temples would put this lamp in any other or better place than the one from which it can illuminate everything at the same time? Aptly indeed is he named by some as the lantern of the universe, by others the mind, by others the ruler. Trismegistus called him the visible God, Sophocles' Electra, the watcher over all things. Thus indeed the Sun as if seated on a royal throne governs his household of Stars as they circle round him. Earth also is by no means cheated of the Moon's attendance, but as Aristotle says in his book *On Animals* the Moon has the closest affinity with the Earth. Meanwhile the Earth conceives from the Sun, and is made pregnant with annual offspring. We find, then, in this arrangement the marvellous symmetry of the universe, and a sure linking together in harmony of the motion and size of the spheres, such as could be perceived in no other way.[1]

---

* An expanded version of the 4th Dyason Memorial Lecture of the Australasian Association for the History, Philosophy, and Social Studies of Science, delivered during the Association's 27th Annual Conference held at Griffith University, Brisbane, July 9-12, 1994.

G. Freeland and A. Corones (eds.), 1543 and All That, 189 – 270

As far as the issue of the origins of Copernicanism is concerned, our text needs to be taken along with references to his precursors made by Copernicus in his Preface addressed to the Pope. The passage reads:

> And first I found in Cicero that Nicetus had supposed that the Earth moved. After that I discovered in Plutarch also that certain others held the same opinion, and I have decided to quote his words here, so that they are generally accessible:
>
> 'Some say that the Earth is at rest, but Philolaos the Pythagorean says that it is carried in a circle round the heavenly fire, slantwise, in the same way as the Sun and Moon. Heraclides of Pontus and Ecphantus the Pythagorean give the Earth motion, not indeed translatory, but like a wheel on its axis, from west to east, about its own centre.'[2]

It should not surprise us that these acknowledgments of predecessors should have sparked off something of a Copernican industry inquiring into whether adherence to some ancient arcane school or other might not have been a major, or even the prime, reason that Copernicus came to advance a heliocentric cosmology. In particular, the distinguished Copernican scholar, Edward Rosen, was moved to pen a series of articles posing the key questions: 'Was Copernicus a Pythagorean?',[3] 'Was Copernicus a Neoplatonist?',[4] 'Was Copernicus a Hermetist?'.[5] Being no disciple of Frances Yates,[6] Rosen's answer to each question was a resounding 'No'. What seems not to have been asked is the question 'Was Copernicus a Christian?'; perhaps because the answer would all too obviously have been 'Yes', and that would have spoiled the game. And yet, as we shall see, there seems good reason for drawing the conclusion that, along with the Ancient schools of philosophy, there are indeed in the passage we have taken as our text very possibly allusions to an ecclesiastical cosmology whose roots can be traced back to Antiquity. And, moreover, the allusions are to be found in those seemingly quaint metaphors of the cosmos as a temple, of the sun as a lamp hanging in the midst of the cosmic temple, and of the sun as a monarch seated on his royal throne governing the heavenly bodies. Before we turn to a detailed examination of these metaphors, we should, however, set in place a bit of scenery so as to provide some context to our discourse.

First, it should be noted that some considerable question mark should be placed over the very assumption that Copernicus' commitment to an heliocentric model was principally, or even in any significant measure, a consequence of his adherence to ancient tradition at all. Rosen's work alone should be sufficient to lead us to exercise caution here. Copernicus was, in fact, as Rosen demonstrates, decidedly cavalier in his references to Ancient sources. In our text, he attributes Sophocles' expression for the sun, 'the all seeing', to *Electra* instead of *Oedipus at Colonus*. And as far as Hermes is concerned, Copernicus even manages to get his name wrong (as is confirmed by his own manuscript) calling him 'Trimegistus'.[7] There is, however, little point in our going over ground covered by others; and certainly the technical astronomical background to *De revolutionibus* lies beyond the scope of this paper. The fact of the matter, though, is that Copernicus became a Copernican primarily for astronomical reasons,[8] not because he was some sort of Renaissance magus translating Renaissance Pythagoreanism, Neoplatonism, Hermetism, or the like into astronomy. It is true, nevertheless, that Copernicus was, in part, a product of a Renaissance humanist culture.

Copernicus liked to portray himself as coming from the backwoods of Europe. The truth of the matter, however, is that the Renaissance had begun to penetrate Poland by Copernicus' time and at the University of Cracow, where he commenced his university studies, he would have enjoyed some exposure to the Renaissance *zeitgeist* then sweeping Europe. However, in 1496 he left Poland on an intellectual pilgrimage to the *fons et origo* of the new movement, Northern Italy, where he completed his education. At Bologna University his work was focussed on law, but he also studied philosophy and some medicine and learned Greek. Here it was that he read Plato and was exposed to Neoplatonism, Pythagoreanism and other such 'isms'. Clearly, he also continued his systematic study of astronomy, which he had commenced back at Cracow. In 1500 he was in Rome, where it seems that he gave a series of lectures on astronomy. He returned to Poland in 1501, where he had been appointed a canon of Varmia, secured for him by his uncle, the bishop, some years earlier. Almost immediately, however, he obtained leave of absence to return to Italy to study medicine at the University of Padua. In 1503 he took out a doctorate in canon law at Ferrara (apparently because it was cheaper to have a doctorate conferred by Ferrara than either Bologna or Padua). He returned to Padua but did not take out a doctorate in medicine. The same year, 1503, he returned to

Poland where he spent the rest of his days as an ecclesiastical administrator at Frauenburg, practising a little medicine on the side and, of course, pursuing his astronomical interests. Canon law was regarded as the most appropriate qualification for budding ecclesiastical administrators, and this would be the obvious reason he took out a doctorate in this field.[9]

There is not the slightest doubt that Copernicus loved Italy and revelled in Renaissance culture. At Padua, medicine and arts (which consisted largely of study of the Aristotelian system) were combined, so he would have been exposed to the critical, empiricist, Averroist Aristotelianism which was so strong in Northern Italy, and which was to make a major contribution to preparing the ground for the Scientific Revolution. Copernicus begins to come into focus. The *quattrocento* produced a breed of learned, cultured, urbane, humanist, eclectic, wordly-wise, and (typically) tolerant pre-Tridentine clerics not unlike, one would surmise, the sophisticated pre-Vatican II Jesuits with whom the likes of G.K. Chesterton, Hilaire Belloc and Evelyn Waugh were wont to rub shoulders closer to our own time. But, as in the case of his modern counterparts, beneath the easily worn accoutrements of a man of his times, who could pass muster in the most sophisticated quarters of society, was the committed churchman. *De revolutionibus* is the product of the man; a work of astronomy, arising out of astronomical and calendrical problems, but embellished by humanist allusions to Hermetism, Neoplatonism, Pythagoreanism and the like. But Copernicus the cleric must also not be neglected (though he usually is). We must not overlook the motivation to glorify God through the revealing of the wonder of the true system of the universe, nor should we close our eyes to possible allusions to the medieval Christian heritage.

1. THE RENAISSANCE CULT OF THE SUN

Even if Copernicanism has its genesis in current astronomical problems, it is certainly the case that Copernicus elicits support from the past, and the development of his own thinking might well have been influenced by Ancient ideas. But there is one glaring omission from the list of names in *De revolutionibus*; that of Aristarchus. This is curious as Aristarchus was the only one of the Ancients to have arrived at the full-blown heliocentric and heliostatic model. Aristarchus is ‘the Ancient Copernicus’, as Copernicus is ‘the Modern

Aristarchus'. Aristarchus, however, is mentioned in the original manuscript preserved in the Library of the University of Cracow, and it would seem that its omission from the printed text was very possibly an accident.[10] But if Copernicus was influenced by the Ancients, could he have been influenced also by writers closer to his own times? There are those who believe that Copernicus could have been inspired by a Renaissance cult of the sun which developed under the umbrella of a major revival, partly in opposition to the Aristotelianism of the later Middle Ages, of Platonism and Neoplatonism.

The Archpriest of this Renaissance cult was the translator of Hermes Trismegistus, Marsilio Ficino.[11] Scriptural warranty for the adulation of the sun was found particularly in the words of Psalm xviii (xix, MT), which in the Prayer Book version reads: 'In them [the heavens] hath been set a tabernacle for the sun' (xix:5) but in the Latin (xviii:5, LXX) *In sole posuit tabernaculum suum* (He has set his tabernacle *in* the sun).[12] Rather a warm spot for it, but it does suggest that the sun is to be identified as the place from which the Almighty governs the cosmos. Much has been made of this cult in relation to Copernicus by virtue of the stress which was placed on the centrality of the sun. The problem is, however, that Ficino, Pico della Mirandola and other devotees had a *very* different understanding of solar centrality than that of Copernicus.[13]

For Copernicus, of course, the sun was at the physical and geometric centre of the universe;[14] but for the Renaissance sun cultists the sun occupied the central position in the sequence of planetary spheres. It was from its position in the midst of the planets (Moon, Mercury and Venus being placed between the Earth and the Sun; Mars, Jupiter and Saturn between the Sun and the fixed stars) that it ruled the cosmos, not from its physical centre. Copernicus could not have been led to geometric heliocentricism from this quarter. As Fernand Hallyn points out,[15] Ficino and others had a linear, axial conception of centrality, not the spherical conception of Copernicus; the sun occupied the mid-point of an axis drawn from the earth through the orbits of the sun and other planets of the geocentric and geostatic model (although they did realise that its position was not mathematically precisely central). Hallyn refers to the frontispiece of Gafurio's *Practica musice*, 'which clearly illustrates the subordination of circular thinking to axial thinking' by linking the spheres by means of a three-headed serpent.[16] However, certainly these Renaissance scholars would have made a contribution to a climate of thought conducive to

Copernicanism by stressing the cardinal importance of the sun within the system of the world, and by bringing questions concerning centrality to the fore.

Also possibly relevant to Copernicus is a debate, to which Hallyn draws attention, over whether the High Altar of a church should be placed at the east end or in the centre. Advocates for the east end argued, stressing the transcendence of God, that the altar should be placed at the periphery of the church as that corresponded to the Empyrean, the most exalted domain of the received cosmological model, from which God ruled the universe.[17] Those arguing for the centre maintained that: 'The right place in the church for God's image is at the center of the family of the faithful, *as* it is right that his cosmic symbol should be at the center of the planetary family'.[18] This debate could well have had some impact on Copernicus, since the period during which most of the churches with a central altar were built, between 1490 and 1530,[19] includes the years that Copernicus was in Italy. But, again, the central altar school of thought adhered to a concept of axial, not spherical, centrality, so the debate as such would not have tended to promote (geometric) heliocentric ruminations. However, the *actual placement* of the High Altar in the physical centre of a church could have promoted such novel speculations if the church itself were thought of as an image or model of the cosmos.

So, just *why* does Copernicus liken the universe to a temple, and the sun to a lamp hanging in that temple? And why, changing the metaphor, does he picture the sun *enthroned* in the geometric centre of the universe? Certainly, as our text reveals, possible sources for such metaphors are to be found amongst the literature of pagan Antiquity and, indeed, in the language used by the Renaissance sun cultists. But could, in addition, there be Ancient or medieval Christian sources? This seems to be the question which has not been properly addressed. In order to arrive at an answer, the primary objective of this paper, we will need to look not so much to Copernicus the Renaissance man of the world as to Copernicus the Polish canon. Although Renaissance thought had started to diffuse into Poland earlier, it was not until the beginning of the sixteenth century that Renaissance art and architecture descended on Cracow, and when it came it came in the form of a blend of Medieval and Renaissance elements.[20] Copernicus would have been closer to the medieval roots of European culture than his contemporaries born and bred in Florence or Venice. Let us, however, turn for the moment from Copernicus to consider these metaphors as they developed within Christian thought, architecture and iconography from early times.

2. THE CHURCH AS COSMOS AND THE COSMOS AS A CHURCH

The notion that a Christian church (or temple, the words are synonymous) can be thought of as an image of the cosmos clearly dates back to early Christianity. Christianity of course inherited the traditions of pagan Classical Antiquity, in both their Greek and Latin manifestations (and to a much lesser extent traditions of local indigenous pagan cultures), as well as the traditions of Judaism. We read in Acts of early tensions between the Greeks and the Hebrews, and of the beginnings of the efforts of the Church, which were to last for centuries, to bring about a synthesis of the two cultures. The success of this creative and expansive integration was (despite subsequent schisms) to transform a Jewish sect into a World Religion, and to yield a distinctive universal Christian culture (albeit one which allowed for local diversity within the constraints of catholic unity). The traditional Christian understanding of the nature of the temple is, thus, not entirely *sui generis*, but a development which has roots tracing back into Jewish and Graeco-Roman culture.

The Latin concept of the temple, *templum*, was in fact a particularly rich and wide-ranging one. But the most fundamental of this cluster of meanings was that of the temple (or temples) of the heavens. This connotation is admirably described by Joseph Rykwert.[21] Rykwert refers to Varro's discussion of *templum*. Varro '... quotes a line of Ennius about Romulus: 'There will be one whom you shall raise to the bright temples of the sky.' He goes on to say: '*Templum* is used in three ways: with reference to nature, to divination, and to resemblance, with reference to nature, in the sky; to divination, on the ground; and to resemblance underground."[22] Divination is the key to the primary link between the *templum* of the heavens and the temple in our modern use of the word. The mode of establishment of the augur's *templum* can be understood in terms of three key technical terms: *conregio, conspicio* and *cortumio*. The three steps (following Rykwert) are:

(1) *Conregio*. The temple of the heavens is fixed in a diagram marked out on the ground with the *lituus*, the augur's wand; and this would include the tracing of easterly-westerly and northerly-southerly axes by means of four movements, forward, backward, and left, right (it would seem that the axes were not determined by strict astronomical means). The axes, of course, divide the diagram into four quarters, each of which is called a *templum*. Local landmarks would then be named, in relation to the diagram, defining the boundaries of the *templa*. In other words, a correspondence is effected between the heavens and an area of the earth's surface; a sacred space is defined. We see here how the Greek word, *temenos*, a sacred enclosure, relates to *templum*.
(2) *Conspicio*. The total scene visible to the augur's eye, falling within the limits indicated by his diagram, is inspected in detail and features considered to be of significance with respect to oracular events noted in relation to the four quarters, or *templa*, determined by the diagram. The significant landscape features and the four *templa* fixed within the diagram are then unified, thus creating a single *templum*. An augur's temple can thus be seen to be a conceptually enclosed three-dimensional volume of space, incorporating both structural elements, notably the axes, derived from observation of the heavens, and oracularly significant landscape features. Having constructed his open-air *templum*, the augur proceeds to pronounce the covenant, the *legem dixit*, the formal statement of what is to be decided and what will count as portents.
(3) *Cortumio*. Any portents observed are interpreted.[23]

Similar processes were followed for defining the layout of a city, camp, etc. and for the placement and ordering of an architecturally constructed temple and its *temenos*. The conception of the Christian temple, the house of God, as a structure which effects the unification of heaven and earth very clearly is, in this respect, in part a descendant of the augur's *templum*.

For the Jews the house of God was of course the Temple in Jerusalem. But the Temple had been preceded by the Tabernacle, and before that we find abundant evidence for an open air *templum* similar to that of the augurs. In the account of Jacob's ladder (Gen. xxviii:10-22) we read how awaking from his dream, in which he saw angels ascending and descending, Jacob said: 'Surely the Lord is in this place: and I did not know it. … How awesome is this place! This is none other than the house of God, and this is the gate of heaven' (vv. 16 and 17). And he set up an altar with the stone he had used as a pillow and poured oil over it. The Tabernacle (and subsequently the Temple) was interpreted in cosmological terms. God gave Moses meticulous instructions for every detail of the Ark, the Tabernacle, the priestly robes and the worship which was to be offered (Ex. xxv-xxxi). It is of the greatest importance to note that the Tabernacle and its furnishings were made by Moses in strict accordance with the divine archetype shown to him on the mountain. God commanded Moses to '... see that you make them after the pattern for them, which is being shown you on the mountain' (Ex. xxv:40). The Christian temple, therefore, being the successor of the Jewish Tabernacle and Temple, could claim as its author the Divine Architect. The cosmological symbolism of the Tabernacle was commented on extensively by Philo and by several Fathers of the Church. Clement of Alexandria notes, for instance, that:

> ... the covering and the veil were variegated with blue, and purple, and scarlet, and linen. And so it was suggested that the nature of the elements contained the revelation of God. For purple is from water, linen from the earth; blue, being dark, is like the air, as scarlet is like fire.

The altar of incense was:

> ... the symbol of the earth placed in the middle of this universe .... And that place intermediate between the inner veil ... and the external court ... was, they say, the middlemost point of heaven and earth.

The seven-branched candlestick, which Clement saw as a symbol of Christ, was interpreted as representing the sun in the midst of the planets. Thus Clement clearly adhered to the concept of axial-centricism:

> The lamp ... was placed to the south of the altar of incense; and by it were shown the motion of the seven planets, that perform their revolutions towards the south. For three branches rose on either side of the lamp, and lights on them; since also the sun, like the lamp, set in the midst of all the planets, dispenses with a kind of divine music the light to those above and to those below.[24]

What form the few churches proper (that is, as opposed to house churches) of the pre-Constantinian period took we do not know. But from the

Constantinian period onwards we find evidence of the application of cosmological thinking in relation to churches. In fact, a cosmological dimension is inherent in the very division of a traditional Christian church, following the layout of the Jewish Temple, into three areas; that of the narthex (corresponding to the vestibule), that of the nave (corresponding to the great hall) and that of the sanctuary (corresponding to the Holy of Holies of the Temple).[25] As the sanctuary is the area within which the offering of the Eucharist is made, it naturally represents heaven. The nave, on the other hand, being the place where the faithful stand, naturally represents earth. The Eucharist makes present the eternal Kingdom of God, thus sweeping up time into eternity, the image of which, according to Plato in the *Timaeus*, is provided by the regular revolutions of the heavenly bodies.[26] The sanctuary, therefore, represents the fulfilment of the *eschaton*; it is the place of the Heavenly Banquet of the Lord. The Eucharist is liturgically outside of time.[27] The nave, in contrast, represents the created world subject to change and time. But the two areas, with their inherent natural symbolism, of course interact, and in early churches this was signified by the actual penetration of the sanctuary into the space of the nave, being separated from the latter only by a low partition.[28] The Holy Mysteries, consummated within the sanctuary, thus work to transfigure, to transform, the temporal order signified by the nave. The nave is not the fallen, unredeemed world which lies beyond the church, but the world 'being saved' (Acts ii:47b). The narthex (or atrium), on the other hand, is a place of transition, a place, in early times, for penitents and catechumens. The narthex signifies the, yet unredeemed, creation 'groaning in travail' in faithful expectation of its salvation in Christ (Rom. viii:19-23). Significantly, the baptistery usually opened off the narthex.[29]

This interpretation, which really is *necessitated* by the theology of the Eucharist and the division of the church into narthex,[30] nave and sanctuary, finds powerful expression in early nave floor mosaics which depict the creation as Paradise *being* restored. Thus in the church built over the traditional site of the miracle of the multiplication of the loaves and fishes at Tabgha, above the Lake of Galilee, we find a late fifth-century/early sixth-century mosaic depicting a marvellous Nilotic scene, with an abundance of exotic birds and plants.[31]

The horizontal division of the body of a church or chapel into nave/earth and sanctuary/heaven came to be matched by a corresponding vertical division. Thus, early on, the starry heavens were commonly depicted in the semi-dome of an eastern apse, as, for example, in the late fifth- to early sixth-century mosaic of the Archbishop's Chapel in Ravenna.[32] Enormously extended possibilities were opened up with the development of the 'cross-and-square' domed Byzantine church. Given that the sphere/circle is symbolic of heaven and the cube/square (or the rectangle in general) of earth, the architecture, by virtue of the cross which it forms, secures the union of the two and thus becomes a visual image of both the Divine Liturgy (Mass) and of the Incarnation itself. In plan view, the four great arches which support the central dome create a square around the circle of the dome, and this then becomes the centre of a Greek (i.e., equal-armed) Cross, which, in its turn, is then usually enclosed (although there are exceptions, such as San Marco in Venice) in a square formed by the outer walls of the church (fig. 1).[33]

It should be noted that the symmetry and proportions of the church reflect those of the human body; the church is, then, also an image of the microcosm. That architectural space can express human proportions is grounded in the concept of *homo quadratus*. Vitruvius argued that the human body yields a square since the height of a person is equal to the distance between fingertips and fingertips when the arms are stretched out straight. In other words, as with a square, the height and breadth of the human body are equal. In fact, the human body, as Vitruvius notes, also yields the circle. If the arms and legs are stretched out, the feet and hands will touch the circumference of a circle drawn with the navel as centre. (There is a well-known drawing of Leonardo which illustrates the point.) A human person is not just a creature moulded from the common clay of the earth (Gen. ii:7), but a being touched by the divine, made in the image and after the likeness of God (Gen. i:26).[34]

The central dome, a natural image of the heavenly vault in itself, was reserved for depiction of the heavenly realm, with Christ at its centre and usually an abundance of angels and/or prophets (though in some cases Apostles rather than prophets). The principal feasts usually occupied the associated squinches, or pendentives, and the surrounding vaulting. Lower down on the walls, or in less significant areas, appeared the saints, who link the Church Militant with the Church Triumphant.[35] In Byzantine churches the lower walls were not covered with iconography but were, at least in the case of grander churches, clad in precious marbles. The rich variety of colours, and the intricate patterns of veins, clearly represented the richness of the fine texture of the created world. Almost nothing in Byzantine churches was devoid of symbolic meaning.[36] In the Slavic Orthodox world, heavenly bodies often adorn the

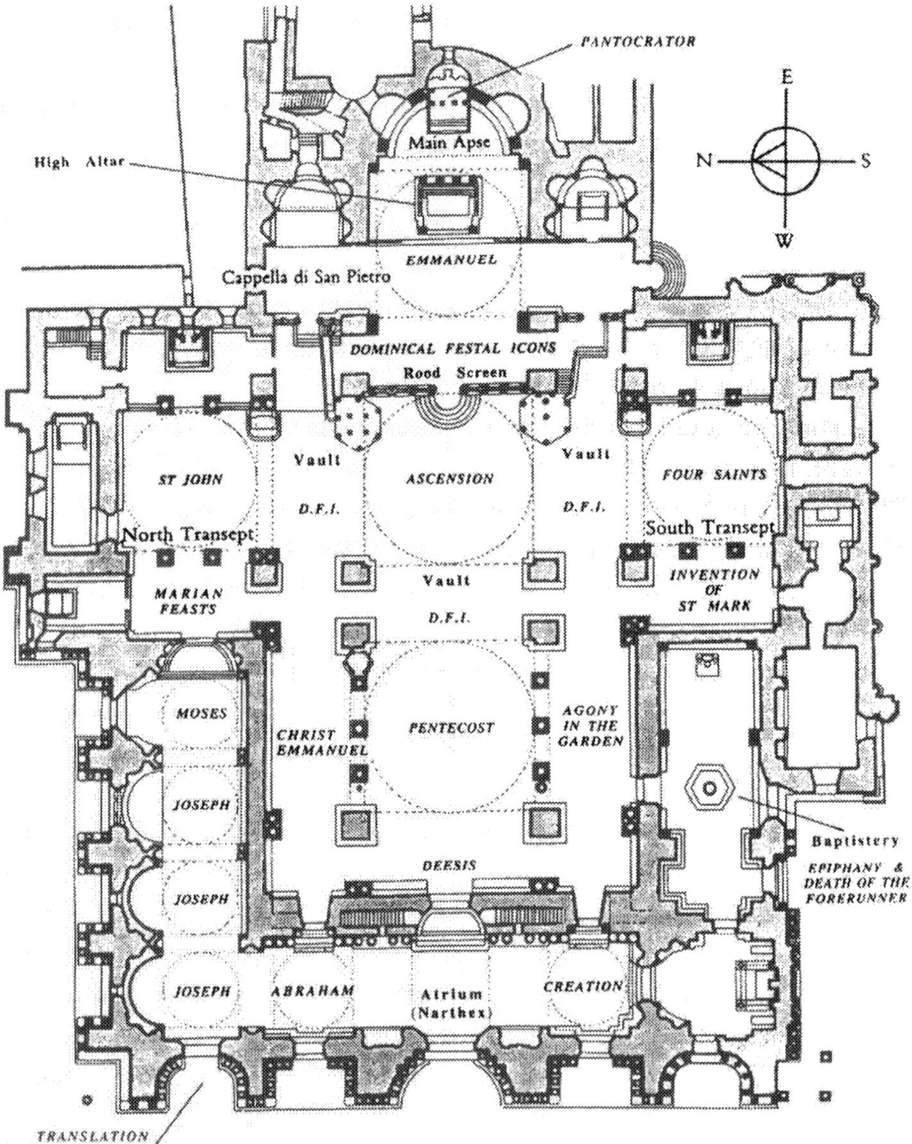

Figure 1. *Ground plan of San Marco, Venice*, with principal iconographical ensembles.

exterior surface of 'onion' domes, as with the Cathedral of the Dormition of the Theotokos at the Trinity Monastery of St Sergius, Zagorsk.[37]

By combining these horizontal and vertical cosmological dimensions—along, of course, with non-cosmological criteria, such as the placing of depictions of Old Testament (OT) types of the Eucharist in or close to the sanctuary—the Byzantine iconographic programme commenced its development. It reached its more or less final form following the restoration of Byzantine sovereignty in Constantinople, after the fateful Fourth Crusade, in 1261. However, even then, there was to be some further development with a late flowering of Byzantine painting and architecture, the so-called Palaeologan Renaissance. Of particular significance was the establishment of Mistra, in the Peleponnese, as a regional Byzantine capital in the fourteenth century. It was principally via Mistra that the Byzantines were to have an impact on the *quattrocento* well before the influx of Greek-speaking scholars from Constantinople, following its fall in 1453 (Mistra itself falling to the Turks in 1460, only to be taken by the Venetians in 1464). In fact, a Platonist philosopher from Mistra, Gemistos Plethon, played a major role in initiating the Renaissance Platonic movement which has so impacted on the historiography of Copernicanism. Indeed, he actually advocated a recension of polytheistic paganism in his *Nomoi*, which was modelled on Plato's *Laws*, and much of which was destroyed by a horrified George Scholarios who edited the work after Plethon's death.[38] For our purposes, however, the important point to be noted is that, even while the *perspectivi* of the *quattrocento* were developing the illusionistic art which was destined to sweep away Byzantine art and its derivatives in the West, there were real contacts with the world of Byzantine art and an understanding of its principles. It would have been well nigh impossible to have studied in Northern Italy during the Renaissance, as did Copernicus, without some direct experience of Italo-Byzantine art, and without the tradition of Byzantine art and architecture making some impact. But let us move on to the Romanesque and Gothic periods.

While there are enormous variations in the degree to which the styles of the various schools of Romanesque art directly borrow from Byzantine art, it is clear that Romanesque art *per se* was in large measure a product of the Byzantine. Nevertheless, often schools of Romanesque art owed a considerable debt to earlier medieval regional styles. There is one Western source of inspiration which stands out from all others, that of the illustrations of manuscripts of Beatus of Liébana's *Commentary on the Apocalypse* (c. 776) dating from the tenth century on.[39] Combine the Beatus illustrations and John's Apocalypse with Byzantine iconography and the Byzantine iconographic programme, add a local art style or two to taste, sprinkle with pagan imagery or pure fantasy, and to all intents and purposes you have the Romanesque.[40]

In Byzantine monasteries, depiction of the Apocalypse was usually to be found in the refectory rather than the church. With the Romanesque, it sometimes informs the whole programme of a church. We frequently first meet it with Christ in Majesty on the tympanum over the door through which we enter the building. The tympanum of what is perhaps the most outstanding of Romanesque porches, that of Moissac (probably between 1115 and 1120), is based on Rev. iv:2b-7 and v:1 and 8 and depicts Christ in Majesty seated on a throne, surrounded by a mandorla against a starry background, flanked by angels and the symbols of the Evangelists. At His feet is the 'sea of glass' and either side and below are the twenty-four elders.[41] But the Christ in Majesty of the Apocalypse also frequently appears in the iconography of the sanctuary or chancel along with other elements taken from the Apocalypse. The frescoes (1120-40) on the barrel-vault of the chancel of the Norman church at Kempley, Gloucestershire, depict a Byzantine *Majestas*, with His feet resting on a sphere, representing the earth, and with the sun and the moon above His head. Surrounding Him are the symbols of the Evangelists, many stars, the seven candlesticks of Revelation and, in the corners, four six-winged seraphim. Around the north and south chancel windows are scenes depicting the towers of a city. It all represents the New Jerusalem, the Kingdom-to-Come which is, nevertheless, made present in the here and now through the celebration of the Mass. Depictions of the Heavenly Jerusalem occur also in Romanesque apse paintings of Christ in Majesty, as with the frescoes in St Gabriel's Chapel, in the crypt of Canterbury Cathedral, and St Michael and All Angels, Copford, Essex (fig. 2).[42]

The distinction between the Old Jerusalem, the Kingdom of this World, subject to time and decay, and the New Jerusalem, the eternal Kingdom of God, thus becomes for the iconography of the Middle Ages essentially an apocalyptic recension of the distinction, given concrete expression in the division of the body of the church into nave and sanctuary, and accentuated in bicameral buildings, between the created spatial world subject to time, and the eternal, heavenly realm.[43] Perhaps the main difference between the two versions of what

Figure 2. *Christ in Majesty*, with signs of the zodiac on the soffit of the arch. Note also the buildings of the Heavenly Jerusalem. Apse of St Michael and All Angels, Copford, Essex.

is essentially the same dogmatic and liturgical distinction is that the Romanesque recension tended to lead to a more pessimistic view of the created order, as is witnessed to by the terrifying torments of the damned, which often enlivened the west wall of the nave,[44] and the extraordinarily explicit ubiquitous grotesque images depicting *luxuria* and *avaritia*.[45]

Iconography of the Apocalypse does, however, in some places appear along the side walls of the nave. This is the case at Clayton in Sussex (eleventh century, possibly pre-Conquest). Although much of the iconography is now missing, it seems likely that the whole programme of the nave was devoted to the Judgment and that its main inspiration was the Apocalypse. But this programme is no exception that proves our rule, for here there is no identification of the nave with the New Jerusalem. What is depicted in the nave is the pilgrimage through life of the blessed *to* the New Jerusalem, depicted at the east (or chancel) end of the north wall, and the damned to hell.[46] This notion of the nave as the place of humanity's pilgrimage to heaven or the New Jerusalem, signified by the chancel, was given expression later in the French

Gothic cathedrals by the great tile labyrinths which were constructed on the floors of their naves.[47] The devoting of the whole of the nave to the Judgment is certainly extraordinary (a tetchy ecclesiastic unhappy with his local flock?), but it does shed light on the apocalyptic understanding of what was to become an extremely common feature of the iconography of British parish churches, the Doom scene painted around the chancel arch on the east wall of the nave. This depicts the cosmic Christ seated in Judgment upon the rainbow with his feet resting on the terrestrial globe. To His right the saved enjoy eternal bliss, to His left the damned are committed to the tortures meted out by Satan and his minions. The message is brutally clear, the only passage from this world (the nave) to heaven (the chancel) is via the Judgment.[48]

The painting of Romanesque and Gothic vaults with heavenly bodies and/or angels must have been extremely common. One very notable survival is the beautiful vaulted ceiling painting (thirteenth century), depicting angels in roundels surrounded by stars, in the Guardian Angels Chapel in Winchester Cathedral.[49] In the crypt chapel of Our Lady Undercroft in Canterbury Cathedral are the remarkable remains of a recently restored fifteenth-century painted vaulted ceiling with heavenly bodies, to which small pieces of mirror set at various angles are attached so that they reflect the flickering candle light.[50] More common, though, are simple designs of stars set against a blue background, sometimes with striking effect as at La Sainte-Chapelle in Paris.[51]

Medieval churches and cathedrals, both Romanesque and Gothic, are remarkable for the enormous variety of their iconography. Common are depictions of the three great annual cycles—the astronomical year, signified by the signs of the zodiac; the agricultural year, signified by the labours of the months; and the liturgical year, signified by the principal feasts (often broken down into distinct Nativity and Passion cycles). The first two annual cycles are often placed around arches, signifying the rise and fall of the year.[52] Extremely common also are depictions of plants—one thinks, for example, of the remarkable botanical carvings of the chapter house of Southwell Minster, 'the leaves of Southwell' (late thirteenth century)[53]—and animals. The plants frequently had symbolic meaning. The animals depicted almost invariably had tropological meanings derived from the bestiaries. Mythical beasts mingled with exotic creatures and common domestic animals. Nor were the human nations and races neglected, including the mythical races—the Sciopods, the Cynocephalics, the Antipodeans—believed to live in remote eastern regions of the Northern Hemisphere or in the Antipodes. As has often been remarked, sculptural programmes often closely correspond to *mappaemundi*, those extraordinary encyclopaedias of the Medieval World, of which the finest surviving example is the Hereford *mappa mundi*.[54] Sometimes, the history of the world from the Creation to the Second Coming, according to the biblical account, is recorded.

So the iconography of a medieval church was by no means limited to strictly religious themes, but encompassed the whole cosmos: the heavens, the cycle of time, the plants, animals and human races, and the daily lives, history, legends, beliefs and morals of the people who created those great buildings.[55] The temple was, then, indeed an *imago mundi*. But if the church is an image of the cosmos then the concept of correspondences demands that it must also be the case that the cosmos is an image of the church. Although the proliferation of altars in later medieval churches blurred to some degree the ancient distinction between nave and sanctuary, it was, as we have seen, the differentiation of these two principal spaces of a church which, given the nature of sacramental theology, *entailed* a cosmological interpretation of the church building.

The *quattrocento* saw the triumph of the new illusionistic perspectival art and the eventual overthrow not only of the Gothic tradition of painting (which owed much to the earlier Romanesque and Byzantine traditions) but of the traditional iconographic programmes. Although the medieval tradition managed in places to linger on into the sixteenth century—one thinks, for example, of the great systematic glass programmes of Fairford and King's College Chapel, Cambridge[56]—the iconoclasm of the Reformers combined with the championing of illusionistic Baroque iconography by the Counter-Reformation to give the *coup de grâce* to traditional Christian iconography in the West. With the Baroque preoccupation with special dramatic illusionistic effects and an intense humanist realism, the new individualistic iconographic programmes largely dispensed with the ancient ecclesiastical cosmology which had been conveyed by Byzantine and Western medieval programmes.

However, if we turn to architecture the situation is somewhat different. As we have seen, the Renaissance saw a debate over the placing of the altar which was grounded in cosmological argument. The church, for architects, was still an image of the cosmos. But it was not only this debate which led to support for central-plan churches. The humanism and the reverence for Classical culture of the Renaissance brought architecture back to something close to the

Byzantine tradition. The ancient principle that the church should be built according to the proportions of the human body was reasserted with vigour. This meant, of course, that a Renaissance church, built as an image of the microcosm, was, *ipso facto*, also an image of the macrocosm. Although transformed into a Latin Cross church by the decision to lengthen the nave, the new St Peter's, Rome, was, as is still clear today, itself designed as a central altar church based on the circle, square and Greek Cross scheme of Byzantine churches.[57]

The most explicit statement known to me of the correspondence between church and cosmos is to be found in *The Church's Mystagogy* by St Maximus the Confessor (580-662).[58] Maximus was not only a great theologian, currently undergoing something of a vogue in certain quarters, but one whose thought transcended the growing divisions of East and West. Although a Greek Father of the Church, Maximus spent a quarter of a century in exile in North Africa, and from there visited Rome, where he is known to have contributed to the Lateran Council of 649. He died in 662 shortly after, it is said, judicial mutilation of his tongue and right hand in Constantinople for refusing to repudiate the orthodox teaching that there are two wills in Christ.[59] The key passage reads:

> ... God's holy church in itself is a symbol of the sensible world as such, since it possesses the divine sanctuary as heaven and the beauty of the nave as earth. Likewise the world is a church since it possesses heaven corresponding to a sanctuary, and for a nave it has the adornment of the earth.[60]

And in another passage, which occurs shortly before the above, he writes:

> ... he [the elder to whom Maximus attributes these ideas] used to speak of God's holy Church as a figure and image of the entire world composed of visible and invisible essences because like it, it contains both unity and diversity. ... In this way the entire world of beings produced by God in creation is divided into a spiritual world filled with intelligible and incorporeal essences and into this sensible and bodily world which is ingeniously woven together of many forms and natures. This is like another sort of Church not of human construction which is wisely revealed in this church which is humanly made, and it has for its sanctuary the higher world assigned to the

> powers above, and for its nave the lower world which is reserved to those who share the life of sense.[61]

So the temple is an image of the cosmos, and the cosmos is an image of the temple. We need look no further for a possible Christian source for Copernicus' allusion to the universe as a temple. It is simply a traditional commonplace.

Maximus in fact goes further and says that the church is also an image of the human person:

> ... holy Church is like a man because for the soul it has the sanctuary, for mind it has the divine altar, and for the body it has the nave. It is thus the image and likeness of man who is created in the image and likeness of God. By means of the nave, representing the body, it proposes moral wisdom, while by means of the sanctuary, representing the soul, it spiritually interprets natural contemplation, and by means of the mind of the divine altar it manifests mystical theology. Conversely, man is a mystical church, because through the nave which is his body he brightens by virtue the ascetic force of the soul by the observance of the commandments in moral wisdom. Through the sanctuary of his soul he conveys to God in natural contemplation through reason the principles of sense purely in spirit cut off from matter. Finally, through the altar of the mind he summons the silence abounding in song in the innermost recesses of the unseen and unknown utterance of divinity by another silence, rich in speech and tone.[62]

And, lastly, Maximus argues that the church represents also just the soul, within which he distinguishes, following patristic tradition (as also in the passage just quoted), an intellectual faculty which he calls the mind (= spiritual intellect, which can roughly be equated with the *nous* of Aristotle) and a vital faculty which he calls the reason (= rational soul or *psyche*), the former corresponding to the sanctuary and the latter to the nave.[63]

So while the church is an image of the universe, the macrocosm, it is equally an image of the human person, the microcosm. The church then mediates between *microcosm* and *cosmos*, between *anthropos* and *macroanthropos*; in other words, it can be regarded as a *mesocosm*. Further, if the church is an image of the human person, the human person is an image of the church: 'Do you not know that you are God's temple ...' (1 Cor. iii:16).[64, 65]

It is certainly the case that the theology of the temple, as also the theology of the icon, was more highly developed in the East than in the West, but there is no doubt that similar conceptions were commonplace in the West also.[66] However, the West did find new, but less systematic, ways of giving expression to the conception of the church as an image of the cosmos, which, along with the liturgy, effected the union of heaven and earth. Romanesque art and architecture, even where they owed more to earlier Western styles than to the Byzantine, were deeply influenced by the principles underlying Eastern architecture and iconographic themes and programmes. But at the birth of the Gothic we still find Abbot Suger of Saint-Denis displaying an understanding of the function of the church as a mediator between earth and heaven, as a *mesocosm*. Thus while defending the lavish use of precious jewels at Saint-Denis he says, in an often quoted passage:

> Thus, when—out of my delight in the beauty of the house of God—the loveliness of the many-colored gems has called me away from external cares, and worthy meditation has induced me to reflect, transferring that which is material to that which is immaterial, on the diversity of the sacred virtues: then it seems to me that I see myself dwelling, as it were, in some strange region of the universe which neither exists entirely in the slime of the earth nor entirely in the purity of Heaven; and that, by the grace of God, I can be transported from this inferior to that higher world in an anagogical manner.[67]

And, of course, with the *quattrocento* we find a return to central-plan churches and an architectural re-affirmation of the correspondence of macrocosm and microcosm.

3. CHRISTOS HELIOS

To conceive of the temple as an image of the cosmos and the cosmos as an image of the temple does not, of course, automatically yield a heliocentric model of the universe. We must formulate also a gloss for the *lamp* in the temple and for the enthroned sun if we are to arrive at Copernicus. In order to prepare to make the connection we must turn to a very early Christian conception, that of Christ as the sun—*Christos Helios/Christus Sol.*

Figure 3. *Christos Helios*, based on the mosaic in the Chapel of the Fisherman, grottoes of St Peter's, Rome.

In the grottoes beneath St Peter's, in the Vatican, is to be found a mid-third-century mosaic on the vault of Mausoleum M of the Julii, the so-called Chapel of the Fisherman (fig. 3). In an octagonal space surrounded by vine is an image of Christ as the sun god riding his chariot, with two rearing white horses. Rays springing from the golden disk of the sun, which creates a halo around Christ's head, form the figure of a cross. In His left hand Christ holds an orb, while the likelihood, given the position of the arm, is that He was giving a blessing with the missing right hand. The mausoleum was originally built late in the second century for the Julii family but later converted as a Christian tomb. The image is hardly open to misinterpretation, but, in any case, a Christian provenance for the mosaic is beyond doubt as the walls carry typical early Christian iconography: Jonah and the whale; Christ as the Good Shepherd; an angel hooking a fish (hence the popular name for the mausoleum).[68] Another early

Roman example is to be found on the vault of the arcosolium in the Tricliniarca Crypt in the Cemetery of Peter and Marcellinus. This takes the form of a painted image of *Christos Helios* in a roundel flanked by scenes of Jonah and the whale, an OT type of the Passion and Resurrection of Christ.[69]

Portraying Christ as the pagan sun god, Sol, might seem strange. But, in fact, pagan myths could be used as types for New Testament (NT) antitypes in precisely the same way as OT stories, for in the use of typology it is the structural identity of the type and antitype which is involved, not the actual content of the stories.[70] A number of other characters from pagan mythology have been used as types of Christ, including Bellerophon, Orpheus,[71] and Theseus. Pagan types are particularly common in early Christian floor mosaics—Orpheus charming the animals, Bellerophon and the Chimaera, and others—doubtless because their Christian meaning would, in an age of persecution, be likely to escape the casual observer (indeed, they seem often to have fooled modern scholars).

There are in fact multifarious linkages between Christ and the sun. In Matthew's account of the Transfiguration, Christ's face is said to have 'shone like the sun, and his garments became white as light' (xvii:2b). Similarly, in the vision of the Son of Man in Revelation, Christ's face is described as being 'like the sun shining in full strength' (i:16b), and his eyes 'like a flame of fire' (i:14b). Then there is the 'woman clothed with the sun' in Rev. xii:1-6. Although this image of the woman primarily referred to the Church, it came to be applied to the Virgin Mary; and the sun, with which she was clothed, was taken to refer to Christ, the 'Sun of Righteousness'. Even more significantly, from the perspective of Copernicus' metaphor, in Rev. xxi:23, Christ (the Lamb) is said to be the *lamp* of the New Jerusalem: 'And the city has no need of sun or moon to shine upon it, for the glory of God is its light, and its lamp is the Lamb'.[72]

In the Old Testament, we find references such as that of 2 Sam. xxiii:4, in which it is said that God 'dawns' on just rulers 'like the morning light, like the sun shining forth upon a cloudless morning'. The really crucial text, however, is the prophecy of the Day of the Lord in Malachi iii and iv, from which Christ's title, the Sun of Righteousness, comes: 'For behold, the day comes, burning like an oven, when all the arrogant and all the evildoers will be stubble… . But for you who fear my name the sun of righteousness shall rise, with healing in its wings' (iv:1a and 2a). For Christian exegetes, the 'sun of righteousness' was understood as a prophecy of the coming of Christ. But apart from specific allusions to the sun in relation to God, both the OT and NT are replete with passages which refer to light or to God's glory. Thus the theme of Christ as the light of the world is introduced right at the beginning of John's Gospel, and, in Jn viii:12, Christ Himself says: 'I am the light of the world; he who follows me will not walk in darkness, but will have the light of life'; while in 1 Jn i:5b we read: 'God is light and in him is no darkness at all'.[73]

But the Scriptures aside, it was easy for the early Christians to associate the familiar sight of sunrise on the eastern horizon with the Resurrection of Christ, and by extension with the anticipated Second Coming of Christ, which would herald the general resurrection of the dead and the restoration of paradise. On the spiritual level, the rising of the sun was a sign of the rising of the Sun of Righteousness in the hearts of the faithful. The rising of the physical sun each day was, therefore, a moment to open the doors of the soul to the Risen Christ.[74] It was a common practice for Christians to be buried facing the east and for them to turn to the east in prayer.[75] Some early basilicas, such as St Peter's, Rome, were orientated with the entrance doors, rather than the sanctuary, facing east (although this did mean that the Pope celebrating Mass across the altar, as has always been the practice in St Peter's, would have faced east). Some major churches, such as Santa Sophia in Constantinople, were orientated on Jerusalem. However, the most common orientation for a church, from very early times, was eastwards, towards the rising sun. On medieval *mappaemundi*, which placed east at the top, the Garden of Eden was located at the extreme east; and on the Hereford *mappa mundi* Christ in Judgment is depicted, outside the map, above the easternmost point.

The facts that the Resurrection happened to occur on the day the pagans dedicated to the sun, and that the Church from the very beginning naturally celebrated that day as the Lord's Day (Rev. i:10) and chose it for the regular weekly gathering for the Eucharist, of course immediately established an association between Christ and the sun.[76] It seems to be very widely believed, however, that the solar symbolism attaching to Christ also derives in significant measure from the choice by the Church in the fourth century of the 25th December as the feast of Christ's Nativity in order to counteract the pagan festivities associated with the winter solstice, kept in the Julian calendar on that day. The Church, it is said, promoted the Nativity of the Sun of Righteousness (*natalis solis iustitiae*) in opposition to the pagan feast of the Nativity of the

Unconquered Sun (*dies natalis solis invicti*) and of the Nativity of Mithras, which was also kept on that day. The pagan solar feast had been instituted by the emperor Aurelian on the 25th December 274 in order to promote a monotheistic cult of the sun which could unify all existing cults. As such, the cult of the Unconquered Sun would, it is said, have been seen as particularly pernicious by the Church. In fact, there seems to be no real evidence for this thesis. The date of Christmas was almost certainly determined by the supposed date of the Annunciation, 25th March (the original Julian date for the equinox), and, in any case, the association of Christ with the sun goes back to earliest Christian times. As far as the Malachi prophecy is concerned, even documentary evidence for the association of the Sun of Righteousness with the Nativity of Christ traces back to 243, thirty-one years before Aurelian's establishment of the *dies natalis solis invicti*.[77] But whatever the origins of the date of Christmas, its celebration on the 25th December could hardly have failed to *strengthen* the association of Christ with the sun.[78]

The association of Christ with the sun is strongly reflected in the worship of both the East and the West, but here we must restrict ourselves to the following few instances. At Vespers in the Byzantine rite, the early hymn at the lighting of the lamps—Φῶς ἱλαρὸν 'O Gladsome light of the immortal Father's holy glory, the heavenly, the holy, the blessed Jesus Christ!'—is still sung.[79] At Mattins, the 'Sun of Righteousness' is invoked twice in the morning prayers recited secretly by the priest and the ancient cry saluting the dawn Δόξα σοι τὸ δείξαντι τὸ φῶς—'Glory to you who has shown us the light'—is met by the ringing of the church bells and the singing of the Great Doxology, the *Gloria in excelsis*.

In both East and West, the symbolism of light plays an extremely prominent role in the liturgy of Easter. In the Byzantine rite, at the end of the Midnight Office, all the lights in the church are extinguished except for a single sanctuary lamp. The celebrant lights a candle from the lamp, and then comes through the Royal Doors elevating the candle high and declaring: 'Come, receive from the unwaning light, and glorify Christ, who rose from the dead'. The congregation then light their candles from the flame of the celebrant's candle and, when they leave the church, carry the lighted candles back to their homes, a potent symbol of the light of the risen Sun of Righteousness spreading through the world and transfiguring the entire creation. In the West, the Paschal Vigil commences with the blessing of the New Fire and the Paschal Candle, a very large beeswax candle which is marked with the Cross, the Greek letters A and Ω and the current Year of Grace. Five grains of incense, signifying the five wounds of Christ, are inserted into the Cross. As the New Fire is carried into the church and through to the sanctuary the deacon pauses three times to proclaim that it is the Light of Christ, *Lumen Christi*. Having elevated the candle onto a large prominently placed candlestick[80] the deacon sings before it the sublime proclamation of Easter, the *Exsultet*.[81] The Paschal candle is an aniconic symbol of the Risen Christ—the Pantocrator, the Lord of the Ages, the Alpha and the Omega, He through whom all things are made, the Sustainer, Governor and Saviour of the cosmos who shall come again in glory to judge the living and the dead, the Light of the World.[82] But it is to the iconic portrayal of Christ Pantocrator to which we must now turn.

The Pantocrator is one of the most important images to have developed in Byzantine iconography. Since no one has seen the Father at any time (Jn i:18), strictly speaking He cannot be portrayed in iconography. However, the New Testament tells us that Christ is the image of the Father,[83] and so, by virtue of the Incarnation, the Father can be portrayed through the icon of the God-man, Jesus Christ. The Pantocrator, the Almighty (the literal meaning of the word), is supremely the explicit icon of Christ as the representative (χαρακτὴρ) of the Father. The Pantocrator is therefore portrayed enthroned like an emperor or judge, a figure of great power and dignity, and sometimes (as at Daphni) also sternness.[84] The essential meaning and significance of the image are provided by the opening words of Hebrews (i:1-4), which also cast a great deal of light on both the Byzantine iconography of Christ and the Eastern doctrine of iconography in general:

> In many and various ways God spoke of old to our fathers by the prophets; but in these last days he has spoken to us by a Son, whom he appointed heir of all things, through whom also he created the world. He reflects the glory of God and bears the very stamp of his nature, upholding the universe by his word of power. When he had made the purification for sins, he sat down at the right hand of the Majesty on high, having become as much superior to angels as the name he has obtained is more excellent than theirs.

So Christ Pantocrator is prophet, judge, high priest and king, enthroned in the heavens: 'The Lord has established his throne in the heavens, and his kingdom rules over all' (Ps. cii (ciii MT):19); 'Heaven is my throne and the earth is my footstool ... ' (Is. lxvi:1a). But He is more than that, for He is also the reflection or radiance (ἀπαύγασμα) of the glory of the Father. In the words of the Nicene Creed, Christ is not only 'true God from true God' but 'light from light'. While we can *distinguish*, we can no more *separate* the radiance of the Son from the glory of the Father than we can *separate* the sunlight which radiates from the sun from the sun itself. The image of the Pantocrator, the cosmic Christ, by its very nature carries with it the powerful metaphor of the sun: Christ the 'Sun of Righteousness'. That this is so is confirmed both by the iconographic details of the Pantocrator and its placement in the church. Christ Pantocrator is traditionally shown either as half body or enthroned. The letters A and Ω may appear either side of the figure and He holds the Book of the Gospels, sometimes closed but frequently open. There is some variability in the text displayed but by far the most commonly found words are those from Jn viii:12, *ego sum lux mundi*, ἐγώ εἰμι τὸ φῶς τοῦ κόσμου—'I am the light of the world'; to which sometimes is added the rest of the verse.[85] While other positions are not unknown,[86] the image of the Pantocrator has two traditional locations. In a domed church it is usually to be found at the apex of the central dome, and this became virtually canonical for the fully developed Byzantine programme *sensu stricto*. The other traditional position is in the central eastern apse (figs. 4 and 5).[87]

The easternmost placement carries most potent associations with the sun. As we have seen, the rising sun was universally seen from earliest times as a metaphor of the Resurrection and of the Second Coming. That the image of Christ at the eastern end of a church typically carries with it specific connotations of Christ as the Light of the World, the Sun of Righteousness, is underscored by the fact that in a few churches the image of the Saviour is replaced by a window. Thus above the later Virgin and Child in the apse of the chapel of S. Zenone in Sta Prassede, Rome (see below), there appears on the east wall a ninth-century Deesis mosaic. But the Theotokos and St John the Baptist, either side, are not turned in supplication towards the Pantocrator, as is the norm with a Deesis ensemble, but towards a rectangular window.[88]

The placement of the Pantocrator in the central dome picks up another aspect of the metaphor of the sun, the centrality of Christ as the spiritual sun within the supercelestial heavens.[89] It is important to note that this concept of centrality is of geometric, spherical centrality, not axial. Christ Pantocrator, the

Figure 4. *Christ Pantocrator*, based on the mosaic in the apse of Pisa Cathedral, by Cimabue and others. See Note 96.

Supercelestial Sun, is, in this position, not (in either sense of the word) a peripatetic sun but the Christ who creates, governs and sustains the whole cosmos from the *centre*,[90] the true *Cosmocrator*. Iconographically, the association with the sun is frequently enhanced by the enclosing of a half-body icon of Christ Pantocrator within a roundel and by the large golden nimbus surrounding Christ's head—itself an image of the risen sun, which had earlier been placed behind the heads of rulers in Asia, and, of course, the pagan sun god.

One of the most dramatic of all images of Christ Pantocrator is that of the Church of the Theotokos Pammakaristos (Fethiye Camii), Constantinople.[91] The dome of this church is ribbed so that the overlying early fourteenth-century mosaic creates the impression of an effulgence of golden sunlight radiating out from the Pantocrator in the central roundel to a circle of prophets around the rim of the dome. Frequently, such images (though not that of Pammakaristos) have angels encircling Christ. As the eye travels around the dome, these heavenly beings are naturally set into motion.[92] Christianity inherited from Judaic angelology the idea that the motions of the heavenly bodies are

Figure 5. *Half figure Pantocrator*, based on the apse mosaic of Cefalù Cathedral, Sicily.

controlled by angelic powers. The angels circling around the supercelestial sun thus represent, or at least readily suggest, the revolving planets.[93] Movement around the centre is also picked up in late Byzantine iconography by the depiction of the Great Entrance of the Divine Liturgy around the central roundel, with angel deacons carrying fans, tapers and thuribles and angel priests bringing the Eucharistic gifts to Christ and bearing the *epitaphion*.[94] The scene represents the Heavenly Liturgy, which is made one with the Divine Liturgy celebrated within the church. The angelic ministers are portrayed as performing a kind of solemn liturgical dance around the Pantocrator at the centre of the dome.[95] To any High Medieval or Renaissance scholar it could hardly fail to suggest the ancient conception of the dance of the heavenly bodies.[96]

This notion can be traced back at least as far as the Pythagorean, Philolaus. In the account given by Aetius, 'ten divine bodies dance' around the fire at the centre of the cosmos, the 'house of Zeus'. Moving from the outer inwards, these are: '... first the sphere of the fixed stars, then the five planets, next the sun, then the moon, then the earth, then the counter-earth ... '.[97] The idea of the heavenly dance is picked up (though of course applied to a geocentric and geostatic model) in the *Epinomis* (a work of the Academy, possibly of Plato himself)[98] and in the pseudo-Aristotelian *De mundo*. In *De mundo*, the motions of the heavenly bodies are pictured as a 'solemn choral dance with all the stars in the same circular orbit revolving unceasingly for all time'. Pseudo-Aristotle expands the analogy:

> The single harmony that is produced by all these as they sing and dance in concert round the heavens has one and the same beginning and one and the same end, in a true sense giving to the whole the name of 'order' (κόσμος) and not 'disorder' (ἀκοσμία). Just as in a chorus at the direction of the leader all the chorus of men, sometimes of women too, join in singing together, creating a single pleasing harmony ... , so also in the case of the god who controls the universe: the note is sounded from on high by him who might well be called the chorus-master; then the stars and the whole heavens move continually.[99]

While Ptolemy in the *Planetary Hypotheses* writes:

> The parts of the planetary orbits are free to undergo translations and rotations in their natural positions in various ways, except that their movement is uniform revolution, like the chain of hands joined in a circle in a dance, or like the circle of men in a tournament who assist each other and join forces without colliding so as not to be a mutual hindrance.[100]

And St Clement of Rome, in his First Epistle to the Corinthians (which only just missed achieving canonical status), wrote (xx):

> The sun, the moon and the dancing stars [ἄστερων τε χοροὶ] according to his appointment circle in harmony within the bounds assigned to them without any swerving aside.[101]

At this juncture, we need to consider another image which is afforded particular prominence in Byzantine churches, the image of the Theotokos with the Christ child. While the placing of the Pantocrator in the eastern apse is common in Italo-Byzantine churches, elsewhere Byzantine churches typically have the Pantocrator in the central dome and the Virgin and Child in the central eastern apse. However, the association of the easternmost point with the Pantocrator and with the sun are not lost by this arrangement. Although sometimes shown standing (as is the case also with some images of Christ Pantocrator), typically the Theotokos is portrayed seated on a royal throne in the same manner as Christ Pantocrator, with the frontally facing Christ child on her lap. In Byzantine iconography, unlike Renaissance and Post-Renaissance Western art, the Christ child is never portrayed with the features of a baby. The Byzantine Christ child is, in fact, the Christ Emmanuel, the eternal Son of the Father, the prototype of humanity, begotten before all ages, whose coming in the flesh to redeem the world was foreseen in shadow by the prophets. The child on the lap or in the arms of the Virgin is thus the cosmic Christ, the Christ Pantocrator of the Apocalypse: "'I am the Alpha and the Omega,' says the Lord God, who is and who was and who is to come, the Almighty [Παντοκράτωρ]" (Rev. i:8).[102] But let us continue our discussion by analysing a specific example of the icon of the Theotokos Enthroned, that in the apse of Hosios Lucas, in Phocis (fig. 6).[103]

Against a golden background, this eleventh-century mosaic depicts the Theotokos seated in the style of the Pantocrator[104] on a purple cushion placed upon an ornate royal throne, with her feet resting on a footstool. On her dark purple *maphorion* (veil) are the three traditional stars, one on the forehead and one on each shoulder, signifying her perpetual virginity *ante partum, in partu et post partum* (before, during and after giving birth). The Christ child, dressed in brilliant gold and with a golden halo, is seated in a full frontal posture on the Virgin's lap giving the blessing with His right hand, while in the left hand He holds a scroll. Visually, it is impossible not to associate Christ with the sun, and the Virgin with the starry heavens. Jesus enthroned upon His mother is indeed the cosmic Christ enthroned in the heavens, but here it is the Virgin who signifies the heavens. We have but to look to the liturgy to confirm the correctness of this interpretation. Although Christ Himself is sometimes referred to as the Morning Star (Venus) rather than the sun (in accordance with Rev. ii:28; xxii:16; and possibly 2 Pet. i:19),[105] more usually it is the Theotokos who is called the Morning Star, who accompanies the rising of her son, the Sun of Righteousness. But the liturgy goes much further than this and sees Mary, who

Figure 6. *Theotokos and Child*, based on the apse mosaic of Hosios Lucas, Phocis.

contained the uncontainable in her ever-virginal womb, the eternal Logos through whom all things are created, as embracing the entire cosmos. Take for example the hymn to the Theotokos sung during the Eucharistic Anaphora (= the Canon of the Mass) of the Liturgy of St Basil (the Divine Liturgy sung at the Vigil of Christmas in the East):

> All creation rejoices in Thee … . Thou art a hallowed temple, and a spiritual paradise [Παράδεισε λογικέ], … whence God had taken flesh and became a little child, He who is from eternity our God. For He made thy womb His throne, and formed thy body to be broader than the heavens … .[106]

If further confirmation is required that the image of the Virgin Enthroned is a close Marian relation of Christ Pantocrator, with its attendant solar symbolism, it is to be found in a mosaic, probably of the eleventh century, in the chapel of S. Zenone in Sta Prassede, Rome. Usually the scroll held by Christ is rolled up but here it is partially open to reveal the words, *Ego sum lux*.[107]

An alternative image to that of the Theotokos Enthroned, to be found particularly in the Slavic Orthodox world, is that of the Virgin of the Sign. This icon derives from the ancient image of the Virgin *Orans*, which, in its classical form, depicts the Theotokos without the Christ child standing with arms outstretched and raised in the ancient posture of prayer (and which also signifies the Cross). The Virgin of the Sign adds the half body of the Christ Emmanuel within a golden clipeus over the Virgin's breast, thus enhancing the imagery of the Sun of Righteousness set within the supercelestial heavens.[108]

The cosmic imagery is even preserved in a late Western development, that of the Coronation of the Virgin. This image has both Christ and the Virgin seated on thrones in the manner of the Pantocrator, with Christ shown in the act of crowning His mother. Like other images of the Pantocrator family, the Coronation of the Virgin is to be found in the eastern apse. Here we have Christ, the Sun of Righteousness, honouring His mother, the Morning Star. The Coronation of the Virgin carries with it nuptial imagery—let it not be forgotten that in the East the sign of the sacrament of matrimony is the crowning of bride and groom by the celebrant—since Mary as a symbol for the Church is the mystic spouse of Christ. So here we have a complex of relationships, since Christ is both bridegroom and son of Mary but also, as Mary is His creature, father. In like manner, Mary is the bride of Christ, but also His mother and daughter. These Mariological insights are, of course, not *sui generis* but proceed from the Christological doctrine of the mystery of the Incarnation. The iconography of the Coronation of the Blessed Virgin raises Mary from the chthonic domain of the caves of the Annunciation and Nativity to the cosmic realm, with its symbols of the supercelestial sun and its attendant herald, the radiant supercelestial planet Venus, the Star of the Dawn.[109]

Other images are sometimes to be found in the semi-dome of the eastern apse. Very well known is the sixth-century mosaic of the Transfiguration in the apse of the Church of the Theotokos in St Catherine's monastery, Mount Sinai. But this also is an image of the cosmic Christ; it is the icon of light *par excellence*. It portrays the Saviour transfigured with uncreated light. Not only did His face shine 'like the sun' (Matt. xvii:2) but even 'his garments became glistening, intensely white, as no fuller on earth could bleach them' (Mk. ix:3). In the Sinai Transfiguration, Christ with glistening garments and brilliant gold nimbus is portrayed within a blue mandorla. Rays of light reach out to each of the surrounding figures of Moses and Elijah, and Peter, James and John.[110, 111, 112]

Figure 7. *Cosmogram*, based on a detail of the apse mosaic of the Transfiguration, Sant' Apollinare in Classe.

In early churches we tend to find a more symbolic treatment of the apse. In the centre of the upper section of a magnificent sixth-century mosaic, in the huge semi-dome of the apse of Sant' Apollinare in Classe, is a large medallion, inscribed within which is a jewelled Cross set against a blue background studded with ninety-nine stars. In the centre of the Cross is the head of Christ, within a circlet of pearls (fig. 7). To one side of the medallion, surrounded by clouds, is Moses and to the other Elijah, while below are three sheep, symbolising Peter, James and John.[113] The mosaic represents the Transfiguration, with Christ symbolised by the Cross. The head of Christ at the centre of a cross which extends across a roundel signifying the cosmos cannot but be an allusion to Christ as the Sun of Righteousness, governing the cosmos and redeeming the whole of creation. Christ, as the spiritual sun, thus brings the number of the heavenly bodies to one hundred, the number of completion.

The Cross set against the backdrop of the starry heavens, with its clear allusion to the concept of *Christos Helios*, is a symbolic device found in a number of locations in addition to Sant' Apollinare in Classe. Thus a golden

Figure 8. *Cosmogram*, based on the mosaic on the barrel-vault of Hagia Sophia, Thessaloniki.

Latin Cross, orientated to the east, is at the centre of the fifth-century mosaic in the dome of the mausoleum of Galla Placidia in Ravenna.[114] On the barrel vault of the eighth-century church of Hagia Sophia, Salonica, is a golden Greek Cross in a roundel set against variously shaded blue concentric rings. On the outermost ring are large golden stars, and from the ends of the Cross, and from its centre along the intercardinals, rays of golden light penetrate the cosmos.[115] This image is of particular interest as, instead of forming a backcloth, the fixed stars are restricted to the outermost ring. Unequivocally, the centre, occupied by the centre of the Cross, is the centre of the cosmos and the outermost ring is the sphere of the fixed stars. The Sun of Righteousness thus casts His golden rays (which Eastern theology would identify as a manifestation of the divine energies) out into the creation, interpenetrating the whole cosmos (fig. 8).[116]

Similarly, on the wall above the eastern apse of the sixth-century church of San Vitale, Ravenna, is a medallion depicting the cosmos borne aloft by two flying angels. At the centre of the 'onion skin' cosmogram, in which the spheres are represented by graded bands of white, blue and green (although,

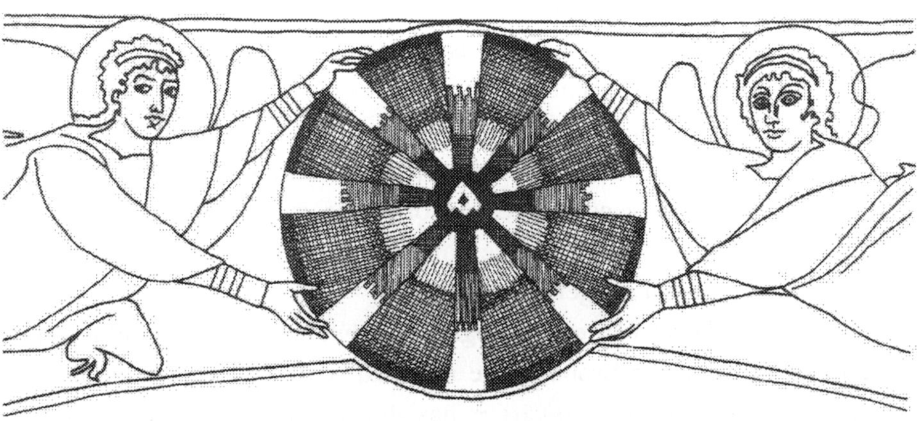

Figure 9. *Cosmogram*, based on the mosaic above the apse in San Vitale, Ravenna.

unlike Salonica, the outermost band is not embellished with stars), is a circle of orange red from which fan out along the cardinals and intercardinals (thus combining the Greek Cross with the X of the XP monogram, the cross of the earth with the observational cross of the heavens) eight orange-red beams of light which mid-way change to white. Inscribed within the central orange red circle is the Greek letter 'A' in white.[117] In this case, there cannot be much doubt that it is the sun which is depicted at the geometric centre of the spiritual cosmos. Although 'A', being comprised of three strokes, can signify the Trinity, the white 'A' surely here identifies the sun as the Sun of Righteousness, He who is the A and the Ω, the Light of the World. Since 'A' signifies the beginning, the cosmogram seems also to carry an allusion to the first act of the creation—*fiat lux*. Perhaps the medallion should be seen as a 'cosmogonogram' as much as a cosmogram (fig. 9).

Another symbol which often is used in place of the Pantocrator is that of the lamb. In the centre of the dome of San Vitale, the Lamb of God, with a brilliant golden nimbus, stands within a roundel backed with gold and silver stars. Four angels hold up the medallion, which is wreathed with leaves and fruits. Each of the angels stands on a globe.[118] That the Lamb of God carried with it, from Apostolic times, solar symbolism is clear from the Apocalypse, where John, as we have seen, says that the Heavenly Jerusalem needs no physical sun or moon since Christ, the Lamb, is its lamp. The physical universe might have been geocentric for Byzantine iconography and its derivatives, but the spiritual cosmos was clearly typically conceived as being heliocentric—just, of course, as the theology was Christocentric. That leaves and fruit should adorn a cosmogram might at first sight seem strange. However, plants, sometimes accompanied by animals, do frequently appear in conjunction with depictions of the supercelestial heavens. The reason, of course, is that Paradise belongs to the supersensible domain, and hence paradisal imagery can rest comfortably alongside supercelestial imagery.[119]

It is to late medieval Western iconography, however, that we should turn in order to locate what are probably the most explicitly heliocentric of all pre-Copernican paintings. Reutersvärd[120] has drawn attention to a number of Swedish examples of vault paintings which depict a central solar emblem, representing the spiritual sun, surrounded by a number of satellites. A fine example, dating from around 1300, from Rone, Gotland, has six such surrounding heavenly bodies.[121] Whether they should be thought of as planets, stars or subsidiary suns is not clear, but there can be no doubt that the painting denotes the supercelestial heavens. He also mentions a number of examples of the fifteenth and sixteenth centuries. At Mörarp, Skåne, the ribs of the vault form an eight-pointed star. At the centre is a boss, clearly representing the spiritual sun, and there are smaller suns or stars on the bosses at the junctions of the vaulting.[122] Most interesting of all such examples is that of the nave vaults at Äspinge, Skåne, which have as their respective keystones a sun, a rosette (an emblem of the sun) and the Sacred Heart surrounded by the crown of thorns. On the first of these vaults, the sun, again at the junction of eight ribs, is depicted with a human face and a number of uneven rays. The central area, which is filled with vegetation, including bunches of grapes, is surrounded by large stars. But, even more interestingly, suspended from the sun by strings are what look to all the world like eight horizontally banded, spherical glass Christmas tree decorations. It is difficult not to interpret these baubles as planets. As Reutersvärd notes, that the central body is the spiritual, not the physical, sun is indicated by the surrounding vine and grapes (emblems of Christ and the Eucharist) and the fact that the vault has been painted from the same Christological perspective as that which has the Sacred Heart at its centre.[123] In England, an equally explicit heliocentric image is presented by an unusual, or even unique, borrowing of the notion of Christ as the spiritual sun

Figure 10. *Christ as the Sun Surrounded by the Seven Acts of Mercy.* All Saints', North Street, York.

in order to symbolise the Seven Acts of Mercy. In a panel in a splendid late medieval fifteenth-century window in All Saints', North Street, York, seven planets, symbolising the Seven Acts, revolve in a circle around a central sun, symbolising Christ; an allegory imposed upon a metaphor, one might say (fig. 10).[124]

Glass opened up new possibilities not realised in other media. Obviously stained glass permitted a much more intimate fusing of the macrocosm without with the iconographic program within—the light of the macrocosm passing directly through the translucent walls of the building to make visible the patterns created by stained and painted pieces of glass. Without the light emanating from the macrocosm, the iconographic program could quite literally not be manifested. In Byzantine churches, the concept of *Christos Helios* is, as we have seen, usually given expression by deployment of the image of Christ Pantocrator at the apex of the central dome and/or in the semi-dome of the central eastern apse. In a Gothic church the image of *Christos Helios* does not need to be crafted iconographically, since the physical surrogate of Christ the Sun is the physical sun fully energising first one expanse of glass then another.

Glass presented new possibilities for the creation of cosmograms. The most notable example is that of that marvellous invention, the rose window. As Patrik Reuterswärd has shown, the rosette was a common Christian cosmic symbol from early times.[125] Sometimes it seemingly signifies Christ as *Christos Helios* or the divine dominion over the cosmos, but at other times it seems to represent the cosmos *per se*, albeit a cosmos infused with the uncreated light of God, with *Christos Helios* at its centre. The rosette, as also the rose window, is a form of *rota*, and, as Reuterswärd notes,[126] God was thought of as ruling the universe from the hub of the cosmic wheel, and so an image of God is traditionally placed at the centre of a rose window. One example of a rose window must suffice here, that of the Lausanne rose (c. 1230). In a *schematum*, consisting of two overlapping squares and eight circles, reminiscent of Cosmati pavements, the rose brings together depictions of the four elements, the four seasons, the four rivers of Paradise, eight monstrous races, the eight winds, the twelve signs of the zodiac and the twelve labours of the months. Also there would originally have been four modes of divination, corresponding to the four elements. Two of these, geomancy and hydromancy, are missing and have been replaced by the sun and the moon, which were almost certainly originally in the centre, the glass of which was replaced in the nineteenth century. What was originally depicted in the centre is a matter of debate. One suggestion is that it had 'the year' in the middle bounded by the sun and moon above, and light and dark below. This, however, is improbable. The sun and moon were frequently placed either side of Christ, particularly of Christ on the Cross. So a more likely possibility would be the Crucifixion in the middle, with the sun and moon (the surviving pieces) above, and the Virgin and St John below. This arrangement would have been appropriate here, as it would signify both that Christ is the centre of the earth and cosmos, and that in Christ heaven and earth, macrocosm and microcosm are as one.[127]

A powerful image of *Christos Helios* to emerge close to Copernicus' time was that of the monstrance. Following the emergence of devotion to the Blessed Sacrament and the institution of the feast of Corpus Christi in the thirteenth century, the monstrance, in which the Host was displayed in a crystal container, made its appearance during the fourteenth century.[128] Originally rectangular, it later became circular, as it appears in Raphael's early sixteenth-century *Disputa del Sacramento* in the Vatican.[129] In this fresco, the monstrance has not yet acquired its modern form, in which gold and/or silver sun's rays surround the Host, although this was a development which became established during the Baroque period. The shape of the *Disputa* monstrance, however, suggests the disk of the sun, and the connection is confirmed by the scene immediately above the monstrance in the painting, in which the Risen Christ is portrayed amidst the clouds against a medallion displaying the golden rays of the risen sun. In ecclesiastical design, the Host itself, with the letters IHS superimposed upon it, came often to be depicted on its own surrounded by sun's rays.[130]

Caution, of course, needs to be exercised in interpreting all appearances of central suns as evidence of heliocentric concepts, whether spiritual or physical. The sun, for example, is to be found at the centre of the face of some pre-Copernican clocks and at the centre of the zodiac.[131] But in such cases it is unlikely that anything more is meant than that the hands (or hand) of the clock record the direction of motion of the (Northern Hemisphere) sun through the heavens during the course of the day, or that the zodiac is comprised of the constellations through which the sun passes during the course of the year. Even in unequivocally ecclesiastical contexts, care needs to be exercised in drawing inferences. For instance, the sun is an heraldic device, and arrangements of suns, stars, etc. sometimes have little or no significance beyond a purely heraldic one. A case in point is that of two magnificent circles of suns, one with a large central sun, within the riotous stellar vaulting of the sanctuary and crossing in Tewkesbury Abbey. For this was not an outpouring of *joie de vivre* on the part of some pubescent Aristarchian, but a victory cry by Yorkists on their triumph over the House of Lancaster at the Battle of Tewkesbury in 1471, the gilded suns being emblems of the House of York.[132] But, of course, even in these cases the placing of the sun in a central position helped to familiarise people with the image of a heliocentric universe.[133]

As we have seen, the concept of *Christus Sol* is of primitive origin, but one of the reasons that the concept of Christ as the supercelestial sun continued within Christian iconography for centuries after *Christus Orpheus, Christus Bellerophon* and the like had passed from the mind of the Church (though *Christus Theseus*, did, as a consequence of the Church labyrinth, stage an appearance during the High Middle Ages) was because of a Christian theology of light. To trace the history of this theology would take us well beyond the requirements of this paper, but its existence should be noted, particularly as its principal early source—the writings of Dionysius the Pseudo-Areopagite—underwent a revival during the Renaissance. These works, deeply influenced by Platonism and Neoplatonism, had been written, probably during the late fifth century, by a writer who assumed the name of St Paul's disciple (Acts xvii:34). They were profoundly influential throughout most of the Middle Ages in both East and West, with the first commentary on them being written by Maximus the Confessor. They were highly valued by Grosseteste[134] and St Thomas Aquinas and underwent something of a revival during the *quattrocento*, perhaps impacting most notably on Nicholas of Cusa. Although their authenticity was sometimes questioned, and indeed even their orthodoxy, great kudos attached to them because of their supposedly Apostolic provenance.[135] For our purposes, though, what needs to be particularly noted is their importance from the point of view of the origins of Gothic art and architecture.[136]

The whole thing was really a chapter of accidents. The local martyr who was the eponymous saint of the Abbey of Saint-Denis came to be confused with the NT Dionysius the Areopagite. So the writings of Dionysius, which first made their way to the Isle de France through a gift of the Pope to Pepin the Short in 785, were assumed to be the work of the Abbey's (and France's) patron, and, not surprisingly, their translation, by no less a scholar than John Scotus Erigena, in the ninth century had a major impact. Most importantly, they fired up the great Abbot Suger to rebuild the Abbey, the resting place of kings, as a temple of light. The central teaching of Dionysius is well summarised by Georges Duby:

> God is light. Every creature stems from that initial, uncreated, creative light. Every creature receives and transmits the divine illumination according to its capacity, that is, according to its rank in the scale of beings … . The universe, born of an irradiance, was a downward-spilling burst of luminosity … . And because every object reflected light to a greater or lesser degree, the initial irradiance brought forth from the depths of the shadow, by means of a continuous chain of reflections, a contrary movement, a movement of reflection back toward the source of its effulgence.[137]

It was, Duby adds: 'This concept [which] held the key to the new art—an art of light, clarity and dazzling radiance. This was to be the art of France, and Suger's abbey church was its prototype'.[138] The introduction of flying buttresses made possible the replacement of masonry by great windows, which in the course of time were to become veritable walls of glass.

For Dionysius, this uncreated light is the creative power which, interpenetrating the whole cosmos, gives form and life to everything within the universe.[139] The ultimate source is the Godhead Itself, what Dionysius calls 'the Good'. Nevertheless, the sun and its light can, Dionysius argues, be used as an *analogy* for the Good:

> And what shall I say concerning the sun's rays considered in themselves? From the Good comes the light which is an image of Goodness; wherefore the Good is described by the name of 'Light,' being the archetype thereof which is revealed in that image. For as the Goodness of the all-transcendent Godhead reaches from the highest and most perfect forms of being unto the lowest, ... and so gives light to all things that can receive It, and creates and vitalises and maintains and perfects them, and is the Measure of the Universe and its Eternity, its Numerical Principle, its Order, its Embracing Power, its Cause and its End: even so this great, all-bright and ever-shining sun, which is the visible image of the Divine Goodness, faintly re-echoing the activity of the Good, illumines all things that can receive its light while retaining the utter simplicity of light, and expands above and below throughout the visible world the beams of its own radiance.[140]

Dionysius speaks of 'the Good', but according to Christian doctrine all things are created by God through the Logos, the second person of the Trinity. Dionysius is thus, by use of this analogy, endorsing the concept of *Christos Helios*.

As far as the creators of the Gothic were concerned, the light of the sun which flooded into their churches was seen as an image of God. The Cistercians, with their puritan proclivities, preferred unmediated light,[141] but the unreformed Benedictine Suger, who was attracted by Cluniac ideas, opted for stained glass. The rays of the sun striking the otherwise dark and inert glass gave the windows form and life, casting moving mosaics of brilliantly coloured light, like the jewels which formed the foundations of the Heavenly Jerusalem (Rev. xxi:19-20),[142] onto pavements and pillars. Again, we see the church as a mediator between earth and heaven, leading the faithful, as they contemplate the image of Christ, His mother, the angels and the saints in glass irradiated with light, ever upwards towards the throne of God. Duby, speaking of Suger's windows, writes: 'They were intended as devices by which to ennoble the divine light, infuse it with ruby- or amethyst-hued iridescence, the colors of the celestial virtues, and thereby lead the viewer's unseeing spirit 'along the paths of anagogical meditation''.[143] The medieval theology of light has long been in eclipse in the West, yet even today tourists sometimes freeze in their steps as they come upon a great window, such as that of *Notre Dame de la belle verrière* (an icon of the Virgin Enthroned) at Chartres, fully illuminated by the rays of the sun.[144] To those who have had such an experience, the words of the medieval lyric might not seem so totally incomprehensible: 'As the sun that shines through glass, so Jesus in his mother was'.[145]

4. THE LAMP IN THE TEMPLE—AN UNSUSPECTED EUREKA EXPERIENCE?

In a recently found letter, which Copernicus wrote to a friend back in Cracow while he was studying at the University of Padua, the University of the Most Serene Republic, he recounts a visit to the Doge's Chapel of San Marco. He tells his friend that he stood at the entrance to the nave transfixed, not knowing for a moment whether he was in heaven or on earth. Above him was the Pentecost dome, with tongues of fire descending like rays of the sun upon the twelve Apostles from the dove of the Holy Spirit, hovering above the unoccupied throne awaiting the return of Christ at the Judgment. On the wall to his left he noted a sublime image of Christ Emmanuel, clothed in radiant gold, with white undergarments glittering with mother of pearl, and surrounded by the starry heavens (fig. 11). As he walked towards the crossing, his guide told him of plans to copy the great image of the Pantocrator badly damaged by fire in 1419 in the eastern apse, and how in the centre of the presbytery dome there was another image of Christ Emmanuel against the backdrop of the star-filled heavens, and with a golden halo like the disk of the sun.

As they approached the crossing, Copernicus' eye was caught by the red-glassed lamp in the middle of the church, its long chain hanging from the apex of the central dome.[146] As he gazed up into the dome he saw the image of the

Figure 11. *Christ Emmanuel*, west arm of the north wall, San Marco, Venice.

ascended Christ, as Pantocrator and as the judge who was to come at the *Parousia*, seated on the rainbow with His feet on the globe of the earth, against a representation of the cosmos, again resplendent with stars. Four angels performed a heavenly dance as they held aloft the cosmogram. On the pendentives he noted the images of the four Evangelists, and beneath them the four rivers of Paradise, and below each river a tree of Paradise (figs. 12 and 13). Set in the pavement beneath was a great square of green-veined marble, which his guide told him was known as the 'sea'. It led him to recall, he says, the sea of glass before the throne of God (Rev. iv:6a) and John's description of the Heavenly Jerusalem in the Apocalypse: 'Then he showed me the river of the water of life, bright as crystal, flowing from the throne of God and of the Lamb through the middle of the street of the city; also, on either side of the river, the tree of life ...' (Rev. xxii:1-2).[147] And then it struck him.

There, in the middle of the church, itself an image of the cosmos, was that glowing red lamp, hanging from the icon of Christ, the Sun of Righteousness, in the centre of the dome. And, of course, if the lamp were an image of the supercelestial sun, then it was also a symbol of the Lamb: '... the city has no need of sun or moon to shine upon it, for the glory of God is its light, and its

Figure 12. *The Ascension Dome and the Lamp*, San Marco, Venice.

lamp is the Lamb' (Rev. xxi:23). Where else, but from the geometric centre, could the Sun of Righteousness shed forth His light to the whole of Creation? But if the Sun of Righteousness be enthroned at the centre of the supercelestial heavens, why, then, should not the physical sun, a symbol of Christ, also occupy the geometric centre of the world? What better place could there be for the sun, situated where it could best illuminate the whole universe? Why should the sun in the physical heavens occupy a different position than the Sun of Righteousness within the spiritual heavens? And then, he says, his head began to spin as he began to catalogue the astronomical advantages which an Aristarchian universe would have over the geocentric model. How much easier to explain the retrograde motions of the planets; how much more simple it would be to have the earth revolve on its axis each day rather than the great sphere of the fixed stars; how much more sense it would make of the differences between the behaviour of Venus and Mercury, on the one hand, and the rest of the planets, on the other; how much more elegant things would be

Figure 13. *The Ascended Christ in Majesty,* detail of the Ascension Dome, San Marco, Venice (without the lamp chain).

if that tiresome Ptolemaic device, the equant, could be eliminated, and if astronomy and the physical structure of the cosmos could be brought into agreement ...

Where else, indeed, Copernicus remarks, could the physical sun be placed in the cosmos, that 'most beautiful of temples', but in the midst of the temple, like the Lamb of God, from where 'it can illuminate everything at the same time'. Just as the Sun of Righteousness is enthroned in the midst of the supercelestial heavens, from whence He bathes the whole creation in divine light, so 'in the middle' of the physical universe must surely be 'the seat of the [physical] Sun', from whence it bathes the universe in its beams, impregnating the earth 'with annual offspring'. Thus the physical sun at the centre of the material cosmos, like the Pantocrator enthroned in the midst of the star-studded supercelestial heavens, 'as if seated on a royal throne governs his household of Stars as they circle round him'. Since God, as the Scriptures tell us, has 'arranged all things by measure and number and weight' (Wis. xi:20b), 'We find, then, in this arrangement the marvellous symmetry of the [physical] universe, and a sure linking together in harmony of the motion and size of the spheres, such as could be perceived in no other way'.[148]

Unfortunately such a letter has not come to light, nor is there the slightest vestige of evidence that such an epistle was ever penned. But could an 'eureka' experience of the kind we have imagined have happened? Yes, surely it is by no means beyond the realms of possibility that such a 'disclosure situation'[149] did occur in San Marco or some other church; although the reader might well feel that two great 'eureka' experiences in the history of science occasioned by a lamp hanging in an Italian church to be well over the odds.[150] Certainly, it would be most improbable that Copernicus could have spent several years in Padua without ever visiting the nearby State church of the Venetian Republic. He might well also have visited other Italo-Byzantine churches, perhaps even those of Ravenna. But whether or not there was such an eureka experience—or perhaps a confirmatory experience which strengthened a growing conviction that Aristarchus was right, or perhaps just an experience, or experiences, which later came into focus—we are unlikely ever to know.

The fact remains, however, that there was a tradition, deeply embedded in Christian art, architecture, poetry and liturgy, which saw in the temple an image of the cosmos and in the macrocosm the image of the church. The metaphor of Christ as the sun was equally well established, and the Apocalypse of St John the Divine gave the stamp of scriptural authority to the image of Christ, the Lamb of God, the Sun of Righteousness of the prophecy of Malachi (Mal. iv:2), as the lamp hanging in the midst of the New Jerusalem (Rev. xxi:23). Moreover, the icon most often placed in the central dome of the church, or sometimes in the central eastern apse, was that of the enthroned Christ Pantocrator, Christ as the only image of the Father (Jn i:18 etc.), 'the Alpha and the Omega, the beginning and the end' (Rev. xxi:6 etc.); an image which carried with it the connotation of Christ as *lux mundi* (Jn viii:12 etc.), the spiritual sun.

So, to the other antecedents listed by Copernicus, surely we must consider adding another, a traditional ecclesiastical cosmology. Other sources can be found for images he employs—the enthroned sun, for instance, could have been borrowed from the sun cultists—but there is just too good an over all fit with the ecclesiastical cosmology for it to be rejected out of hand as a probable influence on Copernicus' thinking, 'eureka' experience or no.

5. THE DEMISE OF AN ECCLESIASTICAL COSMOLOGY

The question that must now be addressed is that if this Antique spiritual cosmology did indeed inform the metaphorical language Copernicus employs in *De revolutionibus*, why did he not explicitly invoke its aid in defence of the heliocentric universe? Surely here was an ally he could call to arms which would impress the Church. There are, however, extremely good reasons why Copernicus would have been unlikely to have embraced any such a strategy. For a start, one of the few things which this man, who was a canon lawyer, estate manager, accountant, physician, astronomer, calendarist, mathematician, Renaissance man of letters, amateur painter[151] and philosopher, was not was a theologian. Churchman he certainly was, but he took the well-established path of the ecclesiastical bureaucrat not that of the theologian or pastor (he was never ordained to the priesthood). His case in his letter to the Pope in *De revolutionibus* is that the *mathematical* (a term which subsumed 'astronomical' at the time) arguments in favour of the heliocentric/heliostatic model over the geocentric/geostatic are such that the theologian (and physicist) should accept the system.[152] And he holds out as a carrot to the Pope the prospect of the availability of more accurate data for the reform of the Julian calendar, the discrepancy of which with respect to the Nicaean vernal equinox (21st March) having become a growing cause for concern.[153] (In the event, the Church greedily devoured the carrot but then, with the Gregorian calendar safely established in lands under Papal sway, unleashed the Church's powerful post-Tridentine quasi-literalist wolves onto the Copernican system.[154])

Apart from any reluctance Copernicus may have had to argue theological points with the theologians, there would have been difficulties in basing a case for heliocentricism on our ecclesiastical cosmology. This cosmology was in reality a bundle of concepts and images, of an intrinsically metaphorical character, which was deeply embedded in Christian poetry, liturgy, iconography and architecture, domains which traded in fluid and imaginative images and were not bound by rigid logical consistency. Although it found substantial support in biblical typology and the Apocalypse of John, and although theologians such as St Maximus the Confessor did from time to time address aspects of it, it was not a cosmology which had been defined by theologians or endorsed, at a formal level, *as* a cosmology of the Church. Belonging preeminently to the realm of the Church's arts and liturgy, and also to the language of mystical experience, this cosmology never resolved itself into a wholly comprehensive and consistent system. Although intrinsically heliocentric and heliostatic, it was, however, not free of geostatic allusions. Perhaps the most obvious one is seen in the enthroned Pantocrator. This image, in part, finds its Scriptural warranty in the text: 'Thus says the Lord: 'Heaven is my throne and the earth is my footstool ..." (Is. lxvi:1). Now, for a heliocentricist this naturally geostatic image presents no real problem; the text simply is making the theological point that God cannot be contained within a human-made building, the verse continuing: 'what is the house which you would build for me, and what is the place of my rest?' The language used is geostatic because that was the language of the time. However, this text does clearly present a geostatic image, and was indeed one of the texts used to attack Copernicanism.

What, however, would have been a much greater problem for Copernicus would have been the axial-heliocentricism defended by the Renaissance sun cultists, and also those who argued for the central placement of the altar. One cannot help feeling that whatever benefit might have accrued to Copernicus, by virtue of the stress laid upon the sun, was outweighed by the muddying of the waters by those who wished to make strong claims for the governorship of the universe by the sun, but retained the physical and geometric centrality of the earth.

For Copernicus to have argued that the concept of *Christos Helios* by its very nature placed Christ at the geometric centre of the spiritual heavens would have been difficult as the iconography was frequently open to alternative interpretations. The placing of an image of Christ at the apex of a central dome as such is, for example, by no means free of ambiguity, for while it is most natural to see this position as an assertion that Christ as the spiritual sun governs His cosmos from the geometric centre, it is also *possible*, however implausible that might be in many instances, to see it as merely depicting the spiritual sun at the zenith above a central earth. On the other hand, where the Pantocrator, Christ Emmanuel, the Lamb or the Cross etc. is placed in the centre of a star-filled or, *a fortiori*, star-bounded roundel, spiritual, spherical heliocentricism does seem to be unambiguously indicated. And, of course, there are cases where an image of the sun itself is depicted at the centre of an array of attendant heavenly bodies. Clearly, however, the iconographic evidence over all is not as clear cut as a Copernicus, fighting from the ropes in any case, could wish for. But the sun cultists would have made successful argument virtually impossible. Interestingly enough, the sun cultists also preferred to remain largely silent about a spiritual cosmology of whose existence they must have been aware; but in their case one can guess that the reason was that they could see that it was fatally contaminated by geometric heliocentricism. Instead they preferred to focus on texts such as that of Ps. xviii:5 (LXX), *In sole posuit tabernaculum suum*, a text, of course, which does not refer to any title specific to Christ: 'Sun of Righteousness', 'the Word', 'Wisdom', 'Son of Man' etc.

It was, however, not only the sun cultists who would have muddied the waters for Copernicus. There was in fact an alternative spiritual cosmology to the ancient version we have been considering, that of Dante's *The Divine Comedy*. Dante obviates the problem of having to reconcile a heliocentric spiritual cosmology with a geocentric physical cosmology by grafting spiritual connotations on to descriptions of the Aristotelian universe (as of course others had done before him). While the sun is a symbol of God, the Trinity Itself is located in the Empyrean. Clearly, had Dante been around in the late fifteenth century he would have favoured the placement of the altar at the east end. Strong advocate of a spiritual interpretation of the cosmos that he was, it is none the less likely that he contributed to the demise of our Antique ecclesiastical cosmology.

There is, however, something of importance which differentiates both Copernicus and the sun cultists from their medieval and Ancient forebears in general. Earlier medieval epistemology was centred on biblical hermeneutics.[155] This traces back to the NT itself, the Fathers in late Antiquity, and, in a number of regards, to Judaic and Classical roots. There were many variants, but the system which proved most influential was that of John Cassian (c. 360-435), which had been influenced, in particular, by the hermeneutics of Clement of Alexandria. This system distinguished four levels of interpretation: the literal, the allegorical, the tropological and anagogical.[156]

Literal needs to be distinguished from literalist. A literalist interpretation assumes that a text has a single clear descriptive meaning; that, for example, the creation stories in Genesis i and ii can be taken to be simple historical descriptions of how God created the cosmos, what He created and when (that is, on which of the six days of creation). The literal meaning in traditional hermeneutics is quite different. Ascertaining the literal meaning of a passage in an OT text—and for the sake of simplicity let us assume it is an OT text—essentially involves establishing what the author meant by what was written. An important part of defining the literal meaning, then, is the determination of the kind of literary text the author has penned. Once a literal meaning has been determined we can proceed to higher hermeneutic levels.

The term 'allegorical level' is decidedly unfortunate as one could well determine that the literal meaning of a given OT passage was allegorical (figurative or metaphorical, etc.) rather than historical, although exegetes varied in their willingness to treat the literal meaning of a text as allegorical. It would be preferable to call this second level the *typological* level, since essentially what it involves, in the case of an OT text, is the finding of a NT antitype for the OT passage, that is a NT passage which is *structurally* identical to the OT 'type'. By virtue of this structural identity, the OT—which is the NT in shadow—is able to deepen one's understanding of a passage. (Often, for a given NT passage there will be a number of OT types.[157]) The tropological meaning is the moral meaning which a passage has; what it means to us, in terms of what we should or should not do in our daily lives; how the passage applies to the human soul. Clearly there can be more than one such meaning. The anagogical meaning is the meaning in terms of final destinies, of the Church Triumphant, of the Last Things—heaven, hell and the judgment. It is the supreme spiritual hermeneutic level.[158]

So a passage seldom (if ever) has just one meaning, and may have a significant number. There is not just one truth but a number of truths of different kinds, all held together by a principle of concordance.[159,160] This biblical hermeneutic was applied—with appropriate adaptation when called for—to almost everything. Thus, to take two well-known examples, Dante applies traditional hermeneutic techniques throughout *The Divine Comedy*, and the bestiaries ascribed, in the tradition of Aesop's *Fables* and the *Physiologi*, moral meanings to every animal they described.

Biblical hermeneutics accustomed medieval humanity to handle different levels of meaning of other kinds as well. Thus, they had little difficulty with the tactic adopted by some Ptolemaic astronomers and others of giving a realist interpretation to Aristotelian cosmology and physics but an instrumentalist interpretation to astronomy. In other words, they accepted that there could be mathematical truths concerning astronomical phenomena—in the sense of geometrical models which generated accurate predictions—which were not compatible with the physical structure of the cosmos according to Aristotelian principles. 'Saving' the appearances of the motion of a planet by means of the most effective mathematical model available involved one kind of truth, a purely instrumental or formal truth; giving an account of the structure of the universe according to Aristotelian metaphysical and physical principles another kind, a realist truth.[161] As Copernicus undoubtedly appreciated, in this case it is difficult to discern any real principle of concordance which binds the two systems together, although Dante handled the problem by simply mapping the epicycles etc. onto the simplified (non-Eudoxan) medieval version of the Aristotelian model.[162,163]

The so-called 'doctrine of double truth', allegedly held by some Averroists in the thirteenth century, clearly did, however, stretch medieval tolerance of multiple truths beyond breaking point. This is the doctrine that conclusions of Aristotelian philosophical reasoning which are inconsistent with faith can be accepted as truths alongside the incompatible truths of theology. That is, a proposition which is true in philosophy and false in theology, or vice versa, can nevertheless be accepted as a truth. So, for example, the propositions 'The cosmos is eternal' and 'The cosmos was created *ex nihilo*' can both be accepted as truths. It seems likely, however, that the 'doctrine of double truth' was but an 'Aunt Sally' set up to attack Averroism, and that no thirteenth-century philosopher did in fact hold any such a doctrine. Averroës' own position seems, indeed, to be in perfect harmony with the medieval hermeneutic epistemology.[164]

Certainly the medieval mentality would have had little difficulty in accommodating both a traditional spiritual heliocentric cosmology and physical Aristotelian geocentric cosmology. Spiritual cosmology was concerned with interpreting the world theologically, by means of a symbolic and metaphorical language, while physical cosmology was concerned with the supposed actual material structure of the world. The relationship between the two sets of descriptions, however, was not, despite the anagogical character of spiritual cosmology, analogous to a literal and an anagogical interpretation of a specific scriptural text, but rather to a typological interpretation of a text. There was a sort of loose typological correspondence between the physical and spiritual universes.

It was, of course, the strong sense of concordance which prevented medieval humanity descending into mindless relativism or the chaos of Feyerabendian Dadaism.[165] It also greatly assisted them in resolving a central paradox of the

Christian faith, that the Kingdom of God to come, secured by Christ through the Paschal Mystery, was none the less—along with all the saving acts of the Lord and the whole of sacred history from the beginning of time with the creation to its ending with the Second Coming—made present, and could be experienced, in the eternal present of an invariably uncertain and often violent life. Time and eternity, Babylon and the New Jerusalem were co-existent.

Now, however, some new mentality was breaking through, and it can be seen in both Copernicus and the sun cultists. What in effect both were endeavouring to do was to collapse epistemic levels into a single level, moreover into a level which was proper to basic physical descriptions of the universe.[166] The sun cultists didn't turn to the traditions of *Christos Helios*, which distinguished the spiritual sun from the physical sun, seeing in the one simply a correspondence of the other, but physically identified their Neoplatonic sun, the seat of God, with the actual physical sun firmly located at the axial centre of the order of planets within a geometrically geocentric system of the world. Copernicus adopted a similar strategy, only in his case allusions to the ancient spiritual cosmology, with its heliostatic connotations, would have seemed a lot more attractive than they would have done to the geometric geocentricist sun cultists. But as we have already suggested, to have made this argument explicit would have invited theological debate, and theological debate was not what he was looking for.

It is possible, however, that Copernicus could see a specific theological danger (one which the sun cultists also courted). As a consequence of collapsing not only astronomy but also spiritual cosmology into a heliocentric and heliostatic physical cosmology, Copernicus could have been seen as identifying the physical sun with the locus of divine control of the cosmos, *Christos Helios*, and not simply in the sense of typological identity—which would preserve the distinctness, the autonomy, of the typical and the antitypical domains—but in the sense of actual fusion. Despite the richly metaphorical language of our text—certainly strongly reminiscent of the earlier poetic extravaganzas of the Neoplatonist sun cultists—it seems hardly likely that Copernicus was tempted to follow the sun cultists' lead and swim out into dangerous pantheistic waters. Nevertheless, given the Western theological climate of the times—and time had moved on from the days of the sun cultists—to suggest that the divine was present within the natural processes of the cosmos as a universal matter of course, rather than intervening, *ad extra*, by infusion of supernatural grace, would have been to court theological objections.

Although at the time of Copernicus' death in 1543 the Council of Trent had yet to gather (it met for its first session on 13th December 1545), Western theology was already well down the track to rejecting the earlier Christian belief, common at one time to both the East and the West, that the divine interpenetrates the material universe. Creator and creation were to be separated; God's transcendence could not be compromised. In Aquinas, grace is a gift, bestowed, as is the human soul, by God by miracle; the Transcendent choosing to act within the material universe, which, in itself, although a wonder of God's handiwork, in a sense even a reflection of the Divinity capable of moving humanity to the worship of its Creator, is devoid of itself of actual penetration by the divine.[167] The issue came to a head in the fourteenth century in the dispute over Hesychasm. The debate led St Gregory Palamas in the East to develop the early patristic distinction between God's essence and energies. The Holy Trinity in Its essence is unknowable, but through Its energies, which interpenetrate the natural world, the divine can be truly experienced. The analogy used by Palamas is the venerable Christian metaphor of the sun; the Godhead, like the disk of the sun, is unknowable in Its essence, but yet, like the rays of the sun, the divine energies vitalise the whole creation and can be directly experienced by humanity. Palamas was appealing to the very spiritual cosmology we have been considering in this paper, as between the physical and the spiritual suns there is a structural, typological (or at the least quasi-typological) identity.[168]

The distinction between God's essence and energies, well supported by patristic texts, has become central to the mainstream theology of the Orthodox Church. Western theologians in general, however, rejected the distinction as they thought that it undermined not only the transcendence of God but the simpleness of the Godhead, by introducing a distinction into the divine nature Itself. This, of course, was denied by the Eastern theologians as the distinction, they claimed, did not affect the divine nature in Itself; the Godhead in Its essence remains simple and indivisible, but the divine energies, the outflowing of God's creative love, interpenetrate the whole natural creation. The East, therefore, was able to re-affirm the sacredness of the creation, whereas the West drove in a wedge between God and the material cosmos, the Creator and the created. Divine action in the world thus became a matter of the imputation of supernatural grace alone; there was no interpenetration of God into the very fabric of creation. Given this shift in Western thinking, Copernicus could well have been fearful of having to confront a charge of pantheism.[169]

St Thomas Aquinas, while accepting traditional Christian hermeneutics, as such, restricted its application to biblical exegesis *sensu stricto*. He also, of course, adhered to a rigorous Aristotelian empiricism: *nihil in intellectu quod non prius fuerit in sensu*. Empiricism, being a philosophical position which imputes ontic privilege to sensory data, tends to look with jaundiced eye on notions of multiple truths. The heavenly domain, the New Jerusalem, was not obliterated by this mode of thinking, but knowledge of its existence rested on faith and it was firmly disentangled from the created world.

The difference between this new Christian way of thinking and the older is well brought out by iconography. Byzantine and Western medieval iconography sought to assist the liturgy in making the eternal kingdom of God present in the here and now. The object was to bring down the New Jerusalem from heaven. There was an integration of earth and heaven, earthly things enjoyed proper transcendental meanings.[170] The Baroque iconography that gave expression to the theology of the Council of Trent, which saw the triumph of Thomism, had another purpose, to lift the faithful up to the heavens. By means of the illusionistic tricks of linear perspective, the architecture of the church was seemingly extended upwards so that it opened out into heaven, another world which lay beyond the earthly.[171] The message was that, by moral and spiritual struggle in this world, the believer could achieve entry to heaven in the next, there to enjoy the Beatific Vision, to experience the Godhead in Its essence.[172]

The trials of the Saints were commonly depicted in medieval Western art, and were by no means unknown in Byzantine, but now there is a new emphasis on the sufferings, and also the ecstasies, of the saints and of Christ and His mother. The new art is designed to appeal to the emotions of the faithful, to encourage them to empathise with the earthly passions of those who intercede for the faithful before the throne of God, and to take up the Cross themselves that they might attain to their heavenly reward which lies beyond the doors of death. Depiction of the Passion of Christ and of the struggles and torments of the saints in the flesh were painted in an increasingly ultra-realistic manner, with the artists routinely working from models acting the part of the holy individual in question. By contrast, Byzantine art, and to a large degree also Western medieval art, sought to portray the deified person. It is not that the individual human features were ignored—traditional likenesses, in many cases authentic, were often strictly reproduced—but the flesh was shown as transfigured. At least with Byzantine art, far from there being any attempt to express emotional states, the aim was to portray dispassion. The heavenly realm was iconographically brought down to earth so that the worshipper would have a strong sense of their earthly worship being joined with the worship of the saints and angels before the throne of God. The iconography was of a piece with traditional hermeneutics, different levels of meaning co-existing within the framework of a building which represented the cosmos as a whole and which mediated between heaven and earth, and thereby worked to achieve their union.

From well before the time of Copernicus' sojourn in Italy the traditional understanding had been in retreat. Largely through changing theological conceptions, on the one hand, and the development of linear perspective, on the other, verism was pushing out the Byzantine and Western medieval tradition of iconography. The earthly realm was no longer depicted *sub specie aeternitatis*, no longer were anagogical connotations intrinsic to images; the heavenly realm was being teased apart from the mundane, connected with it only by supernatural interventions. As far as the heavenly realm was concerned, we can detect two conflicting tendencies. On the one hand, figures of saints in heaven were depicted with the same insistent verism as the same saints were depicted during their life on earth. On the other hand, the heavens were often reduced to absurd fantasy, with ludicrous winged cherubs and the like aimlessly besporting themselves, and saints squatting on clouds. Allegory and fantasy flourish as never before in the art of the Renaissance, but this is not the allegory of the allegorical/typological level but of the literal level, the 'text' itself. Symbolism also flourishes, but becomes individualistic and hence often indecipherable, and ceases to belong to a traditional cosmic iconographic language. As in many other departments of Renaissance culture, what we are seeing is a strong tendency to collapse the Ancient and medieval hermeneutic-epistemic levels into a single epistemological, and ontologically privileged, level. What won't collapse is banished to an essentially new category, a category which involves the re-drawing of the boundaries of the supernatural.[173] In short, we have commenced our journey into the modern 'enlightened', scientific (also, unfortunately, scientistic), positivistic age.

Paradoxically, although the spiritual cosmology might have played a role in the genesis of the Copernican theory, Copernicus, by fusing together spiritual and physical cosmology, and by collapsing astronomy and physics into a unified physical cosmological system, was hammering one of the final nails into the coffin of the Antique ecclesiastical cosmology in the West. The new empiricism of which Copernicus was an advocate did not sit well with the ancient concept of the correspondence of macrocosm and microcosm or with the medieval acceptance of levels of truths generated by traditional hermeneutics and its derivatives. The coherence of the many-levelled medieval world view was under siege from all sides. Natural philosophers and astronomers such as Copernicus were beginning to demand not only that scientific models, laws and theories should be solidly grounded in observable phenomena, but that physical theory and operations for 'saving' the phenomena should be in agreement; indeed, that physics and cosmology, on the one hand, and mathematics and astronomy, on the other, should be merged into a single empirical enterprise.[174]

At the same time, the Reformation brought about a new emphasis on the literal meaning of biblical texts at the expense of the higher hermeneutic levels. (The anagogical level was, indeed, often discarded entirely.) Since the Protestants maintained that individual believers could interpret the Bible for themselves by virtue of a God-given inner light, without the mediation of the authority of the Church or the insights of traditional exegesis, a door was in fact opened to literalism. In other words, what mattered was the plain grammatical sense of a text, rather than what the human author might have meant by what they wrote. For literalists, indeed, the human author virtually disappears altogether, every sentence being treated as the record of the unmediated voice of God.[175] However, while Copernicus and the new breed of biblical interpreters might have been alike in their revolt against medieval multiple truths, this fact was far from being a recipe for harmony, since the new exegetes of course maintained that the plain meaning of a number of texts of Scripture entailed the truth of the geocentric and geostatic understanding of the cosmos.[176]

The Renaissance, the Reformation, the Counter-Reformation and the Proto-Scientific Revolution all conspired, in their own ways, to bring an end to the Antique spiritual cosmology. No longer did iconographic programmes (where iconography was permitted at all) present the church, save in a very vestigial fashion,[177] as an image of the cosmos, although some memory of spiritual cosmology did survive in architecture. The concept of *Christos Helios* , as such, was not discarded, but it was detached from its ancient context in spiritual cosmology. In Roman Catholicism, the sun, as a symbol of Christ, actually burst out into greater brilliancy than ever with the splendid Baroque monstrances which surrounded the Host with golden and silver rays.[178] But despite such isolated echoes of ancient ideas,[179] the spiritual cosmology as such, which once it had shared with the East, was almost totally lost even from the memory of the Western Church. But if Copernicanism compromised the old heliocentric spiritual cosmology, it was destined to obliterate the alternative geocentric spiritual interpretation that had been built onto the Aristotelian-Ptolemaic model, most notably by Dante in the *Commedia*. Well could Donne lament, in an oft misquoted passage from the *First Anniversary*, the passing of a venerable tradition which had comprehended the physical, tropological and anagogical domains within a comprehensive Christocentric cosmological belief system. Until Sir Isaac Newton was to bring the sacred back into the System of the World for a season late in the seventeenth century, it did indeed seem as if the new cosmology had stripped away all meaning, that: ''Tis all in pieces, all cohaerence gone'.[180] But Newton's partially resacralised cosmology could only have brought slight comfort to those who felt as did John Donne. The truth be told, Newton's was but a sadly reduced and highly sanitised vision, which failed to encompass the rich range of hermeneutic interpretations that had so characterised the ancient ecclesiastical cosmology and that of Dante's *The Divine Comedy*.[181]

School of Science and Technology Studies, University of New South Wales

NOTES

[1] Copernicus, trans. A.M. Duncan, *On the Revolutions of the Heavenly Spheres* (Newton Abbot, 1976) p. 50 (I, 10). For illustration, see Martin Kemp's paper in this volume, fig. 11.

[2] *Ibid.*, pp. 25-6. The translator observes in a note to this passage that 'Nicetus' should in fact read 'Hicetas', Hicetas being an obscure Pythagorean. Cf. Cicero, *Quaestiones academicae*, IV, 29. He also notes that '... some people interpret Plato's rather obscure account of the universe in *Timaeus* as meaning that the earth revolves'. There is even evidence for thinking that Plato might, late in life, have favoured the hypothesis advanced by Heraclides that Venus and Mercury orbit around the sun. See H. Blumenberg, trans. R.M. Wallace, *The Genesis of the Copernican World* (Cambridge, Mass., 1987) p. 233. The quotation is taken from Plutarch, *De placitis philosophorum*, III, 13.

[3] *Isis* 53 (1962) 504-8.

[4] *Journal of the History of Ideas* 44 (1983) 667-9.

[5] In R.H. Stuewer (ed.), *Historical and Philosophical Perspectives of Science* (New York, 1970) 163-71.

[6] The publication of such works of Frances Yates as *Giordano Bruno and the Hermetic Tradition* (London, 1964) led to something of a 'silly season' in the history of science when the most extravagant claims were made for the influence of the Hermetic and magical traditions on the origins of the Scientific Revolution in general and the mechanical philosophy in particular. This is not to say, of course, that the occult and arcane traditions are totally irrelevant to the understanding of the development of modern science, or that there is little of value in the writings of Frances Yates. For an assessment of the historiographical issues surrounding the Yates' thesis, and also the internalist-externalist debate more generally, see Mary Hesse's essay, 'Hermeticism and historiography: An apology for the internal history of science', in Stuewer, *Perspectives,* 134-62.
[7] Rosen, 'Was Copernicus a hermetist?', p. 166 (see Note 5).
[8] Blumenberg suggests that the starting point of the train of thought which led Copernicus to his system was the fact that on the geocentric hypothesis there was no satisfactory way of deciding between the inferior and superior conjunctions of Venus and Mercury (that is whether the two planets had their orbits between the Moon and the Sun or between Sun and Mars). For Blumenberg's argument see *Genesis,* pp. 231-6.
[9] A convenient summary of the life of Copernicus, on which I have drawn, is provided by A. Koyré, trans. R.E.W. Maddison, *The Astronomical Revolution: Copernicus-Kepler-Borelli* (Paris, 1973) pp. 18-25.
[10] A likely explanation for the omission of Aristarchus has been suggested by Owen Gingerich. Gingerich points out that there is mention of Aristarchus in the original manuscript in the University of Cracow Library. Some changes, however, were made just prior to a copy of this manuscript being delivered to the Nuremberg printer, Petreius: 'The original Book One, containing Copernicus's cosmological assumptions, was merged with the original Book Two, which provided a short manual of trigonometry, complete with tables. As a result, an interesting passage about ancient precursors of the heliocentric view was removed and reworked into a new preface and dedication to Pope Paul III. In the process, the name of Aristarchus ... was eliminated, perhaps inadvertently'. 'A fresh look at Copernicus', in *Encyclopedia Britannica, The Great Ideas Today,* 1973 (Chicago, 1973) 154-78, pp. 156-7.
[11] On the sun cult, see F. Hallyn, trans. D.M. Leslie, *The Poetic Structure of the World: Copernicus and Kepler* (New York, 1990) ch. 5.
[12] There are, unfortunately, two different systems for numbering the Psalms. Greek and Latin texts use the numbering of the Septuagint (LXX), while many vernacular versions use that of the Hebrew Masoretic Text (MT). Except here, Biblical translations are from the Revised Standard Version.
[13] A point overlooked by H.P. Nebelsick, in his otherwise estimable study, *Circles of God: Theology and Science from the Greeks to Copernicus* (Edinburgh, 1985). See, e.g., p. 215.
[14] Strictly speaking, the sun wasn't quite at the centre of the Copernican universe, but rather the centre was that of the centre of the earth's orbit, which itself revolved around a movable eccentric circling the sun.
[15] Hallyn, *Poetic Structure,* pp. 132-3.
[16] *Ibid.,* p. 132 and fig. 5.
[17] *Ibid.*, pp. 136-9.
[18] *Ibid.,* p. 137.
[19] *Ibid.,* p. 138.
[20] Although there is Orthodox influence, the medieval art and architecture of Poland are predominantly Romanesque and Gothic. The fifteenth century was, in fact, the great age of the Gothic in Poland. Cracow Cathedral is basically Romanesque, but with Gothic and Renaissance additions. The Royal Castle, Wawel, was rebuilt by Sigismund I in the early sixteenth century in a vibrant blend of Polish Gothic and Italian Renaissance styles. It might not perhaps be going too far to say that Copernicus, like his king, brought the Italian Renaissance back to Poland to blend

it with the late medieval Polish culture which was still in the ascendancy at the commencement of the sixteenth century. Certainly, *De revolutionibus* strikes many as a blend of the two cultures. See A. Rhodes, *Art Treasures of Eastern Europe* (New York, 1972) pp. 130-71.

[21] *The Idea of a Town: The Anthropology of Urban Form in Rome, Italy and the Ancient World*, 2nd edn (Cambridge, Mass., 1988) particularly pp. 44-65 and 88-96.

[22] *Ibid.*, p. 45 (*On the Latin Language*, 7:6).

[23] Rykwert, *loc. cit.*

[24] *Stromata, or Miscellanies*, 5:6, in A. Roberts and J. Donaldson (eds), *The Ante-Nicene Fathers*, ii, rev. edn (Grand Rapids, Michigan, 1989) p. 452.

[25] On the architecture of the Temple, see, e.g., G.A. Barrois, *Jesus Christ and the Temple* (Crestwood, NY, 1980) pp. 54-61.

[26] '... he [the Demiurge] determined to make a moving image of eternity, and so when he ordered the heavens he made in that which we call time an eternal moving image of the eternity which remains for ever at one'. Plato, trans. H.D.P. Lee, *Timaeus* (Harmondsworth, 1965) pp. 50-1.

[27] There is, therefore, unlike the Hours, which each have a proper time of the day for their celebration, no specific time for the celebration of the Eucharist. As Alexander Schmemann puts it: 'The Eucharist is the actualisation of one, single, unrepeatable event, and the essence of the Sacrament consists first of all in the possibility of the conquest of time, i.e. the manifestation and realisation ... of a past event in all its supra-temporal, eternal reality and effectiveness. No matter when the Liturgy is celebrated ... it is essentially independent of the day or hour; it is not determined or restricted by them'. This is not to say that rules have not been devised from time to time to prescribe when the Eucharist should be celebrated on different days. During the Western Middle Ages, for example, it was generally laid down that public Mass—as opposed to private Masses, which could be celebrated at any time—should be celebrated after Terce on feast days and after Nones on fast days. See A. Schmemann, trans. A.E. Moorhouse, *Introduction to Liturgical Theology*, 3rd edn (Crestwood, NY, 1986) pp. 41-5 (the quotation is on p. 43) and J.A. Jungmann, trans. F.A. Brunner, *The Mass of the Roman Rite: Its Origins and Development* (Missarum Sollemnia), revised and abridged edn (London, 1959) pp. 176-9.

[28] Something of this arrangement survives in the Church of S. Clemente, Rome. However, in this case it is the choir, with its two ambones, which projects into the space of the nave, enclosed within a low screen. The altar beneath its *baldacchino* is situated behind another low screen just forward of the eastern apse. Although the present church dates from the beginning of the twelfth century, some of the marble used for the choir and chancel screens came from the sixth-century arrangement within the early basilica which lies beneath the later building. Cf. R. Beny and P. Gunn, *The Churches of Rome* (New York, 1981) pp. 48-50 and pl., p. 92.

[29] A term sometimes used in the West for the narthex of monastic churches is that of 'Galilee'. It was so called because the angel at the empty tomb instructed the women to tell Peter and the disciples that Christ had risen and was going before them to Galilee (Mk. xvi:6, Matt. xxviii:7). Thus solemn processions from the High Altar to the narthex were seen as symbolic of the Risen Christ going before His disciples into Galilee.

[30] During the High Middle Ages, except in large monastic churches supplied with a Galilee, the narthex was most often replaced by a porch or porches, or even dispensed with entirely.

[31] Strangely, one of the finest quality pavements of this kind is to be found in the Australian War Memorial in Canberra. This is the Shellal Mosaic, of 561/2, discovered by Anzac troops during the First World War. It depicts splendid birds and animals, mostly in pairs, surrounded by a vine, from which spring not only bunches of grapes but heads of grain, thus signifying the bread and the wine of the Eucharist. Of the gifts of the earth God has given humanity, the faithful offer the bread and the wine, which in the sanctuary are transformed in to the food of angels. Further Eucharistic symbolism is found running up the spine of the mosaic and in its borders. Cf. P. Henderson, 'The Shellel Mosaic: A reappraisal', *Journal of the Australian War Memorial*, 12 (1988) 35-44.

[32] Illustrated in G. Bustacchini, *Ravenna: Capital of Mosaic* (Bologna, 1990) pl., p. 82. Sublime renderings of the starry heavens are to be found on the fifth-century mosaics of the vault and of the dome of the Mausoleum of Galla Placidia in Ravenna; see *ibid.*, pls, pp. 12, 15 and 22.
[33] In three dimensions (ignoring the dome, apses, etc.) it will be noted that a three-dimensional cross will be formed. The volume of the building is thus interpenetrated by the sign of the Cross. The Cross of the Crucifixion unites heaven and earth. This is achieved by virtue of the fact that the vertical beam, pointing upwards, signifies heaven and eternity, but also, since the Cross is planted in the earth, that the eternal came down to earth and was incarnate in the person of the God-man, Jesus Christ. The horizontal beam, while it signifies the created world and ordinary historical time, signifies also the firmament which divides the cosmos into an upper, heavenly realm and a lower, earthly realm. It is worth remarking that a cross within a circle is the traditional astrological/astronomical symbol for the earth (diag. *i* below), signifying the four quarters of the earth. Considered three-dimensionally, it represents the sphere of the earth, the orb (as in the orb of a monarch). On to the two-dimensional diagram (which can represent also the four quarters of the heavens) we can superimpose a representation of the four quarters of the heavens as they are actually observed. This is obtained through humanity's common experience of the night sky. An observer always experiences the night sky, wherever they might be on the surface of the earth, as if they were at the centre of the universe. An observer (other than one at the poles or the equator) turning towards the observable celestial pole will observe stars close to the pole continually wheeling through the sky, neither rising nor setting. Turning to the east, stars and planets will be observed rising from the horizon, and, turning to the west, they will be observed setting. In the direction opposite the pole, the observer will see, towards the east, heavenly bodies rising, ascending to only a slight elevation above the horizon, and then setting towards the west. The sky thus divides into four observationally very different quarters defined by the intercardinals rather than the cardinals (diag. *ii*). If we combine the cross of the earth and the observational cross of the heavens, diags. *i* and *ii*, we arrive at an eightfold division (diag. *iii*). As the device combines the letter *Chi* with the Greek Cross, and as the upright of the Cross can signify the letter *Rho* (as in the case with an eleventh-century tympanum at Jaca Cathedral, which has the loop of the *Rho* depicted on the upright) it can constitute a Christogram. (See photograph in 'The forgotten symbols of God', in P. Reuterswärd, *The Visible and Invisble in Art: Essays in the History of Art* (Vienna, 1991) 57-136, ill. 62.) From this diagram we can, of course, readily construct an octagon (diag. *iv*). The octagon is a figure which is common in church architecture; in particular, being the favoured form for baptisteries (as with the Baptisteries of the Orthodox and the Arians at Ravenna; see Bustacchini, *ibid.*, pls, pp. 86-91 and 97-100). There are some churches with an octagonal basic plan, including San Vitale, in Ravenna. The octagon symbolises the eight days, as does an eight-pointed star (diag. *v*) or rosette, which in fact also often signify the sun (see below). From very early times Christians regarded the Lord's Day, Sunday, as not only the first day of the week but also the eighth day; the New Day that dawned with the rising of the Sun of Righteousness from the tomb which ushered in the eternal Kingdom of God. Within the liturgical year, the whole fifty days from Easter Sunday to the Sunday of Pentecost inclusive (a week of weeks plus one day) was regarded as but a single day and outside of ordinary time; the eighth day, the day of the Resurrection. The great fifty days (Gk. πεντηκοστή) was thus a foretaste of eternity within time. Cf. J. Danielou, *The Bible and the Liturgy* (Ann Arbor, 1956) particularly ch. 16.

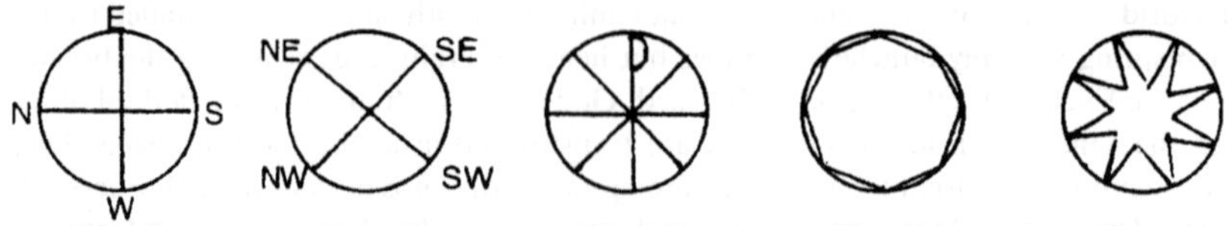

Diag. i Diag. ii Diag. iii Diag. iv Diag. v

It should be noted that a six-pointed floral-like star often appears in the same contexts as the eight-pointed star. Perhaps, in contrast to the eighth day signified by the eight-pointed star, the six-pointed star signifies the six days of creation, and time as opposed to eternity. Either star can signify Christ, through whom both the heavens and the earth were made, as the spiritual sun.

[34] Vitruvius, trans. M.H. Morgan, *The Ten Books of Architecture* (New York, 1960) pp. 72-3 (iii, 1, 3). See also, U. Eco, trans. H. Bredin, *Art and Beauty in the Middle Ages* (New Haven and London, 1986) pp. 35-6.

[35] A useful summary of the development of Byzantine ecclesiastical architecture and the iconographic programmes, along with their symbolic interpretation and links with liturgy, is provided by H. Wybrew, *The Orthodox Liturgy: The Development of the Eucharistic Liturgy of The Byzantine Rite* (London, 1989). See also, T.F. Mathews, *The Early Churches of Constantinople: Architecture and Liturgy* (Philadelphia, 1971) and G. Freeland, 'Time, architecture and the Byzantine iconographic programme', *Phronema* 4 (1989) 75-88.

[36] On aesthetics see G. Mathew, *Byzantine Aesthetics* (London, 1963).

[37] Colour plates, pp. 78-9, 80-2, 84-5 in G. Galitzine, *Imperial Splendour: Palaces and Monasteries of Old Russia* (London, 1991). Note that the crosses above the domes have a sunburst at their centre. Good sunburst crosses can also be seen above the dome of the Church of the Holy Trinity in the Monastery of the Holy Ghost, Novgorod, *ibid.* pl., p. 25. Other notable examples of Russian onion domes depicting the starry heavens include those of the Yuriev Monastery, Novgorod, *ibid.*, pl., pp. 30-1, and the Church of the Theotokos of Kazan, Kolomenskoe, *ibid.*, pls, pp. 125, 126-7 (also on dust jacket).

[38] See R. Webb, 'The *Nomoi* of Gemistos Plethon', *Journal of the Warburg and Courtauld Institutes* 52 (1989) 214-19. On Plethon and his context within Renaissance philosophy, see B.P. Copenhaver and C.B. Schmitt, *History of Western Philosophy, 3: Renaissance Philosophy* (Oxford, 1992) particularly pp. 87-90 and 140-2.

[39] See J. Williams, *Early Spanish Manuscript Illuminations* (London, 1977) pp. 24-8, and pls 12-30, 32-3, 36-40, with accompanying commentary. The importance which the early medieval Spanish Church came to attach to the Apocalypse of John is borne out by the decision of the Fourth Council of Toledo (632) that: 'The Apocalypse is a canonical book and should be read in the church from Easter to Pentecost; whoever objects may be excommunicated' (quoted by Williams, p. 24). By contrast, the Eastern Orthodox Church, although it regards Revelation as a canonical book, never reads it within the liturgy. The earliest extant illustrated manuscript of Beatus' *Commentary* (c. 940) is the Morgan Beatus Manuscript, in the Pierpont Morgan Library, New York (M. 644). The illustrations from this manuscript have been reproduced in full in J. Williams and B.A. Shailor (eds), *A Spanish Apocalypse: The Morgan Beatus Manuscript* (New York, 1991).

[40] The influence of Byzantine and other Eastern Christian art and architecture on the Romanesque has tended to be stressed by more recent writers. See V.I. Atroshenko and J. Collins, *The Origins of the Romanesque: Near Eastern Influenes on European Art 4th-12th Centuries* (London, 1985).

[41] For a detailed analysis see M. Schapiro, *The Romanesque Sculpture of Moissac* (New York, 1985).

[42] Demus, on the other hand, thinks that the buildings depicted at Kempley represent Bethlehem and Jerusalem respectively. See O. Demus, trans. M. Whittall, *Romanesque Mural Painting* (New York, 1970) p. 509. See also pl. 233. For St Gabriel's Chapel and Copford (sadly over-painted during the Victorian era, but largely authentic) see D.I. Hill, *Canterbury Cathedral* (London, 1986) pp. 163 (with illustration)-165, and Demus, *Romanesque,* p. 509 and pls 234 and 235 (colour), 236 and 237; A.J. Wright, *St Michael and All Angels, Copford* (Copford, 1993), with colour plates but not paginated. The New Jerusalem descending from heaven was also symbolised by great chandeliers taking the form of a hanging wheel rim depicting the walls, towers and gates of the Heavenly City together with prickets for a multitude of candles. Émile Mâle describes the chandelier at Aachen as follows: 'This city of light, the inscription tells us, is

the heavenly Jerusalem. The joys of the soul promised to the elect appear between the battlements, next to the figures of apostles and prophets who guard the Holy City' (*Religious Art from the Twelfth to the Eighteenth Century*, repr. (Princeton, New Jersey, 1982) p. 61). There is a significant example also in St Michael's, Hildesheim, illustrated in J. Evans (ed.), *The Flowering of the Middle Ages*, repr. (London, 1985) pp. 38-9. Also illustrated (p. 39) is the only remnant, one of its feet, of an enormous seven-branched candlestick, several times the height of a person, from St Remi at Rheims. Such candlesticks, of course, derive from those in the Jewish Temple. A reconstruction of the St Remi candlestick is also illustrated.

[43] One likely reason for this iconographic development was the popularity in the West not only of the Apocalypse but also of Augustine's *De civitate Dei.* As in the NT (apart from the Apocalypse) the Temple is always spiritually identified either with the individual person or with the whole community of the faithful (e.g., Eph. ii:19-22) rather than with the church building, so Augustine's City of God denotes the Saints and good angels in heaven and the righteous on earth. Those who comprise the City of God will only be revealed at the end of the world, and it is only then that the City will be complete. See, for example, his homily on Psalm cxxi (cxxii MT), *Augustine of Hippo, Selected Writings*, trans. M.T. Clark (Ramsey, NJ, 1984) p. 234. However, as Augustine's two cities were identified with Babylon and the Heavenly Jerusalem of the Apocalypse, Augustinian ideas could indeed have contributed to the popularity of depicting the chancel as the Celestial City. In a Byzantine church an image of the Heavenly Jerusalem is certainly formed, with the Pantocrator in the dome with the angels, prophets and evangelists, the saints on the walls and the people gathered for the Divine Liturgy below. However, the church building is not portrayed in the literal way as the New Jerusalem, with walls, towers, basilicas and the like, that one finds in the Romanesque church. The church building tended to be seen more as an image of the Tabernacle or the Temple of Jerusalem, on the one hand, and as an earthly image of the supercelestial heavens, on the other. Thus we find St Germanus (Patriarch of Constantinople, 715-30) writing: 'The church is an earthly heaven in which the super-celestial [ἐπουράνιος] God dwells and walks about'. *St Germanus of Constantinople on the Divine Liturgy,* Gk text with trans. by P. Meyendorff (Crestwood, NY, 1984) p. 57 (*Ecclesiastical History and Mystical Contemplation*, 1). The distinction between the images of the New Jerusalem and of the supercelestial heavens might be subtle but it is none the less significant. A Byzantine church, or a close relative, is the Temple in the midst of a sacred city.

[44] An excellent example is the Ladder of Salvation at Chaldon. See I. Nairn and N. Pevsner, rev. B. Cherry, 2nd edn. *The Buildings of England: Surrey* (Harmondsworth, 1971) pp. 140-1. In later English medieval churches the torments of the damned are usually located as part of the Doom painting around the chancel arch, but the west, the place of the setting sun—and hence the position in opposition to that of Christ, the rising sun (see below)—retained its associations with Satan and hell. In the late medieval Gothic church of Fairford, Gloucestershire, the Doom appears in (heavily restored) glass of c. 1500 in the great west window. As, of course, this window is fully illuminated only around sunset, its position is of considerable symbolic significance. See H. Wayment, *The Stained Glass of the Church of St Mary, Fairford, Gloucestershire* (London, 1984) pp. 55-8, pl. xxii.

[45] See A. Weir and J. Jerman, *Images of Lust: Sexual Carvings on Medieval Churches* (London, 1986).

[46] See E. Baker, 'The wall paintings of Clayton Church, Sussex', in G.R. Grove, *Clayton: A Guide to the Church of St John the Baptist,* repr. (Clayton, 1966) and Demus, *Romanesque,* pp. 507-8, pl. 230 and fig. 38.

[47] See P.R. Doob, *The Idea of the Labyrinth: From Classical Antiquity through the Middle Ages* (Ithaca, NY, 1990) ch. 5, and G. Freeland, 'Unravelling the labyrinth?', *Parergon* 30 (1981) 13-28.

[48] Fragments of many such Dooms have survived, but there is an impressive restored example in the Church of St Thomas of Canterbury, Salisbury (c. 1475). E.C. Rouse, *Medieval Wall*

Paintings, 4th edn (Princes Risborough, Bucks., 1991) pp. 57-60, pl. 66.
[49] *Ibid.,* pl. 18 and Demus, *Romanesque,* pl. 239.
[50] Hill, *Canterbury Cathedral,* p. 165 and fig. p. 164. There are two other such ceilings known to me. Although the fifteenth-century sculptured Deep Chapel (today's Blessed Sacrament Chapel) in Cahors Cathedral, Southern France, was severely mutilated during the wars of religion, mirroring can still be seen on large stars on its ceiling. The beautifully restored fifteenth-century ceiling of John Baret's Chantry Chapel in St Mary's, Bury St Edmunds, has a multitude of small stars with each star being comprised of a globule of mirror glass surrounded by golden rays. For details and colour illustration see the church guide, *St Mary's, Bury St Edmunds* (Bury St Edmunds, 1993) not paginated.
[51] Also worthy of special note are the hammer-beam nave roofs, particularly associated with East Anglia, which have numerous carved angels, often playing musical instruments or carrying shields with the instruments of the Passion. Amongst several truly outstanding examples that of the double hammer-beam roof of March perhaps deserves special mention. Of peculiar interest is the mid-fifteenth-century nave roof of St Mary's, Bury St Edmunds (see the church guide, *ibid.*). St Mary's also has an outstanding painted wagon roof in the chancel. The cornice comprises angels holding scrolls with twenty-two verses of the *Te Deum.*
[52] Even the Italo-Byzantine chapel of the Doges, San Marco, has an interesting carved Romanesque arch to the central doorway of the west front. Scenes include the labours of the months and the signs of the zodiac. These cycles also, of course, occur in many situations other than around arches. For example, the two cycles are arranged above one another around the rare twelfth century lead font of St Augustine's, Brookland, Kent.
[53] Made famous by Nicholas Pevsner in his King Penguin, *Leaves of Southwell,* published in 1945. See also N. Summers, *The Chapter House Southwell Minster* (Derby, 1994).
[54] For useful overviews of *mappaemundi,* see J. Wogan-Browne, 'Reading the world: The Hereford *mappa mundi*', *Parergon* NS 9 (1991) 117-35 and D. Woodward, 'Medieval *mappaemundi*' in J.B. Harley and D. Woodward (eds), *The History of Cartography,* vol. 1 (Chicago, 1987) ch. 18.
[55] Iconographic programmes were always considerably more variable in the West than in the East. With the rise of the Gothic, the major iconographic programme inside the church was often that developed in stained glass. The devaluing of wall painting programmes was often underscored by the abandon with which sections were destroyed by the insertion of additional or enlarged windows in older churches. Sculpture retained its importance, with new possibilities opened up by developments in the sculpture of the human figure. With great Gothic cathedrals, such as Chartres, we find a major programme in glass within the cathedral matched with a major sculptural programme on the exterior. Medieval churches also often had major programmes in tapestry or embroidery; but of these *very* little survives.
[56] See Wayment, *Fairford,* and H. Wayment, *The Windows of Kings' College Chapel, Cambridge: A Description and Commentary* (London, 1972).
[57] See Beny and Gunn, *Churches of Rome,* pp. 244-55.
[58] Though it is the fullest, it is by no means the only explicit statement. Another example is provided by a Syriac hymn which gives a description of the Cathedral of Santa Sophia at Edessa, following its rebuilding shortly after the old church had been destroyed by floods in 524/525:

4. Indeed, it is an admirable thing that in its smallness it should resemble the great world,
   Not in size, but in type ... ;
5. Its ceiling is stretched like the heavens—without columns, vaulted and closed—
   And furthermore, it is adorned with golden mosaic as the firmament is with shining stars.
6. Its high dome is comparable to the heaven of heavens;
   ...
7. Its great, splendid arches represent the four sides of the world;
   They also resemble, by virtue of their variegated colors, the glorious rainbow

of the clouds [Gen. ix:8-17].
8. Other arches [probably the squinches] are set all round like some outcrops of rocks on the top of a mountain; ...

Text in C. Mango (ed.), *The Art of the Byzantine Empire 312-1453: Sources and Documents* (Toronto, 1986) 57-60, p. 58. The reference to outcrops of rocks on a mountain top is interesting. From this it would seem that the central space beneath the dome was thought of as the High Place, upon which was situated the altar of the Lord. This thought might also provide a gloss on the choice of the Ascension—which according to Acts i:12 and tradition occurred on the Mount of Olives—for the central dome found in a number of Byzantine churches.

[59] See Introduction by J. Pelikan in *Maximus Confessor: Selected Writings,* trans. and ed. G.C. Berthold (Mahwah, NJ, 1985) 1-13.

[60] *Ibid., The Church's Mystagogy: In Which are Explained the Symbolism of Certain Rites Performed in the Divine Synaxis,* ch. 3., p. 189.

[61] *Ibid.,* ch. 2, p. 188. It should be noted that the three modes to which Maximus refers in this passage bear comparison with the three traditional levels of biblical hermeneutics which were usually added to the literal/historical level—the allegorical, tropological and anagogical (see below).

[62] *Ibid.,* ch. 4, pp. 189-90.

[63] *Ibid.,* ch. 5, pp. 190-5 (particularly first and last paras). To the three modes of interpretation supplied by Maximus, St Symeon of Thessalonica, whose analysis was much influenced by *The Church's Mystagogy,* adds a fourth, a correspondence of the three divisions of the church with the three hypostases of the Trinity. Symeon's commentary bears witness to the continuity of the traditional understanding of the symbolism of the Christian temple through to the fifteenth century. See H. Simmons, 'The influence of Pseudo-Dionysius in the liturgical writings of St Symeon of Thessalonica', *Phronema* 5 (1990) 33-9, p. 38.

[64] By this St Paul means, of course, that the Christian, like the Temple, is a dwelling place of God; not that the human person and the church building (which can hardly be said to exist in St Paul's day) are images the one of the other. However, such verses as this would have suggested to the builders of churches, properly speaking, that the temple should reflect the division of parts and faculties, the proportions and symmetries of the microcosm; the human person, who is made in the image of God and in whom God chooses to dwell.

[65] For a very recent discussion of the cosmic theology of St Maximus, see A. Louth, *Maximus the Confessor* (London, 1996) ch. 5, particularly pp. 74-7 for the *Mystagogy.* See also L. Thunberg, *Man and the Cosmos: The Vision of St Maximus the Confessor* (Crestwood, NY, 1985) particularly ch. 6.

[66] That medieval Western churches were indeed images of the cosmos has not been widely appreciated. A few scholars, including M.D. Anderson in *The Imagery of British Churches* (London, 1955), have, however, noted that this was the case: 'Medieval symbolism was all-embracing and scholarly disposition of subjects could make the humblest village church portray the entire Universe of space and the ultimate limits of time. Its roof, and particularly that of the chancel, represented the sky and therefore the glories of Heaven were painted upon it ...' (p. 67; see also following pages).

[67] E. Panofsky (ed. and trans.), *Abbot Suger, on the Abbey Church of St-Denis and its Art Treasures,* 2nd edn, ed. G. Panofsky-Soergel (Princeton, NJ, 1979) pp. 63 and 65.

[68] W. Dorigo, trans. J. Cleugh and J. Warrington, *Late Roman Painting: A Study of Pictorial Records 30 BC—AD 500* (London, 1971) pp. 110-15 and pl. 79. For full page colour illustration, see C. Bertelli (ed.), *Mosaics* (New York, 1989) pl., p. 53.

[69] Dorigo, *ibid.,* p. 111 and pl. 80.

[70] For discussion of the structural character of typology, see the essays contained in E. Leach and D.A. Aycock, *Structuralist Interpretations of Biblical Myth* (Cambridge, 1983).

[71] Orpheus was an obvious candidate to become a type of Christ as he descended into hell in

search of his wife Eurydice. But he was seen not just as a type but as a prophet who, like the Sibyls and many of the Ancient philosophers, could be shown to have affirmed the unity of God and that this one God had created the cosmos; even, according to some sources, *ex nihilo*. See, e.g., Lactantius in *The Divine Institutes* I, 5, trans. W. Fletcher, in Roberts and Donaldson, *Ante-Nicene Fathers*, vii, p. 13.

[72] It should be noted that, in John v, Christ says of John the Baptist: 'He was a burning and shining lamp, and you were willing to rejoice for a while in his light' (v. 35). In this passage (vv. 30-47) Christ teaches that John bore 'witness to the truth' (33), but that His own 'testimony' is from God and 'is greater than that of John' (36). Christ does not call Himself a 'lamp', but the inference certainly is that He is the new lamp which will outshine John (see Jn i:6-9).

[73] Christ says to His disciples in the Sermon on the Mount: 'You are the light of the world. A city set on a hill cannot be hid. Nor do men light a lamp and put it under a bushel, but on a stand, and it gives light to all in the house. Let your light so shine before men, that they may see your good works and give glory to your Father who is in heaven' (Matt. v:14-16). It is, of course, in so far as they become true images of Christ, who is the only image of the Father, that Christians become the 'light of the world'. (See also Phil. ii:15.) And, at the end, Christ promises, 'the righteous will shine like the sun in the kingdom of their Father' (Matt. xiii:43).

[74] The idea is to be found in the Second Epistle of Peter: 'And we have the prophetic word made more sure. You will do well to pay attention to this as to a lamp shining in a dark place, until the day dawns and the morning star rises in your hearts' (i:19). Unfortunately this whole passage (vv. 16-21) is difficult to interpret. Presumably, we are to take 'the morning star', as is usually the case, to refer to Venus, but are we to interpret it as a symbol of Christ? Possibly. However, as Peter is talking about 'the prophetic word' may be we should interpret 'the morning star' as referring to the word of the New Dispensation, the fulfilment of the prophetic word; in which case, it would seem that Christ is to be identified with the sun, the dawn of which brings with it the word rising in the hearts of the faithful like the Morning Star.

[75] 'Praying toward the East is handed down by the holy apostles, as is everything else. This is because the comprehensible sun of righteousness, Christ our God, appeared on earth in that region of the East where the perceptible sun rises, as the prophet says: 'Orient is his name' (Zech. 6:12); and 'Bow before the Lord, all the earth, who ascended to the heaven of heavens in the East' (cf Ps 67:34 [LXX]); ... and again, 'The feet of the Lord shall stand upon the Mount of Olives in the East' (Zech. 14:4). The prophets also speak thus because of our fervent hope of receiving again the paradise of Eden, as well as the dawn of the brightness of the second coming of Christ our God, from the East', Germanus, *Ecclesiastical History* II, in Meyendorff, *Divine Liturgy*, pp. 63 and 65.

[76] On the famous Creation tapestry from Gerona Cathedral (c. 1100), *Dies Solis* is depicted by *Christos Helios* in a cart-like chariot, with wheels whose spokes form a cross, drawn by four fiery steeds. Colour plate in G. Holmes (ed.), *The Oxford Illustrated History of Medieval Europe* (Oxford, 1988) facing p. 177.

[77] The Ancient mentality led Christians to the belief that the conception of Christ and the Crucifixion would have taken place on the same day of the year. The two popular early dates for the Crucifixion were 25th March and 6th April. The feast of the Epiphany, which included a commemoration of the Nativity, came to be celebrated on the 6th January (still the date on which the Armenians celebrate Christmas), nine months after 6th April. However, the *zeitgeist* moved in favour of the Northern Hemisphere vernal equinox, 25th March, and this suggested 25th December (the winter solstice) for the celebration of Christ's birth. This line of thinking coincided with another, concerning the conception of John the Baptist, which according to Lk. i:24-6,36 occurred six months before that of Christ. It was believed that the annunciation to Zechariah (Lk. i:5-25) had occurred while he was serving in the Temple during the festivals of the Jewish month Tisri. This led to the choice of the autumnal equinox, 24th September, for the feast of the Conception of John (though in the East, where the feast is still celebrated, it was

fixed on the 23rd, the day which marked the beginning of the civil year until the fifth century). This determination, of course, pointed to the 25th December for the feast of the Nativity of Christ, together with the 25th March for the Annunciation and 24th June (the summer solstice) for the Nativity of the Forerunner. Thus were the four principal astronomical points of the tropical year duly boxed, and a structure put into place which was to play a large part in the determination of the cycle of fixed feasts. We know from Chrysostom that it was believed that 25th December was the actual birthday of Christ, and that this was the supreme reason why the feast of the Nativity should be celebrated on that day, not any coincidence with the pagan feasts. However, that is not to say that the solar symbolism was not seen as significant; indeed, the most important days of the solar year had played a crucial role in the determination of the 25th December as the day of Christ's Nativity. For full details see the following work, on which the above account of the origins of Christmas has been largely based: T.J. Talley, *The Origins of the Liturgical Year,* 2nd edn (Collegeville, Minnesota, 1991) pt 2, particularly pp. 85-103. The solar basis for many major feasts has made easy the assimilation of numerous folk observances into the Christian Year, from the Yule Log to leaping over bonfires on St John's Eve. In Russia, John the Baptist was even sometimes called 'St John the Solstice', so closely was he associated with the solar year. But in Orthodox lands there is also a very strong link between the Prophet Elias (Elijah), who enjoys a major cultus, and the sun, with many solar folk practices associated with his day (20th July), including in parts of Eastern Europe the sacrifice of a red cockerel, a symbol of the sun. This cult seems to owe its origin, firstly, to the similarity of the prophet's name and 'Helios' (or 'Gelyos') and, secondly, to the story of the assumption of Elias into heaven (2 Ki. ii:1-12). The popular icon of the 'Fiery Ascension of the Prophet Elijah' is very closely related to the early Christian Roman image of *Christus Sol,* with the prophet shown ascending to heaven in a fiery chariot drawn by two fiery steeds. Sometimes the biblical whirlwind becomes a veritable seething fireball. Sometimes, also, the chariot itself it reduced to a single great wheel; the wheel being a common symbol for the sun. See V.N. Zalesskaya and Y.A. Piatnitsky, 'The sun in Byzantine and Russian art', in M. Singh (ed.), *The Sun in Myth and Art* (London, 1993) 250-65, pp. 260-2 and pls 378 and 382.

[78] Testimony to the importance attached to the rising sun comes from Pope Leo I (440-61), who castigates members of his flock who turn and bow to the rising sun when they have mounted the steps leading up to St Peter's Basilica (Talley, *Origins*, pp. 100-1). It is difficult to believe that the Pope could have been unaware of the universal early Christian practice of turning to the east in prayer. He seems to have suspected that some of the congregation weren't just honouring the creator of the sun, and certainly he feared that such practices could cause recent converts to fall back into the worship of the pagan gods. He might have had some genuine grounds for concern, as we learn from St Cyril of Alexandria that many fourth-century converts in Jerusalem were ex-Manicheans who identified Christ with the sun, and St Ambrose, in the same century, had to warn his flock against identifying Christ and the sun in a literal sense (see Reutersward, 'Forgotten symbols', p. 63).

[79] This ceremony can be traced back to the Jewish rite of the lighting of the evening lamp (Lev. 24:1-4). Early Christian practice was for the deacon to bring in a lamp from outside into the congregation. For Christians, of course, the lamp symbolised Christ. From at least the fourth century in the Church of the Holy Sepulchre the lamp was brought out from the tomb of Christ itself, and this inspired the Eastern practice of conveying the lamp from the sanctuary (N. Uspensky, trans. P. Lazor, *Evening Worship in the Orthodox Church* (Crestwood, NY, 1985) p. 31; see also pp. 14-15).

[80] Sometimes Paschal candlesticks were very tall, intricately carved permanent furnishings. There are several notable stone candlesticks in Rome, including the fine Romanesque example, of 1180, in S. Paolo fuori le Mura (Beny and Gunn, *Churches of Rome,* p. 87). Perhaps the most impressive of all such candlesticks is that in Hildesheim Cathedral, of c. 1020. This takes the form of a 3.2m bronze column with a continuous spiral band, in imitation of Trajan's column in Rome, with narrative scenes from the life of Christ, ending with the Triumphal Entry (discussed

in D.J. Bernstein, *The Mystery of the Bayeaux Tapestry* (London, 1986) pp. 98-9, and pl. 60).

[81] So called from its opening word. The *Exsultet* is a great paean of praise to God for the light of the Resurrected Saviour which, on this holy night, banishes all the powers of darkness. There are some differences between the Tridentine rite and the New Order, but these are not of any great significance.

[82] The Eastern and Western Paschal ceremonies described here do not in fact have the same origin. The Paschal Candle is essentially a baptismal candle, since the Paschal Vigil was the normal occasion for the administration of baptism in early times. The Orthodox ceremony, which occurs at a later point in the Paschal Vigil immediately prior to the singing of Mattins of Easter Sunday, echoes the ceremony of the Holy Fire performed in the Church of the Holy Sepulchre in Jerusalem. All lights having been extinguished, the Greek Patriarch enters the Holy Sepulchre itself alone, and then shortly emerges from it with two bundles of flaming candles. The engendering of the Holy Fire is still today widely believed in the East to be miraculous. A well-known, and highly sceptical, description of the event, first published in 1849, was recorded by the Hon. Robert Curzon in 1834. On this occasion, a panic occurred among the thousands of worshippers trying to leave the church which left well over three hundred dead. R. Curzon, *Visits to Monasteries in the Levant,* repr. (London, 1955) ch. 16.

[83] Jn i:18, xii:45, xiv:9b; Col. i:15; Heb. i:3.

[84] Was, as has been suggested, the Byzantine image of Christ Pantocrator modelled on Phidias' Wonder of the World, the statue of the enthroned Zeus at Olympia? Certainly Phidias' Zeus was readily available to iconographers for a while (until destoyed during a riot) after it had been removed to Constantinople. There is a Byzantine story concerning the depiction of Christ as Zeus which traces back to Theodorus Lector (sixth century): 'At the time of Gennadius [Patriarch of Constantinople, 458-71] was withered the hand of a painter who dared to paint the Saviour in the likeness of Zeus. Gennadius healed him by means of a prayer. The author [i.e. Theodorus] says that the other form of Christ, viz. the one with short, frizzy hair, is the more authentic' (*Hist. eccles.* I, 15; PG 86, 173). Theodorus' tale is also recorded by St John Damascene (PG 86, 221). (Quoted as in Mango, *Sources and Documents,* pp. 40-1). There is good evidence that the Byzantine image of Christ very probably derives from one or more actual likenesses of Christ Himself. The evidence canot be gone into here, but the interested reader should consult: L. Ouspensky, *Theology of the Icon* (Crestwood, NY, 1978) particularly pp. 46-9 and 59-71. It would seem, however, that there was a perceived resemblance between the image of the Byzantine Christ and Phidias' Zeus. But perhaps, although Christ did not owe His iconographic facial features to Zeus, nevertheless the traditional, and perhaps genuine, likeness of the Saviour, portrayed as Pantocrator, did *authenticate* Phidias' Zeus as an iconographic type of Christ. And it is worth noting that the wheels and other sun symbols associated with the iconography of Jupiter were destined to be inherited by Christ. It is also worthy of note that many early paintings and sculptures of Christ, such as paintings in the catacombs, are symbolic in character and make no claim to portraiture; for example, images of Christ as the Good Shepherd. In many such works the depiction of Christ seems to owe a debt to pagan images of Apollo. From the perspective of the present paper, this is perhaps of some interest as Apollo was, of course, a solar deity.

[85] The whole verse can be found, amongst other examples, in the great Italo-Byzantine churches of Norman Sicily. In the mosaics of the main apses of Cefalù Cathedral, the Palatine Chapel at Palermo and the Cathedral of Monreale, Christ Pantocrator is depicted with the open Book of the Gospels displaying the verse in both Greek and Latin. (The first two mosaics are dated by inscription; Cefalù to 1148 and the Palatine Chapel to 1143.) In the Palatine Chapel the half-figure of the Pantocrator also appears at the apex of the cupola, the surrounding inscription in Greek being taken from Is. lxvi:1—'Thus says the Lord: 'Heaven is my throne and the earth is my footstool". The Martorana at Palermo also has the Pantocrator in the cupola. Although Christ holds a closed Book of the Gospels, the surrounding inscription is again Jv viii:12, in Greek. For

full details of the mosaics of these churches, see O. Demus, *The Mosaics of Norman Sicily* (New York, 1988). Colour illustrations appear in Bertelli, *Mosaics*, pls, pp. 188-91.
[86] For example, at Torcello, which is a basilican church, the Virgin occupies the main apse, so the Pantocrator has been placed in the semi-dome of the south apse.
[87] The Pantocrator, in a few instances, in fact occurs in both positions. The central apse is the favoured position for Christ Pantocrator in Italo-Byzantine churches, such as San Marco, Venice, and the Sicilian churches mentioned in Note 85 above. In Romanesque churches the Pantocrator, in the form of Christ in Majesty of the Apocalypse, is also the favoured image for a central apse. Perhaps the most splendid of all surviving examples of Romanesque apse frescoes is that from San Clemente, Tahull, of 1123 (now in the Museo de Arte de Cataluña, Barcelona) which has the open Gospel Book with *Ego sum lux mundi* (Demus, *Romanesque*, pl. 208, with detail in colour p. 15). Images of Christ in Majesty (usually carved) are also frequently found on tympana.
[88] P. Reutersward, 'Windows of divine light', repr. in *The Visible and Invisible*, 45-56, p. 49 and ill. 37.
[89] A term which seems to be synonymous with 'supersensible heavens', 'heaven of heavens', or simply 'supernature'; the realm of God and the heavenly hosts. On a crude level it has been thought of as an actual location outside of the created cosmos contained within the sphere of the fixed stars, the Empyrean. It can, of course, in fact no more be located in space than it can in time. So, at a more sophisticated theological level, it connotes the spiritual (in the sense of anagogical) level of being which, though outside of time and space, none the less relates to the historical spatial cosmos through a system of correspondences. At the centre of the supercelestial heavens is of course the Godhead Itself. Since the source of the Godhead, the Father, can only be imaged by the Son, iconographically He is represented by Christ Pantocrator (equally *Christos Helios*). Supercelestial things can be imaged by earthly things by virtue of correspondences, in a manner analogous to the relationship between antitype and type. So the 'supercelestial (or supersensible) heavens' essentially denotes the representation of spiritual powers by means of images of sensible bodies. Indeed, it is only by means of material images that we can picture spiritual realities.
[90] Following the maxim 'As in heaven, so on earth', we should expect to find a correspondence between the Pantocrator, enthroned in the centre of the heavens, and the location of the saving acts of the incarnate Christ on earth; and so we do. In Christian thought, Jerusalem was considered to be at the centre of the earth (and is so placed on *mappaemundi*). In the Catholicon of the Church of the Holy Sepulchre is an *omphalos* (a navel stone) which marks the exact centre of the earth according to Christian tradition. According to tradition, within metres of this spot Christ was nailed to the cross, was crucified, died, was anointed, buried in the tomb belonging to Joseph of Arimathea, and rose from the dead. It was the *omphalos* at the sanctuary of Apollo at Delphi which marked the real centre of the pagan world; and there is an interesting association, which shows up in early Christian iconography, between Christ and the solar deity, Apollo (see Note 84). The *omphalos* of the Church of the Holy Sepulchre thus proclaims that Jerusalem is the true centre of the earth, not Delphi, and Christ, the Sun of Righteousness, the new Apollo.
[91] Colour plate in Bertelli, *Mosaics*, p. 156. A similar effect is achieved by the dome mosaic of the Church of the Saviour in Chora, Constantinople.
[92] As in the Palatine Chapel, Palermo (Bertelli, *Mosaics*, pl., p. 191). Angels are prominent in mosaics of a variant of the usual Pantocrator image, Christ of the Ascension/Parousia.
[93] Cosmati pavements—reflecting, as it were, in semiprecious stones the sublime beauty of the supercelestial heavens above—pick up the cosmic symbolism. The basic Cosmati design, capable of considerable elaboration, is that of a large central roundel and four encircling roundels. The central roundel is often a striking piece of porphyry and, as Reutersward observes, was known as a *rota* (wheel). The use of the term *rota* in itself might suggest that we are to link the central roundel with *Christos Helios* (the wheel being, as noted above, a symbol of the sun), but that this is the case is confirmed by the fact that the central roundel is sometimes marked by a cross.

Reutersward notes that this is the case in the chapel of S. Silvestro in SS. Quattro Coronati, Rome and a pavement at Reggio di Calabria. See Reutersward, 'Forgotten symbols', pp. 117-20; and also Notes 105 and 142 below.
[94] The epitaphion is a large richly embroidered cloth bearing the icon of Christ laid out for burial. Today it is carried in procession only at Vespers and Mattins of Holy Saturday. In late Byzantine times the epitaphion, which was used for covering the bread and wine prepared for the Eucharist, was carried at the Great Entrance of at least major celebrations of the Divine Liturgy (Mass). (See *Phronema* reference, Note 95).
[95] An excellent example is that of the dome of the Catholicon of the monastery of Stavronikita on Mount Athos, painted by the great iconographer Theophanes the Cretan and his son, Symeon, in 1545-46. For full details and colour plates see M. Chatzidakis, trans. A. Doumas and J. Vulliamy, *The Cretan Painter Theophanis: The First Phase of his Art in the Wall-Paintings of the Holy Monastery of Stavronikita* (Mount Athos, 1986). A theological commentary on the paintings by Archimandrite Vasileios of Stavronikita included in the volume has been reprinted in *Phronema,* 4 (1989) 23-44. It is worth noting that the way in which not only angels but also Apostles are portrayed around the Pantocrator (as on Ascension domes, or the Epiphany domes of the Orthodox and Arian Baptisteries in Ravenna) is often reminiscent of the round dances performed in early Christian liturgy; as is attested to by numerous patristic references to dance. These dances were, in general, symbolic both of the worship of the angels before the throne of God and of the harmony of the heavens. According to the Apocryphal Acts of John, the Apostles performed a ring dance around Christ, at the centre, before His betrayal. A liturgical dance, now performed three times a year, survives at the Cathedral of Seville; although this seems to recall David's dance before the Ark of the Lord (2 Sam. 6:14-16) rather than the ring dances. It is known that dances were performed at Easter on, or around, the labyrinth in several cathedrals during the High Middle Ages. A description of the dance at Auxerre has survived but unfortunately it isn't detailed enough to reveal the actual choreography. But it is known that a ball, called the *pilota*, passed back and forth from the canons to the dean. The dance clearly signified the Harrowing of Hell by Christ, the antitype of Theseus: His threading of the path through the entangling confusion of turnings and blind alleys of hell; His conquest of the Minotaur/Satan at the centre; and His glorious egress from the labyrinth as the Resurrected Saviour. As we have seen, the sun was universally taken as a symbol of the Paschal Mystery from the beginnings of Christianity. The Gospels record that at the Crucifixion the sun was darkened (Matt. xxvii:45; Mk xv:33; Lk. xxiii:44) in cosmic empathy with the dying of the Sun of Righteousness on the Cross. Laid to rest in the tomb just before the Sabbath/Passover began with the setting of the sun, the Sun of Righteousness descended into hell, and then, having vanquished the powers of darkness, rose from the dead, and at the dawn of Easter Sunday appeared to Mary Magdalene at the sepulchre. The cosmic reverberations of the Lord's Passover are typically signified by the placing of the sun and the moon either side of Christ on icons of the Crucifixion. Does, then, the *pilota* in the Auxerre labyrinth dance represent, as several commentators have believed, the sun as an image of the Paschal Victim? It is possible, but, as Penelope Doob argues, it is more likely that the *pilota* represents the ball of thread which Ariadne gave to Theseus (and also, possibly, the ball of pitch which Theseus was traditionally thought to have carried to thrust down the gullet of the Minotaur). Even if this is the case, it would seem extremely likely that the dance did, in part, symbolise the restoration of the harmony of the heavenly spheres effected by the egress from the labyrinth/Resurrection from Hades of Theseus/Christ. Since the participants in liturgical ring dances represented both the angels ('We who in a figure represent the Cherubim ... ' as says the hymn sung at the Great Entrance of the Orthodox Liturgy) and the heavenly bodies, the divine centre around which they danced clearly represented *Christos Helios.* Although, sadly, liturgical dances *sensu stricto* died out in the Byzantine East as they did in the West (except for Seville), three-fold dance-like circlings around a centre take place during several rites. For instance, at the conclusion of the Orthodox Marriage

Office, in a ceremony significantly known as the 'Dance of Isaiah', the celebrant leads the bride and groom around in a circle three times. It is interesting that the rubrics direct that such movements be made 'against the sun' (i.e., in the opposite direction to the sun). This could well reflect a desire to avoid confusion with pagan dances honouring the sun as a deity, but it also means that those performing such a ceremony cannot be representing the physical sun. But there seems good reason for concluding that they represent the heavenly powers executing their eternal ring dance around the Supercelestial Sun. For Auxerre and Christian labyrinth dances, see Doob, *Labyrinth,* pp. 123-8. For the history of Christian liturgical dancing in general see E.L. Backman, *Religious Dances in the Christian Church and in Popular Medicine* (London, 1952). For the Acts of John, see M.R. James (ed.), *The Apocryphal New Testament* (Oxford, 1926).

[96] Both traditional iconographic programmes—one thinks of Giotto's Arena Chapel in Padua—and the image of the Byzantine Pantocrator survived into the Proto-Renaissance period. Dominating the central axis of Pisa Cathedral is the massive image of Christ Pantocrator in the semi-dome of the apse (fig. 4). Enthroned and giving the blessing with His right hand, He holds the Gospel Book open displaying the familiar words of Jn. viii:12. Commenced by Francesco de Simone, the mosaic was completed by Cimabue and Vincino da Pistoia (colour plate in Bertelli, *Mosaics*, p. 211). The Pisa Pantocrator is essentially traditional, but the High Renaissance Pantocrator in the underground chapel of Sant' Elena, in Santa Croce in Gerusalemme, Rome—although an undeniable descendant of the Byzantine half-figure Pantocrator, complete with open book and *Ego sum lux mundi A et Ω*—has an un-Byzantine Noeltide cheerfulness and is surrounded by a crèche of absurd winged cherubs, some playing musical instruments (Bertelli, pls, pp. 246 and 247). The mosaic was probably created by Baldassare Peruzzi.

[97] G.S. Kirk, J.E. Raven and M. Schofield, 2nd edn, *The Presocratic Philosophers* (Cambridge, 1983) p. 343 (Dk 44A16 = Aetius 2.7.7) as quoted in L.C. Taub, *Ptolemy's Universe: The Natural Philosophical and Ethical Foundations of Ptolemy's Astronomy* (Chicago, 1993) p. 145.

[98] See Taub, *ibid.*, p. 143-4.

[99] Aristotle ed. and trans. E.S. Forster and D.J. Furley, *On Sophistical Refutations; On Coming-to-be and Passing Away; On the Cosmos* (Cambridge, 1955) 399a 12-21, as quoted in Taub, *ibid.*, p. 143.

[100] Ptolemy, *Planetary Hypotheses,* 2.12, trans. in S. Sambursky, *The Physical World of Late Antiquity* (Princeton, 1962) p. 145, as quoted in Taub, *ibid.*, p. 144.

[101] Trans. J.B. Lightfoot, *The Apostolic Fathers,* pt I, ii (London, 1890), as quoted in Nebelsick, *Circles of God,* p. 89.

[102] Typically, this verse appears as an inscription on the central dome of a Byzantine church depicting Christ Pantocrator.

[103] Colour plate in Bertelli, *Mosaics*, p. 129.

[104] Kostas Papaioannou writes of the gold backgrounds:

> The element in which the image of the icon or mosaic bathes is a substance beyond all natural colouring: the gold ground. The midnight blue of Galla Placidia still belongs to the world of the senses; but the glowing gold—the only colour never encountered in nature—strips space, matter and bodies of everything which might suggest extent, weight, the hazards of an existence riveted to the earth. Everything the Neo-Platonists and the great Orthodox scholars said about the 'subtle light' of the intelligible world and the illumination of obscure matter by the divine ray, will find monumental expression in the manner by which the icons and mosaics realise the transfiguration of colours through their gold grounds. The gold strewn with russet which unifies the space of the church has a double function. Through its agency, the light becomes co-extensive with the pictorial space: the light is no longer natural but 'pneumatic' light, which surrounds the figures like a nimbus of sanctity and projects them with complete lucidity into the midst of the worshippers. (Trans. J. Sondheimer, *Byzantine and Russian Painting* (New York, 1965) pp. 62-3.)

[105] Above the chancel arch of the fourteenth-century Abbey Church of Pomposa, near Ferrara, is a medallion within which is a depiction of the Pole Star, from which issue a multitude of rays. Beneath Christ in Majesty in the apse appears an inscription which identifies Christ with the star: *astro polis* Χρς *nitet cum carne beata.* This unusual (unique?) association of Christ with the Pole clearly relates Christ with the *axis mundi;* about which the motion of the entire universe is sustained. Polaris is also an appropriate symbol for Christ as He is the unwaning light, the star which never sets. See Reuterswärd, 'Windows of divine light', p. 52 and ill. 41. Reuterswärd also notes that the principal roundel of the pavement at Pomposa is marked with an eight-pointed star, obviously signifying here Christ as Polaris ('Forgotten symbols', p. 119).

[106] To which we might add a gobbet from the hymnody of the Christmas liturgy of the Byzantine rite:

> O most blessed womb of the Maid of God, thou hast been shown forth as spiritually vaster than the heavens. For Him whom heaven cannot contain, thou dost carry and hold within thee. (Mother Mary and K. Ware trans., *The Festal Menaion,* repr. (London, 1977) pp. 218-9.)

The Byzantine liturgy of the Feast of the Epiphany, or Theophany, which commemorates the baptism of Christ in the Jordan, is, like Christmas, particularly rich in allusions to Christ as the Supercelestial Sun:

> Today the Sun that never sets has risen and the world is filled with splendour by the light of the Lord. ... Today Paradise has been offered to men and the Sun of Righteousness shines down upon us. ... Today the blinding mist of the world is dispersed by the Epiphany of our God. ... The waters saw Thee, O God, the waters saw Thee and were afraid. The Jordan turned back, seeing the fire of the Godhead descending bodily and entering its stream. (From the prayer for the Great Blessing of the Waters, *ibid.,* pp. 354-5.)

And, obviously, such allusions are generously distributed through the liturgy for the Transfiguration. Patristic commentaries on the Transfiguration lay great stress on Christ as the Sun: St Ephraem the Syrian and Pseudo-Leo observe that there were two suns visible to the Apostles on Mount Tabor, the physical sun and the face of Christ; St Ambrose speaks of Christ as the 'divine sun'; St Augustine says Christ 'is the Sun foretold by the prophets', and so on. Quoted in J. Miziołek, '*Transfiguratio Domini* in the apse of Mount Sinai and the symbolism of light', *Journal of the Warburg and Courtauld Institutes* 53 (1990) 42-60, pp. 49-51. See Miziołek for further details.

[107] Illustrated in E. Norman, *The House of God: Church Architecture, Style and History* (London, 1990) p. 29.

[108] A notable example is that in the Church of the Holy Apostles, Peć, Yugoslavia. Colour plate in G. Frere-Cook (ed.), *Art and Architecture of Christianity* (Cleveland, 1972) p. 69.

[109] The apse mosaic of Santa Maria in Trastevere, Rome (1140-43) reveals an intermediate stage of the iconographic development from Christ Pantocrator to the Coronation of the Virgin. Christ and His Mother are shown enthroned side by side, but Christ is not crowning the Virgin (Bertelli, *Mosaics*, pl., pp. 186-7). The fully developed iconography is very characteristic of the Gothic period. A fine example, of 1295, is that in Santa Maria Maggiore, Rome, by Jacopo Torriti. Christ and the Virgin are enthroned within a star-studded medallion, with the sun beneath Christ and the moon beneath His mother (*ibid.,* pl., pp. 220-1). The Coronation occurs in Florence Cathedral on the tympanum over the central door into the nave. It is by Gaddo Gaddi, who worked on the Baptistery mosaics around the turn of the fourteenth century. As with the Romanesque Christ in Majesty, the figures are flanked by angels and symbols of the Evangelists (*ibid.,* pl., pp. 212-13).

[110] Colour plate in *ibid.,* pp. 120-1. For a full discussion of this mosaic see Miziołek, '*Transfiguratio Domini*'. Miziołek argues that the focus of the mosaic, which is 'the earliest known example of a Transfiguration with rays' (p. 44), is that of Christ 'as the Sun of Justice' (p. 42). He does not think that the iconography reveals the direct influence of depictions of Helios,

etc. The rays, he suggests, 'derive from representations of the sun as a circle with eight rays or an eight-pointed star—representations which date back as far as the art of the Ancient East' (p. 46). Evidence also exists for the Christian use of the eight-pointed star within a circle as a representation of the sun (see p. 47). Miziołek notes the interesting detail about the Sinai mosaic that: 'Each witness to the vision is lit by one of the rays which changes the colour of his clothing in the area of its impact' (p. 43). It should, however, be noted that the eight-pointed star certainly does not always signify the sun. The star of Bethlehem is sometimes so depicted, as are, on occasion, stars in general.

[111] In the church of the Saviour in Chora (Kariye Camii), Constantinople, there is an *Anastasis* (Resurrection). (Colour plate in R. Browning, *The Byzantine Empire* (London, 1980) p. 174.) Although it is placed in the apse of the Parecclesion (where services for the dead were held), not the main church, the beautifully delicate fourteenth-century fresco depicts Christ in a fashion which closely resembles that of the Transfiguration. Christ is portrayed in brilliant white garments and with a golden nimbus within a star-studded mandorla divided into layers of subtle light blues and white. It is difficult not to conclude that this uncommon treatment, of what in the West is called the Harrowing of Hell, was suggested by the decision to place this particular icon in the apse of the Parecclesion, the easterly direction in which Christians have turned since earliest times to greet the rising of the Sun of Righteousness from the sepulchre, and to which they look in anticipation of the Second Coming when Christ will return to raise the dead from their tombs.

[112] Patrik Reutersward notes ('Windows of divine light', pp. 46-7) that on the eastern wall of the church, over the apse, there is an *Agnus Dei* with two adjacent arched windows above, which are 'separated from each other by a column which has a cross hewn into its capital'. Over the arched windows there is a circular window. On the outside of the church, separating the arched windows, is a column in relief with a relief sun disk marked with a cross immediately above it (ill. 34). Similarly, on the west front of the church, again on the central axis, are two arched windows separated by an ornate column, with another window immediately above. In this case, the upper window takes the form of a small Greek Cross (ill. 35). These arrangements of windows, as Reutersward notes, clearly amplify in symbols the meaning of the Transfiguration in the apse. We can in fact go further than this, for a column can be used as a symbol of Christ as the cornerstone (Matt. xxi:42; Mk xii:10; Lk. xx:17; Acts iv:11; I Pet. ii:6), as Reutersward shows elsewhere ('Forgotten symbols', pp. 106-10). The sun disk, then, tells us that Christ, signified by the column, is the Sun of Righteousness, the Light of the World. So, clearly, the column is Christ of the Transfiguration suffused with uncreated light. But what is the meaning of the round uppermost window at the east end? In the west gallery of the Abbey Church at Gurk, Carinthia, which Reutersward discusses ('Windows' pp. 50-2 and ill. 38), we find a similar arrangement of two arched windows with a circular window above. However, in this case the windows are surrounded by an outstanding mid-thirteenth-century mural of the Transfiguration. Between the two arched windows is painted the Transfigured Saviour and to their respective far sides Moses and Elijah. The three Apostles are depicted beneath these principal figures. Above Christ's head is a circular window, with the Father in a medallion portrayed immediately above it. Cut off by the two arched windows is a figure-of-eight band of light which surrounds the mandorla of Christ, the circular window and the God the Father medallion. Even without the aid of an inscription, there could be little doubt that the circular window represents the 'bright cloud' of which the Gospels speak: 'He was still speaking, when lo, a bright cloud overshadowed them, and a voice from the cloud said, 'This is my beloved Son, with whom I am well pleased ... " (Matt. xvii:5). However, as the Father has this verse on the scroll He holds, all vestige of uncertainty is removed. It needs only to be added to Reutersward's account that, as the cloud is interpreted by the Fathers as a manifestation of the Holy Spirit, the figure-of-eight incorporates all three hypostases of the Trinity. Can there be much doubt that the circular window at Sinai likewise symbolises the bright cloud of the Transfiguration?

[113] Bertelli, *Mosaics,* pl., pp. 86-7.
[114] Colour plate in Bustacchini, *Ravenna*, p. 15.
[115] D. Nalpandis (ed.), *Thessaloniki and its Monuments* (Thessaloniki, 1985), pp. 81-3 and colour pl. 14.
[116] Another interesting three-dimensional cosmogram, which to some degree appears to prove our rule, is a fifth-century mosaic in the baptistery of Albenga, Italy. It shows three differently shaded blue nested concentric spheres, each marked with the XP monogram and A and Ω. See Norman, *House of God,* colour pl. 1 and legend p. 17. Perhaps this, or a similar cosmogram, inspired Dante's image of the Trinity in *Il Paradiso* (xxxiii, 115-20):

> That light supreme ...
> ... showed to me three spheres, which bare
> Three hues distinct, and occupied one space;
>
> The first mirrored the next, as though it were
> Rainbow from rainbow, and the third sacred flame
> Breathed equally from each of the first pair.

D.L. Sayers and B. Reynolds (trans.), *The Comedy of Dante Alighieri the Florentine: Cantica III Paradise* (Harmondsworth, 1962), p. 346.
[117] Bustaccini, *Ravenna,* pl., pp. 50-1.
[118] *Ibid.,* pl., p. 33. In the East, such uses of the Lamb were abandoned following the insistence of the Trullan Synod of 692 (Canon LXXXII) that the symbol of the Lamb was not to be substituted for the image of Christ Himself.
[119] One of the most extraordinary examples of a scheme which brings together both kinds of imagery does, however, appear as somewhat odd. This is the Romanesque pavement at Die, near Valence, which depicts Paradise with the four rivers of Eden and numerous paradisal plants and creatures (including a mermaid), but with the spiritual sun, at the centre, surrounded by scattered heavenly bodies superimposed upon it. (See Reutersward, 'Forgotten symbols', p. 127 and ill. 141.)
[120] Reutersward, 'Forgotten symbols', pp. 128-31.
[121] *Ibid.,* p. 121 and ill. 142.
[122] *Ibid.,* p. 131 and ill. 144.
[123] *Loc. cit.* and ill. 145.
[124] For a brief description of the window, see P. Cowen, *A Guide to Stained Glass in Britain* (London, 1985) p. 212. It is interesting to note that while three of the seven planets resemble the central sun the other four resemble daisy-like flowers. The issue cannot be entered into in detail here, suffice is to note that depiction of heavenly bodies, and not just the sun, as flowers is common—compare the barrel-vault of the Mausoleum of Galla Placidia (Note 32). The correspondence between the starry heavens and the wildflowers bespeckling the fields, common lands and byways of Europe is not hard to discern. That the plants appear to be daisies seems to establish a relationship with the sun. 'Daisy' is derived from 'day's eye'. The most common explanation is that the flower was so called because it opens at sunrise and closes at sunset. However, J.C.J. Metford's entry under 'daisy' in his *Dictionary of Christian Lore and Legend* (London, 1983) reads: 'A flower which symbolises innocence; by popular etymology, 'the day's eye' or the sun; and Jesus, 'the Sun of Righteousness". The All Saints' window might lend some support to this latter interpretation. Either way, the daisy has an association with the sun.
[125] 'Forgotten symbols', particularly pp. 65-76 and (for rose windows) 120-2.
[126] *Ibid.,* p. 121 (though, here, he is following H.J. Dow, 'The rose window', *Journal of the Warburg and Courtauld Institutes,* 20 (1957) 248-97).
[127] See P. Cowen, *Rose Windows* (London, 1979) pp. 84-5, 128-31 and pls 38, 39, 68 and 69.

[128] Jungmann, *Roman Rite,* pp. 90-2.
[129] See Hallyn, *Poetic Structure,* pp. 132-3. For colour illustration see D.R. de Campos (ed.), *Art Treasures of the Vatican: Architecture Painting Sculpture* (New York, 1974) pls 60 and 61.
[130] A device particularly associated with the Jesuits. It is interesting that, in the Transfiguration in the central light of the east window in the Corpus Christi chapel at Fairford, golden rays of light flow out from a gold Host, bearing the letters IHC, over the breast of the transfigured Saviour. No image could make clearer the fact of the extension of the symbolism of *Christos Helios* to the Host which occurred in the West (Wayment, *Fairford*, pp. 11 and 37, pl. x).
[131] The image of Christ encircled by the zodiac is known in both East and West. For example, a tenth-century zodiacal chart depicts Christ, with His head surrounded by sun's rays and with an eight-pointed star to His left and another embroidered on His cloak, in a roundel at the centre of the circle of the signs of the zodiac. (Illustrated in T. Goldstein, *Dawn of Modern Science: From the Arabs to Leonardo da Vinci* (Boston, 1980) p. 143.)
[132] See illustrations in B. Purefoy, *Tewkesbury Abbey* (London, 1981) pp. 12, 14 and 15. It is perhaps worth noting that the emblem of the sun was often borrowed by Christian rulers long before the Sun King, Louis XIV. There is an interesting example in the centre of the fifteenth-century ceiling of the Sala delle Arti Liberali in the Borgia Apartments in the Vatican, depicting the arms of the infamous Borgia Pope, Alexander VI, in a medallion surrounded by golden sun's rays. De Campos, *Vatican,* pl. 42.
[133] I am grateful to Keith Hutchison, a contributor to this volume, for drawing my attention to numerous examples of pre-Copernican heliocentric images in the course of a profusely illustrated (but unpublished) lecture.
[134] Grosseteste was perhaps the greatest High Medieval Western exponent of the metaphysics of light. A point of particular interest is the stress he placed on the sun. See J. McEvoy, 'The sun as *res* and *signum:* Grosseteste's Commentary on Ecclestiasticus ch. 43, vv.1-5', *Recherches de Théologie Ancienne et Mediévale,* 41 (1974) 38-91. For Grosseteste, 'the sun is made the chief activating cause of all natural phenomena' (p. 54). As McEvoy notes: 'Grosseteste's universe is heliocentric in every respect save the cosmological' *(loc.cit.)*. On the Western medieval aesthetics of light, see Eco, *Art and Beauty,* ch. 4.
[135] For the general theological context of Dionysius' theology, see J. Pelikan's Introduction, 'The odyssey of Dionysian spirituality', J. Leclercq's Introduction, 'Influence and noninfluence of Dionysius in the Western Middle Ages', and K. Froehlich's Introduction, 'Pseudo-Dionysius and the reformation of the sixteenth-century' to *Pseudo-Dionysius: The Complete Works,* trans. C. Luibheid (New York, 1987) pp. 11-24, 25-32 and 33-46.
[136] It should be noted that Suger and his architects were, in a sense, re-inventing the wheel. Anthemius of Tralles and Isidore of Miletus might have used very different means to achieve their end in building Justinian's Santa Sophia in Constantinople, but the intent was the same and the measure of success comparable. Although the Pantocrator also occurred elsewhere in the church, including the dome, significantly the late ninth-century mosaic of the enthroned Christ Pantocrator with the open book and the words, 'Peace be with you, I am the light of the world', occurs over the imperial doorway from the narthex into the nave. Prostrating himself before Christ is an emperor, probably Leo VI who undertook much work in the cathedral (see J. Beckwith, *Early Christian and Byzantine Art,* 2nd edn (Harmondsworth, 1979) pp. 190-1 and pl. 158; colour in Bertelli, *Mosaics,* pp. 132-3). Placed where it is, the inscription clearly proclaims, *inter alia*, that the whole church is an expression of those words ἐγώ εἰμι τὸ φῶς τοῦ κόσμου. Certainly that is the case. Procopius wrote in the sixth century of Justinian's building:

> The sun's light and its shining rays fill the temple. One would say the space is not lit by the sun from without, but that the source of light is to be found within, such is the abundance of light ... So light is the construction, the dome seems not to rest on a solid structure, but to cover the space with a sphere of gold suspended in the sky [i.e., on the circlet of light formed by the forty windows beneath the dome] ... The scintillations of the light forbid the spectator's gaze to linger

on the details; each one attracts the eye and leads it on to the next. The circular motion of one's gaze reproduces itself to infinity since the spectator is never so much as allowed to choose what he prefers from the whole ensemble ... The spirit rises towards God and floats in the air, certain that He is not far away, but loves to stay close to those whom He has chosen. (*The Building of Justinian,* as quoted in Papaioannou, *Byzantine and Russian Painting,* p. 106.)

[137] G. Duby, trans. E. Levieux and B. Thompson, *The Age of the Cathedrals: Art and Society 980-1420* (London, 1981) pp. 99-100. For Duby's helpful discussion of Dionysius, Suger and Saint-Denis, see pp. 97-109.

[138] *Ibid.,* p. 100.

[139] In contrast, in the West during the late medieval period the creative light was considered to be corporeal and itself created (see below). Distinctions were commonly made between the words *lux, lumen, color* and *splendor. Lux* is light at its source, the light created by God on the first day; *lumen* is light diffused through space, and is normally invisible; *color* is light made visible on the surface of bodies it strikes or in bodies through which it passes; *splendor* is light reflected from surfaces such as gold leaf or mosaic, or highly polished materials. See P. Hills, *The Light of Early Italian Painting* (New Haven, 1987) pp. 11-12 and Eco, *Art and Beauty,* p. 50.

[140] *The Divine Names,* 4, in Dionysius the Areopagite, trans. C.E. Rolt, *The Divine Names and the Mystical Theology,* 2nd edn (London, 1940) pp. 91-2.

[141] The Cistercians did, however, unbend sufficiently to allow the use of *grisaille.*

[142] These verses of the Apocalypse must have been very much in the minds of the creators and patrons of the great *opus sectile* Cosmati pavements with their semiprecious stones. The finest (at least surviving) example is that before the High Altar of Westminster Abbey (1268). Semi-precious stones in this fine pavement include Egyptian purple porphyry and Greek green porphyry, rescued from Antique buildings, and a remarkable rare Egyptian onyx marble, which forms the central roundel. R. Foster, *Patterns of Thought: The Hidden Meaning of the Great Pavement of Westminster Abbey* (London, 1991) pp. 35-8 and 76-7.

[143] Duby, *Cathedrals,* p. 106.

[144] It is significant that this much loved window is on the south-east side of the Cathedral (Chartres is, unusually, orientated towards the north-east rather than towards the east). As the sun in the Northern Hemisphere north of the Tropic of Cancer is always in the southern sky when it crosses the meridian, windows are sometimes deliberately placed on the sunny south or the shadier north side for symbolic purposes. The south, through the association of Christ with the sun, can, like the east, by symbolically regarded as the region of God, light and goodness, whereas the north, like the west, the domain of Satan, darkness and evil. The north is also associated with the OT Church and the south with the New. At Fairford, which retains a complete programme of glass, these ideas find their most dramatic expression in the clerestory windows. The south clerestory has twelve figures of saints, as 'Champions of the Faith', in the main lights, and angel singers and musicians in the tracery above. The north clerestory has twelve figures of 'Persecutors of the Faith', opposing the 'Champions', in the main lights (including Annas, Caiaphas, Judas Iscariot, Herod and Nero) and a splendid menagerie of devils in the tracery lights (Wayment, *Fairford,* pp. 71-84). The windows along the north and south sides of the nave have saints and prophets. The south side has saints of the Church of the NT. Although the saints of the New Dispensation spill across into the westernmost window of the north side of the nave, the other four windows of the north side are devoted to the OT, with three windows of prophets and one, the easternmost, with OT types of the Blessed Virgin Mary (to whom the church is dedicated). There is some doubt as to the precise original arrangement of the nave windows, but it seems clear that each of the prophets was paired with an Apostle on the south side. The scheme derives from the tradition of associating each of the twelve clauses of the Apostles' Creed with one of the twelve Apostles, who are supposed to have gathered to formulate the Creed. Each clause was then paired with a text from a different OT prophet. The prophets foresaw in shadow (corresponding to the northern skies) what the Apostles

comprehend in the full light of the sun (the southern skies). The twelve prophets and twelve Apostles constitute an Apostles' Creed sequence, and so each prophet must have been paired with his companion Apostle on the opposite wall. Both Apostles and prophets have scrolls with the appropriate texts (*ibid.*, pp. 63-71; Anderson, *Imagery*, pp. 70-2, 152, 189-90). Even a Romanesque church—many of which are far lighter than their reputation would suggest—might make deliberate use of sunlight for symbolic purposes. An interesting case in point is that of Vézelay. Here the architecture is so perfectly attuned to the annual movements of the sun that at mid-day at the time of the summer solstice the sun's rays shining through the south clerestory windows fall exactly along the central line of the nave and exactly in the middle of each bay, while at the time of the winter solstice they fall exactly on the north clerestory capitals, and at the equinoxes the abacuses (for the link between these astronomical times and the Christian feasts see Note 77). The exactitude of the correspondences established between the sun and the architecture of the church does indeed 'bear witness to the builders' desire to establish a relationship between the total structure and the cosmos'. (H. Delautre and J. Greal, *La Madeleine de Vézelay* (Lyon, 1985) p. 27.) It is interesting that John James has detected an allusion to *Christos Helios* in the placing of the signs of the zodiac in mediallions around Christ in the Pentecost tympanum above the central door from the Galilee into the nave at Vézelay. In particular, four signs seem to frame Christ 'as though He were the sun': Aries and Scorpio, the signs for the equinoxes, either side; Capricorn, for the winter solstice, near the bottom; and near the top, in a half-medallion, 'a crane with bent neck, since ancient times the symbol for the summer solstice'. (*The Traveller's Key to Medieval France: A Guide to the Sacred Architecture of Medieval France* (New York, 1986), p. 294.)

[145] Quoted in Cowen, *Stained Glass*, p. 9.

[146] However, the lamp, it would seem, would have been the predecessor of that which hangs in this position today. Perhaps this is the place to remind ourselves that some churches actually have an architectural lantern over the crossing. The most impressive of all such structures has to be the fourteenth-century lantern above the octagon in Ely Cathedral. The wooden vaulting forms the eight-pointed star of *Christos Helios*. At the apex is a half-figure portrayal of the cosmic Christ surrounded by angels and stars. Light floods into the building through eight large windows. It is indeed an image of Christ as the lamp of the Heavenly Jerusalem. (Photograph in W. Swann, *The Gothic Cathedral*, repr. (London, 1981) pl. 246.) The Chapel of the Constable in Burgos Cathedral has a remarkable lantern by Simón de Colonia, 1482-94. At the apex of the lantern vaulting, which takes the form of an eight-pointed star, is a window which also takes the form of an eight-pointed star. (Photograph in A. Shaver-Crandell, *The Cambridge Introduction to Art: The Middle Ages* (Cambridge, 1982) p. 103.)

[147] The Italo-Byzantine church of San Marco, rebuilt in the eleventh century, was modelled on the Apostoleion in Constantinople (which had been built by Justinian but was destroyed in 1459). It is in the form of a Greek Cross with a central dome over the crossing and another dome over each of the four arms (see fig. 1 above). Unlike the Apostoleion which, interestingly, had a central altar, the High Altar of San Marco is in the presbytery. The western half of the church is boxed by the very extensive atrium (narthex) with six small domes, a chapel and the baptistery. The earlier mosaics belong to the twelfth and thirteenth centuries. While all the key elements of the programme are represented by medieval mosaics, the scheme was not completed until the Renaissance. For full details see O. Demus, *The Mosaics of San Marco in Venice*, 4 vols (Chicago, 1984). There is also an excellent one-volume abridged edition which is adequate for most purposes: O. Demus, ed. H.L. Kessler, *The Mosaic Decoration of San Marco, Venice* (Chicago, 1988). This is not an area we can pursue in this paper, but it does need to be appreciated that the concept of 'cosmos' is a temporal as well as a spatial one. San Marco is most unusual in having an inscription in the Capella di San Pietro referring to the temporal meaning of the iconographic programme. Unfortunately, this inscription presents difficulties both of translation and interpretation. In the view of the present writer, which differs in certain key

respects from that of Demus, the inscription reflects the Augustinian doctrine (*Confessions*, 11 *passim*) that the present, the past and the future all inhere in the present, as the present of the present, the present of the past and the present of the future. The inscription asserts, on my reading, that in Christ, whose incarnation is the centre, the spiritual (not temporal) mid-point, of salvation history, the past, the present and the future are all brought together into what Mircea Eliade calls the 'eternal present'. The past, present and future are not teased apart and assigned to discrete iconographic assemblages on the E-W axis as Demus suggests, but are brought together in each of the iconographic units, and most noticeably by the iconography of the three principal domes. Christian spiritual cosmology is a cosmology of the eternal present. See Demus, i, ch. 18 and my paper 'Time, architecture', pp. 80-4.

[148] The quotations are, of course, taken from our text.

[149] I believe I owe the term to the late Anglican philosopher of science, Bishop I.T. Ramsey.

[150] The bronze chandelier in Pisa Cathedral pointed out to tourists as the one Galileo supposedly observed swinging in 1582, and which led him to formulate the law of the isochronism of the pendulum, was, sadly, not installed until 1587.

[151] A point which is not irrelevant to the argument of this paper. See Martin Kemp's article in this volume.

[152] Copernicus, *Revolutions*, pp. 23-7. 'Mathematics', he says, 'is written for mathematicians' (p. 27) and mathematicians should be allowed to determine mathematical (i.e., astronomical) questions.

[153] *Ibid.*, p. 27.

[154] See Fr Barry Brundell's paper in this volume.

[155] During the later Middle Ages a second tradition arose, that of Scholastic epistemology. High Medieval Scholastic philosophy was centred around such great debates as the nature of universals, fought out between nominalists and realists. But the fact that attempts made by Abelard and others to reconcile the nominalist and realist positions met with some degree of success indicates that at bottom both nominalists and realists shared much the same empiricist principles. During the Middle Ages there seems to have been very little friction between the hermeneutic and Scholastic traditions. The main reason for this was probably the fact that they largely operated in different contexts. The hermeneutic tradition belonged to the contexts of spirituality and mystical theology, textual interpretation (not just biblical interpretation), iconography, literature, and popular culture in general, whereas the Scholastic epistemology saturated dogmatic and systematic theology, and academic philosophy and logic. However, in hindsight one can see that Scholastic epistemology was preparing the groundwork for the naturalistic empiricism of the High Renaissance with its rejection of multiple truths.

[156] For a recent study of the early development of Christian hermeneutics, see D.S. Dockery, *Biblical Interpretation Then and Now: Contemporary Hermeneutics in the Light of the Early Church* (Grand Rapids, Michigan, 1992). Cassian explicitly related the four levels to the virtues (cf. *ibid.*, p. 158). Nicholas of Lyra, in the fourteenth century, encapsulated the levels of Cassian in a famous mnemonic: 'Littera gesta docet, (the letter teaches facts) / Quid credas allegoria, (allegory what one should believe) / Moralis quid agas, (tropology what one should do) / Quo tendas anagogia (anagogy where one should aspire)' (as in *ibid.*, p. 159).

[157] So called 'theological' windows, based on illustrated works such as the *Biblia Pauperum,* which displayed OT types in conjunction with their corresponding NT antitypes, were popular during the Middle Ages. There are remains of a set of such theological windows, displaying two types for each antitype, in Canterbury Cathedral. It is typology which underlies the remarkable glass programme of King's College Chapel, Cambridge (see Wayment, *King's College*).

[158] Often drawing out an anagogical meaning involves the allegorisation of the OT passage in the light of its NT antitype (another reason for avoiding the expression 'allegorical level'). Tropological and anagogical meanings might be determined by interpretation of the OT passage directly, particularly in the case of the former, or via a NT antitype.

[159] And, of course, the principle of concordance was also an aesthetic principle, as Umberto Eco notes: 'The poetics of the cathedrals was ruled by an aesthetic principle, the principle of concordance' (*Art and Beauty,* p. 62).

[160] None of this is to say that early and medieval Christianity lacked any conception of *the* Truth. Far from it. For traditional Christianity, *the* truth is not to be sought in some ultimate empirical ground of reality or ontically privileged level of empirical discourse. 'What is *Truth?* said jesting *Pilate*, and would not stay for an answer' (Francis Bacon's *Essay*, 'Of Truth'). Had Pilate waited for an answer, Christ might have given it him: '... I am the way, and the truth and the life; no one comes to the Father, but by me' (Jn xiv:6). In this world, we but 'see in a mirror dimly' (I Cor. xiii:12) like seers practising catoptromancy. It is like the blind men and the elephant. We gather partial truths; we see things first from this view point and then from that. Seldom does even the mystic see things for an instant in their wholeness with the eye of God. Hermeneutically speaking, *the* Truth is to be found not so much through the aggregation of partial truths, exegeses at different hermeneutic levels, but in the principle of concordance which holds all the partial truths together. Amongst the windows which the Abbot Suger ordered for the Abbey of Saint-Denis was one he called his 'anagogical window'. For Suger (following the Pseudo-Areopagite) the *anagogicus mos* is that which lifts the soul from the material to the immaterial. The panels of the window are not typological like the theological windows of Canterbury and elsewhere. Rather, they comment on the source of the principle of concordance upon which the *anagogicus mos* itself rests, and that source is Christ. Thus one panel of the window shows a mill being worked by St Paul (there is another version of the 'Mystic Mill' on a contemporaneous capital in the nave of the Abbey of La Madeleine, Vézelay). From the grain brought by the prophets of the OT, the mill discards the husks and grinds out the flour from which is fashioned the living bread of the New Dispensation. The mill is Christ. (*De administratione,* xxxiv, trans. Panofsky, *Abbot Suger,* p. 75, and also Panofsky's Introduction, pp. 20-6; colour pl. in G. Duby, *Medieval Art,* ii: *Europe of the Cathedrals 1140-1280* (Geneva, 1995) p. 27.) Seeing the presence of God in the hermeneutic process itself is perhaps not altogether unlike seventeenth- and eighteenth-century deists or late twentieth-century cosmologists identifying God as the embodiment of the laws of physics, or those who see the justification of science in terms not of verification or falsifiability but in the coherence, or mutual compatibility, of accepted scientific theories.

[161] A brilliantly clear statement of the instrumentalist position—which owes a debt to medieval nominalists such as William of Ockham—is provided by Osiander in his brief Preface, 'To the Reader on the Hypotheses in this Work', added to the beginning of *De revolutionibus.* This manifesto of instrumentalism, designed to fend off theological objections, was almost certainly added without Copernicus' approval and certainly, despite some somewhat ambiguous comments in his own letter to Paul III, flies in the face of his realist convictions. *Revolutions,* pp. 22-3. See also the paper by Anthony Corones in this volume.

[162] ''Twas once believed that the fair Cyprian [=Venus], whirled / Radiant in the third epicycle, shed / Love's madness on the yet unransomed world' (*Paradiso*, viii, 1-3, trans. Sayers and Reynolds, *Comedy,* p. 115).

[163] Certainly, the origin of the medieval principle of concordance lies in biblical hermeneutics. But biblical hermeneutics and its derivatives did not by any means exhaust the arsenal of medieval techniques for effecting the concordance of different levels of meaning (or being). Other techniques included those based on number or geometrical figures and patterns; that is, concordance effected by showing that two or more different sets of descriptions or entities can be expressed by the same set of numbers or by the same (or within the same) geometrical *schematum.* A number of interesting examples of the latter are discussed in Foster, *Patterns of Thought,* ch. 6. James Franklin in his paper in this volume refers to a number of diagrams or mnemonic devices which were used to effect a concordance between various sets of descriptions or items.

[164] See entries under 'Averroës' and 'Averroism' in P. Edwards (ed.), *The Encyclopedia of Philosophy* (New York, 1967).

[165] See P.K. Feyerabend, *Against Method: Outline of an Anarchistic Theory of Knowledge,* repr. (London, 1978). In a footnote (12, p. 12) to which we are directed by a footnote to the title—a Feyerabendian touch in itself—the author says that he has come to prefer 'Dadaistic' to 'Anarchistic'. Instead of the Anarchist slogan 'Anything goes' Feyerabend decided he preferred a Dadaist slogan, 'Don't take anything seriously'.

[166] That the rejection of medieval multiple truths in favour of a single empirical truth played a significant role in the laying of foundations for the new science of the sixteenth and seventeenth centuries has, of course, been widely acknowledged. Jacob Bronowski in his paper, 'Copernicus as a humanist', in O. Gingerich (ed.), *The Nature of Scientific Discovery: A Symposium Commemorating the 500th Anniversary of the Birth of Nicolaus Copernicus* (Washington, 1975) 170-88, termed this High Renaissance epistemological stance the 'doctrine of single truth', and he recognised Copernicus as an adherent: 'Copernicus' attitude implies that the single truth exists in nature and as nature, and cannot be established or overturned by any authority other than the study of nature herself' (pp. 186-7). For Bronowski, a decisive turning point was Lorenzo Valla's exposure, by means of what Bronowski calls 'literary archaeology', of the Donation of Constantine and other documents as fakes in 1439. Bronowski argues that had the Church drawn the appropriate 'lesson from the scandal ... , there might have been no reason for it to part company from science. Instead, Valla was long persecuted, and 150 years later Cardinal Robert Bellarmine still castigated him as *praecursor Lutheri* ... ' (*loc. cit.*). But equally it could be said that had the Church itself not flirted with the 'doctrine of single truth', but remained steadfast to the medieval acceptance of multiple truths, it could easily have accommodated the new science. The doctrine of single truth might have taken a somewhat different form than it did in the case of Copernicus, but it is all too obvious that many a Protestant and Counter-Reformation scholar, not excluding Bellarmine, had become dangerously ensnared by the new mentality (see Fr Brundell's paper). Biblical literalism, in its way, is as much a manifestation of the doctrine of single truth as scientific physicalism.

[167] '... into those which he [God] moves to follow the eternal supernatural good does he infuse certain supernatural forms or qualities, whereby they may be moved by him gently and promptly to seek the eternal good. And so the gift of grace is a certain quality' (*Summa Theologica,* II, I, cx, 2c; as quoted in E.L. Mascall, Via Media: *An Essay in Theological Synthesis* (London, 1956) p. 150). Ch. 4 of Mascall provides a useful discussion, from a position generally sympathetic to St Thomas, of the knotty theological issues touched on here. We are more concerned with the consequences of St Thomas's theology, and of the way in which he was read, than what his intentions might have been or, for that matter, what the truest reading of his writings might happen to be. These are matters for Thomist scholars. Clearly it was not his intention to open up gulfs between God and humanity, humanity and the natural world, mind and body, and so on, but it is difficult not to sheet home some measure of the responsibility for such disastrous dualisms to Scholasticism.

[168] It is interesting that Kepler turned to spiritual cosmology to supply an apt analogy for his own Copernican cosmology. For Kepler, however, the sun does not correspond to Christ or the Trinity as such but to the Father, ' ... in the sphere, which is the image of God the Creator and Archetype of the world ... there are three regions, symbols of the three persons of the Holy Trinity—the centre, a symbol of the Father; the surface, of the Son; and the intermediate space, of the Holy Ghost. So, too, just as many principal parts of the world have been made—the different parts in the different regions of the sphere: the sun in the centre, the sphere of the fixed stars on the surface, and lastly the planetary system in the region intermediate between the sun and the fixed stars' (*Epitome of Copernican Astronomy* IV, trans. C.G. Wallis, repr. in M.K. Munitz (ed.), *Theories of the Universe: From Babylonian Myth to Modern Science* (New York, 1957) p. 198). For Kepler, the sun is the new *primum mobile* from which issues the *anima motrex,* corresponding to the Holy Spirit, which sweeps the planets around in their orbits.

[169] For an analysis from the Orthodox standpoint, see Vladimir Lossky's essay, 'The theology of

light in the thought of St Gregory Palamas', in V. Lossky, J.H. Erickson and T.E. Bird (eds), *In the Image and Likeness of God* (Crestwood, NY, 1974) 45-69. An interesting selection of papers on Palamism, written from several different perspectives, will be found in *Eastern Churches Review*, 9 (1977) 1-71. See in particular the conflicting views of the Thomist, Dom Illtyd Trethowan, in 'Irrationality in theology and the Palamite distinction', 19-26, and of the Palamite, Archimandrite Kallistos Ware, in 'The debate about Palamism', 45-63. It should be noted that since the Second Vatican Council Palamite theology has begun to take a hold in certain Catholic theological circles. A most interesting article calling for a change in direction in Catholic theology is that of G.F. Pollard, 'Existential reactions against Scholasticism', in a highly seminal collection of Vatican II era essays written by Catholics: M. de la Bedoyere (ed.), *Objections to Roman Catholicism* (London, 1964) 141-64. For a highly critical (and rabidly anti-scientific) Orthodox analysis of the consequences of the theology of St Thomas, and the Augustinian tradition of the West in general, see P. Sherrard, *The Rape of Man and Nature: An Enquiry into the Origin and Consequences of Modern Science* (Ipswich, Suffolk, 1987) particularly chs 2 and 4.

[170] As Maximus the Confessor puts it: '... the whole spiritual world seems mystically imprinted on the whole sensible world in symbolic forms' (*Mystagogia,* 2, Berthold (trans.), *Maximus Confessor,* p. 189).

[171] Long after Copernicus' time, the crowning achievement was perhaps that of the incredible *trompe l'œil* ceiling, the *Entrance of St Ignatius into Paradise* (c. 1685), by the Jesuit, Fr Andrea Pozzo, in the church of S. Ignazio di Loyola in Rome. So strict is the perspective that the painting can only be seen without distortion from one precise spot marked on the nave floor beneath. Norman, *House of God,* colour pl. 45.

[172] The Hesychasts brought God very near, as perception of the Uncreated Light—which, for them, was an experience of the Godhead Itself, though not Its essence—could be achieved by anyone by dint of following the spiritual methods they advocated. For the West, the Beatific Vision was something which could—save, at least according to some theologians, in the most exceptional circumstances, such as that of St Paul's conversion on the road to Damascus (Acts xxii:6-16)—only be experienced after death.

[173] Again, this is a topic which cannot be discussed adequately in this paper. However, what, in a nutshell, seems to have occurred was that, while the earlier term 'supernatural' applied exclusively to the heavenly realm, or 'supernature', it came to be extended to cover a range of terrestrial phenomena which didn't seem to fit into the tightening sub-categories of the natural. The most notorious example is that of the phenomena believed during the High Renaissance and Early Modern periods to be caused by witchcraft. (By contrast, belief in the existence of witches was proclaimed heretical by canon law during the Middle Ages.) It is difficult to escape the conclusion that the way was smoothed for this paradoxical new category, which we might call the 'terrestrial supernatural', by the Thomist concept of 'created grace' or the 'created supernatural' (see Mascall, Via Media, pp. 148-65). If the saints could perform miracles by the power of created grace bestowed by God, why could supernatural acts not be performed by witches, empowered by a special gift of created 'anti-grace' by Satan? The supernatural was brought down from heaven to become a category of the created terrestrial order alongside the 'natural'. Some of the phenomena were, of course, attributed to the direct acts of demons, who as fallen angels, although cast down to earth (Rev. xii:9), properly belonged to the domain of supernature rather than nature. However, the construction of an elaborate demonology was another undesirable product of Scholasticism.

[174] Possibly the most deeply felt objection Copernicus had to the received astronomy was that its models did not combine to form a consistent and coherent physical cosmology. The models devised to save the phenomena were independent of one another, and, moreover, there were alternative models for one and the same phenomenon. In his Preface to *De revolutionibus* addressed to the Pope, Copernicus likens the activities of the Ptolemaic astronomers to '...

someone including in a picture hands, feet, head and other limbs from different places, well painted indeed, but not modelled from the same body, and not in the least matching each other, so that a monster would be produced from them rather than a man' (trans. Duncan, *Revolutions*, p. 25).

[175] It should be noted, however, that mainstream Protestantism eventually shied away from such an extreme position.

[176] See R.S. Westman, 'The Copernicans and the Churches', in D.C. Lindberg and R.L. Numbers (eds), *God and Nature: Historical Essays on the Encounter between Christianity and Science* (Berkeley, 1986) 76-113, pp. 89-93.

[177] One could argue that the Pozzo ceiling in S. Ignazio (Note 171) presents an image of the cosmos since in its lower reaches it comes down to the earth with representations of the four known continents (a common feature of Baroque and Rococo paintings), signifying the missionary successes of the Society of Jesus. However, this and similar examples are really simply large-scale cosmological paintings placed in a church, rather than part of a deliberate programme to depict the church itself as an image of the cosmos.

[178] The Host and the monstrance became expressions of the triumphalism of the Counter-Reformation Catholic Church. At Loreta, Prague, is a gilded silver sun-burst monstrance which is embellished with more than 6000 diamonds. Colour illustration in Frere-Cook, *Art and Architecture*, p. 81.

[179] One more recent echo of *Christos Helios* and of Christ as the lamp is that of the Pre-Raphaelite painter Holman Hunt's, *The Light of the World*, painted between 1851 and 1853; an enormously popular work, of which the original is in Keble College, Oxford. A late copy (1904), much of which was executed by another hand and which is now in St Paul's Cathedral, London, toured the Empire, including Australia, and drew large crowds, being seen by around seven million people. For full details see J. Maas, *Holman Hunt and the Light of the World* (London, 1984). Christ is shown knocking on a closed door, which symbolises the door of the heart: 'Behold, I stand at the door and knock; if anyone hears my voice and opens the door, I will come in to him and eat with him, and he with me' (Rev. iii:20). Set at night, the scene portrays the Saviour illuminated by the light of the moon and of a lamp which He holds in His left hand, recalling Psalm cxviii(cxix MT):105, 'Thy word is a lamp to my feet and a light to my path'. Millais in a letter to Hunt records the reaction of the Bishop of Oxford, Samuel Wilberforce (the self-same 'Soapy Sam' who crossed swords with Darwin's Bulldog, T.H. Huxley). 'He', reports Millais, 'rather disliked the title for the Christ picture and suggested instead 'The Midnight Visit', thinking 'the Light of the World' should not come from the church but from the Body of Christ ...' (see *ibid.*, p. 60). Soapy Sam was of course right (as in truth he often was). What the Bishop, it would seem, realised, and what escaped others, was that 'the Light of the World' of Hunt's picture must denote not the image of Christ but that of *the lamp*, which was the principal source of light. Light from the lantern is cast both through its glass windows and through small perforations scattered over its domed top. Some of these perforations are explicitly star-shaped and the total effect is to create an image of the starry firmament arching over the central light. The lantern is, thus, an image of the spiritual sun at the centre of the supercelestial heavens. However, it is at the same time clearly an image of the church; hence the reason why the Bishop saw that the Light of the World was not radiating from the body of the Saviour, as it does, for example, in icons of the Transfiguration, but from a symbol of the church, the lantern. (In fact, with its ecclesiastical-looking windows and its dome, the lantern bears a distinct resemblance to a church.) The title of the picture (which derives from Johanine texts such as Jn viii:12 and xii:46) is therefore inappropriate, Hunt's Christ being a *bearer* of light to the closed door of the human heart, rather than the Light itself. The phenomenal public interest in the picture is also a matter of some interest. As a painting, it has had many critics and few would consider it to be a world masterpiece. The deep fascination of the public for the work seems, rather, to derive from the extremely powerful archetypal image it presents. It is as if memory of the Ancient spiritual cosmology, long relegated to the lumber room of the Western subconscious, had suddenly been

brought to the surface by Hunt's canvas.
[180] The passage from Donne's *First Anniversary* (II. 205-9) is correctly quoted in J.A. Mazzeo, *Nature and the Cosmos: Essays in the History of Ideas* (New York, 1977) p. 78.
[181] I am grateful to all who have commented on this or the original version of the paper. A very special word of thanks, however, to my student Valentino Migotto for supplying me with several valuable references and to my wife Jill for putting the entire manuscript onto disk for me.

ANTHONY CORONES

COPERNICUS, PRINTING AND THE POLITICS OF KNOWLEDGE

In his preface to *De revolutionibus,* Copernicus confessed (perhaps rhetorically) that he had 'hesitated for a long time whether to bring my treatise, written to demonstrate its [the Earth's] motion, into the light of day, or whether it would not be better to follow the example of the Pythagoreans and certain others, who used to pass on the mysteries of their philosophy merely to their relatives and friends, not in writing but by personal contact'.[1] His reticence was scarcely surprising given his open admission of the fact 'that the judgments of many centuries had reinforced the opinion that the Earth is placed motionless in the middle of the heaven'.[2] I shall argue, however, that it was not really the weight of contrary opinion which worried Copernicus, since clearly he believed that he had the better arguments; rather, it was the act of going into print.

Having said this, let me add an immediate disclaimer: this is not a paper about the 'impact' of printing on the 'Scientific Revolution', nor would I be inclined to attempt such a project. Not only are so-called 'big picture' stories rather suspect at the moment,[3] but scholars are busy dismantling the elaborate historical and philosophical construction known as the 'Scientific Revolution'; it is even being argued that 'science' was not invented until the late eighteenth century.[4] The notion of the 'Printing Revolution' has also come under fire.[5] While I am somewhat sceptical about 'big picture' stories, there *is* however an interesting 'little' story to tell about Copernicus and printing. Stripped to its bare bones, the argument of this paper is that Copernicus was cautious about printing his treatise on the motion of the earth because the act of printing was politically and epistemologically loaded; and that while he seemed to understand enough about the politics of knowledge at the time to attempt to control the reception of his work, he failed to grasp the ways in which that politics was being affected by printing. The argument is primarily philosophical, and deals with historiographical issues.

G. Freeland and A. Corones (eds.), 1543 and All That, 271 – 289

Despite some notable exceptions,[6] printing does not generally feature in histories of science.[7] There is a good reason for this. In so far as historians of science have tended to focus on the content of science,[8] printing is presumed to be a mere tool[9]—convenient for the communication of ideas, but not in itself a factor in the conceptual development of science. On this view it is not obvious that printing even counts as a positive adjunct to that development, since it could just as easily serve the spread of 'non-progressive' ideas and theories as 'progressive' ones. Why, then, take it into account?

The most direct response to such a position is to question the assumption of the neutrality of printing. Why restrict printing to being a mere carrier of ideas? Could it not also affect the content of science?[10] The most strident versions of this view are, of course, forms of technological determinism; but while technological determinism has a certain reductive charm, it is not enjoying much favour among historians at the moment. In particular, the noted historian of printing, Elizabeth Eisenstein, makes a point of distancing herself from such a thesis: 'The very idea of exploring the effects produced by any particular innovation arouses suspicion that one favours a monocausal interpretation or that one is prone to reductionism and technological determinism'.[11]

Eisenstein's monumental book, *The Printing Press as an Agent of Change,* is unquestionably the most important exploration of the impact of printing on the development of science. It is, nonetheless, a very guarded and well-hedged exploration. Commenting on the title of her book, Eisenstein stresses that 'it refers to *an* agent of change not to *the* agent, let alone *the only* agent of change'.[12] For all that, however, she does think that printing constituted a profound shift in communications, and that this shift has been largely neglected by historians. Her revisionist thesis is stated as follows:

> As *an* agent of change, printing altered methods of data collection, storage and retrieval systems and communications networks used by learned communities throughout Europe. It warrants special attention because it had special effects. In this book I am trying to describe these effects and to suggest how they may be related to other concurrent developments. The notion that these other developments could ever be *reduced* to *nothing but* a communications shift strikes me as absurd. The way they were reoriented by such a shift, however, seems worth bringing out.[13]

Despite the criticism which Eisenstein has received for claiming that the shift from script to print precipitates a 'revolutionary' rupture in European history,[14] I think she is right to emphasise the discontinuities. In any case she does not ignore the continuities, though she is critical of the deep-rooted historical conventions of 'gradualism and continuity'.[15] As opposing historiographical strategies, continuist and discontinuist metanarratives should not blind us to the fact that both are concerned with change, and with how best to characterise the changes being studied. So long as both strategies are in use, historians will be kept on their toes.

To talk of printing as an 'agent' of change invites causal analysis (one speaks, for example, of 'chemical agents' which cause particular reactions to occur; or of human agents, typically responsible for doing things). Indeed, what is significant about printing for Eisenstein is that it had *special* effects—causal language again. It might seem, then, that the task at hand is to determine in what sense printing caused certain things to happen, and to weave this in with other causes and their effects in order to arrive at the full story. In her conclusion to *The Printing Press as an Agent of Change,* however, Eisenstein makes a rather striking claim which indicates that there is more to it than that; that printing is not just one cause among others:

> One cannot treat printing as just one among many elements in a complex causal nexus for the communications shift transformed the nature of the causal nexus itself. It is of special historical significance because it produced fundamental alterations in prevailing patterns of continuity and change.[16]

It's not that there weren't books before the invention of printing, but the way in which the production, marketing and distribution of books was affected by printing (at least, in the European context—the story was not the same in China) made a difference. A more subtle position on the question of whether printing affected the content of science is required, then, than that suggested by the opposition presented earlier between the claim that printing is neutral with respect to content, and the technologically determinist position which assumes that the medium is the message. The 'middle way' would be to establish that there is an indirect, but important, sense in which printing affected the content of science—namely, by showing how printing affected the deliberative and problem-solving activities of scientists.

One strategy is to show that printing increased the cognitive resources available to scientists. Another is to show how those increased resources affected the problem-solving activities of scientists. Yet a third strategy, one which is geared especially to discontinuity in scientific thought, is to claim that students 'who took full advantage of technical texts which served as silent instructors were less likely to defer to traditional authorities and more receptive to innovating trends'.[17] These are the basic strategies adopted by Eisenstein, and she employs them across a wide range of sciences. So far as Copernicus is concerned, however, these strategies are realised in the following way. Firstly, as 'a post-print astronomer, Copernicus had an opportunity to survey a wider range of records and to use more reference guides than had any astronomer before him';[18] and being freed from the need to slavishly copy texts, Copernicus was in a position to do more with more. Secondly, not only did access to many records enable him 'to tackle certain technical problems relating to long-term cycles that had remained out of the reach of astronomers who were served by scribes',[19] but the availability of different texts exposed him to a diversity of views (and possible problem solutions, though Eisenstein does not make this point) about celestial motions. Thirdly, the multiplicity of conflicting models, inconsistently used, was a goad to innovation.[20]

Interesting as these points are, however, they do not go very far. Certainly, printing gave Copernicus access to a great deal, and supplied him with cognitive resources; but this tends to limit the consideration of printing to largely 'internal' matters—that is, to the data available to Copernicus, the problem-solutions and mathematical models to hand, and the 'state' of the field of astronomy. This analysis is exemplified by Eisenstein's telling remark that 'Perhaps the most significant contribution made by Copernicus was not so much in hitting on the 'right' theory as in producing a fully worked out *alternative* theory and thus confronting the next generation with a problem to be solved rather than a solution to be learned'.[21] The language of 'alternatives', problems to be solved and 'contributions' suggests an evolutionary or developmental model of scientific progress focussed on the problematic of the field. This is well and good as far as it goes, but it does not necessarily owe anything to printing—access to the great library of Alexandria in antiquity could have served similar ends just as well. Further, despite Eisenstein's leaning towards 'internalist' historiography, she has little to say about the specifics of Copernicus' theory, and how printing bears on those specifics. Finally, and most especially, she has nothing to say about Copernicus' cautious attitude to printing.[22]

In the Introduction to this paper I adverted to Copernicus' caution about printing his treatise on the motion of the earth, and suggested that this owed something to political and epistemological considerations. Copernicus himself provides us with a significant clue to the role of such considerations in his appeal to the Pythagorean tradition of philosophical secrecy. What he seems to approve of in this tradition is a certain kind of epistemic elitism whereby what can be said publicly is constrained by an assessment of the audience; that is, certain things (philosophical mysteries) should only be passed on to certain people; and then 'not in writing but by personal contact'.[23] The reason why writing (and by extension printing) was mistrusted was because there was no control over possible readers. By restricting the communication of ideas to personally 'screened' listeners, one could better control the distribution and reception of those ideas. According to Copernicus, the Pythagoreans resorted to this strategy:

> ... not as some think from a certain jealousy of communicating their doctrines, but so that their greatest splendours, discovered by the devoted research of great men, should not be exposed to the contempt of those who either find it irksome to waste effort on anything learned, unless it is profitable, or if they are stirred by exhortations and examples of others to a high-minded enthusiasm for philosophy, are nevertheless so dull-witted that among philosophers they are like drones among bees. Accordingly as I thought it over, the contempt which I had to fear because of the novelty and absurdity of my opinion had almost driven me to suspend completely the work which I had begun.[24]

The rhetorical nature of this Pythagorean reflection by Copernicus should not be allowed to mask the seriousness of his concerns about printing *De revolutionibus.* It was one thing personally to inform friends and 'devoted great men' about this theory concerning the motion of the earth' but it was quite another to print a book and put it out into the public domain where just anybody could read it. No wonder Copernicus feared a contemptuous response. That same fear had crippled him some three decades earlier—Copernicus was not even prepared to put his name or give a title to the first brief sketch of his heliocentric theory. The *Commentariolus*, as it came to be called much later, was not printed by Copernicus; and he only dared to distribute a few handwritten copies to friends.[25]

Of course, given the fact that *De revolutionibus* was printed, how was it that his reservations had been overcome? Copernicus lays the responsibility on his friends, claiming that despite *his* resistance, they insisted on publication. Copernicus tells us that Tiedemann Giese, Bishop of Kulm, 'often urged me, and demanded of me, sometimes with reproaches as well, to issue this book ... after I have kept it suppressed and hidden not just for nine years but almost four times nine years already'.[26] The astronomer (and teacher of Kepler) Michael Maestlin (1550-1631), being steeped in the classics, noted the allusion to Horace's *Ars poetica* in Copernicus' claim to have suppressed the book. In recommending that a poet not publish work until nine years had passed, Horace shrewdly observed that 'What you have not published you can destroy; for the word once set forth can never come back'.[27] The manuscript of *De revolutionibus* was not, however, lying in a locked closet for thirty-six years. The book was in fact not completed until just before its publication.[28] Why feign Horatian prudence? Although Copernicus' humanist contemporaries may have appreciated the literary conceit for its own sake, Copernicus uses it as a rhetorical device to heighten the responsibility of his friends for the publication of the book. Indeed, by the end of this passage in his preface to *De revolutionibus*, Copernicus says that 'I was induced by their persuasion ... eventually *to allow my friends to publish this work* [my emphasis], as they had long been asking me'.[29]

The repeated emphasis on 'friends' is indicative of Copernicus' anxiety to secure a sympathetic *hearing* for his 'absurd doctrine'; what he couldn't secure was a sympathetic readership for the book. The shift from speakers/hearers to writers/readers is one which predates printing, and was problematised by Plato in a way which brought epistemological issues to the fore. These issues claim Copernicus' attention in ways which are heightened by the advent of printing. In *Phaedrus* (274b-275b), Plato spins a story about the Egyptian invention of writing, and the then king's reservations about it. Those (Platonic) reservations concern the art of memory and the danger of forgetfulness if people rely on writing instead of practising the mnemonic arts. This leads on to a distinction between true wisdom and the semblance of it:

> ... it is no true wisdom that you offer your disciples [that is, those of the god Theuth, who is alleged to have invented writing], but only it semblance; for by telling them many things without teaching them you will make them seem to know much, while for the most part they know nothing, and as men filled, not with wisdom, but with the conceit of wisdom, they will be a burden to their fellows.[30]

This criticism repeats Plato's reservations about imitation or mimesis in general: imitations or semblances are *not* the real thing; direct knowledge of the Forms is not equivalent to representations of them. Combine this with the claim that it is people who know, not written words, and you have the basic elements of Plato's attack. Despite the apparent absurdity of comparing written words to people, Plato makes a serious point about communication and understanding through a parody of the master/disciple relationship. Speaking about written words, the character Socrates says that 'they seem to talk to you as though they were intelligent, but if you ask them anything about what they say, from a desire to be instructed, they go on telling you just the same thing forever'.[31] The seeming intelligence of the written word is a semblance because it is dissociated from its genuine source—the embodied master of 'wisdom' who has a cultivated and trained intellect. Those seeking wisdom should seek genuine instruction from such a master. To imagine that the written word can substitute for such a source is simply delusion. In the circumstance of discipleship to a book, one cannot be an authentic disciple because there is no possibility of correction or edification, no training or growth.

Copernicus, unfortunately, had no disciples, at least, not until the very end of his life, when Georg Joachim Rheticus sought him out and requested instruction in the 'absurd doctrine'. With Rheticus, Copernicus was able to instruct in ways of which Plato would have approved; and Copernicus was fortunate to have Rheticus asking for the kind of clarification which tightened up certain details of the system—a good example of the master/disciple relation, and one which, very likely, was crucial to the printing of *De revolutionibus*. Were it not for Rheticus, the book may not have been published at all.[32] There is something of a gulf, however, between the cosy Pythagorean relationship shared by Copernicus and Rheticus,[33] and the relationship between potential readers and the book, divorced as such readers would be from direct instruction by the master. Indeed, Rheticus was, perhaps, a better prepared and more capable student than Copernicus could have hoped for. Who knows, however, who might read the book? To pick up Plato's critique again, Socrates argues that 'once a thing is put in writing, the composition ... drifts all over the place, getting into the hands not only of those who understand it, but equally of those who have no business with it; it doesn't know how to address the right people, and not address the wrong'.[34] Plato manages thereby to highlight the sorts of Pythagorean reservations to which Copernicus confesses in his preface. So long as Copernicus controls communication of the work, it is safe from the 'wrong' people. Once in print, epistemic promiscuity and irresponsibility are unavoidable. The written word, as Plato so quaintly puts it, is 'unable to defend or help itself'.[35]

Given such difficulties, a person might, like Socrates, decide not to write at all. And Plato himself never 'spoke' directly in his writings, but resorted to ironic and dialogical modalities. Copernicus, however, did have something explicit to say, and had to say it in a way which would not allow him to 'take back the word'. He found it necessary, therefore, to build in the requisite help and defences. He tried, in other words, to control the reception of the text by defining how it was to be read and who was fit to read it. This task fell largely to the preface, which is a representation of the epistemic and political circumstances within which Copernicus located himself, and which he felt obliged to address. The preface has, of course, been subject to endless comment and interpretation, in support of various historiographical strategies (whiggish, revolutionary, rational reconstructionist, etc.[36]); I follow Robert Westman in seeing in the preface an attempt to shape 'the terms of its own interpretation'.[37] From the perspective being developed here, however, Westman's analysis of the rhetoric and politics of the preface misses the significance of printing. Further, in so far as Westman believes that Copernicus disavowed strict demonstrative ideals in favour of a Horatian aesthetic of 'fittingness' to convince his intended audience of the truth of his system, he underplays the likelihood that Copernicus would have been happy with strict demonstration could he but have found it. In its absence, a weaker 'demonstration' (whereby art imitates nature, and the beauty and *symmetria* of his system are turned into epistemic virtues) was called for. No matter how well judged and historically situated the humanist rhetorical strategies of the preface were (and Westman mounts a convincing case), they failed nonetheless. Like Westman, I see these strategies in social and political terms; but not without taking account of printing.

Let me refer the reader then to Westman's paper for details on Copernicus' intended reformist ecclesiastical audience. For the purposes of this paper, however, the consideration of printing suggests that regardless of authorial intentions, readers cannot be made to interpret a text as intended. Nor can they be forced to adopt new criteria for the assessment of a theory, even if they accept these criteria in another area (poetry/art). Nor does the adoption of recognised rhetorical strategies assure successful persuasion (far more likely, this would work the other way: 'Oh, I see what he's up to. Amusing. But hardly convincing!'). Certainly, it is important to understand Copernicus' strategy, and his targeting of a particular audience. But what else is he doing in this way of proceeding if not adopting a Pythagorean approach; that is, Copernicus uses print as if it is a form of personal dialogue, a dialogue he is having with 'devoted great men', with mathematically informed ecclesiasts whom he assumes are interested in reform; men like Nicolaus Schönberg, Cardinal of Capua, 'famous in every kind of learning',[38] as Copernicus describes him in his preface. Cardinal Schönberg had, in effect, offered Copernicus his patronage in his letter of November 1, 1536, in which he begged Copernicus 'to communicate your discovery to enthusiasts [that is, mathematicians and astronomers]', and declared himself as 'one who is zealous of your reputation and desires to do justice to your merits'.[39] This kind of talk no doubt appealed to Copernicus' Pythagorean sensibilities; but unfortunately, Copernicus could not comply with the Cardinal's request, since the work was not yet complete. And the Cardinal died the following year. Nevertheless, Copernicus had the Cardinal's letter printed in *De revolutionibus* in a prominent position: immediately preceding his own preface. Thus was the dead Cardinal enlisted as an ally in a conversation which Copernicus constructed between himself and like-minded people.

Seen in this light, the dedication of the work to Pope Paul III is not just a piece of flattery by Copernicus, but an appeal to an epistemic authority to endorse and guard the conversation, and to close off the circle of conversation to those unfit to participate:

> ... I have preferred to dedicate these my labour to your Holiness rather than to anyone else because ... you are considered as distinguished by the authority of your office and by your love of all learning and even of mathematics, so

> that by your influence and judgement you can restrain the stings of false accusers ...[40]

Copernicus might well insist that 'Mathematics is written for mathematicians',[41] and imagine that these fellow mathematicians will think his work 'of value to the ecclesiastical Commonwealth over which your Holiness now holds dominion';[42] but this also means that Copernicus was pursuing a divisive strategy within the Catholic Church. With the Reformation under way, he could not have picked a worse time to do so. Copernicus, in other words, plays the game of the politics of knowledge. The more firmly historical studies situate his work 'within the local circumstances of its production',[43] as Westman puts it, the more these local social and political factors come to the fore. The more also we see the self-conscious way in which Copernicus engages these factors as he judges them. It is hardly surprising that people misjudge them more often than they get it right. No matter how 'gently Horatian and Erasmian'[44] Copernicus thought he was being, *De revolutionibus* was taken to be a threat to the established order. Andreas Osiander, it turns out, was a more astute judge in this regard than Copernicus.

Like Copernicus, Osiander made a point of focussing on an intended audience in his prefatory letter to *De revolutionibus*; but he took that audience to have already settled on the 'right' epistemic order:

> I have no doubt that certain learned men, now that the novelty of the hypotheses in this work has been widely reported—for its establishes that the Earth moves, and indeed that the Sun is motionless in the middle of the universe—are extremely shocked, and think that the scholarly disciplines, rightly established once and for all, should not be upset.[45]

Unlike Copernicus, therefore, he does not attempt to reform that order, but to gain acceptance for Copernicus' work within it. Thus, divine revelation remains the sole source of truth, and astronomy remains a hypothetical enterprise. So long as Copernicus' hypotheses save the phenomena, Osiander is certain that 'learned' men 'will find that the author of this work has committed nothing which deserves censure'.[46] This passage highlights the fact that scholars could be censured for going outside the established epistemic order (as Galileo was to learn in the next century). The primary imperative in this order seems to have been the humiliation of human reason; the primary epistemic sin, pride. Osiander warns that no one should 'expect from astronomy ... anything certain, since it cannot produce any such thing, in case he seizes on things constructed for any other purpose as true, he departs from this discipline more foolish than he came to it'.[47] Despite the apparent humility of Copernicus' preface, there is nevertheless a certain arrogance in some passages: men of learning are portrayed as 'seeking out the truth in all things'; and 'opinions which are totally incorrect should be avoided'.[48] One might invoke the grace of God in such undertakings, but that does not disguise the intent. The strategies of persuasion, therefore, are markedly different.

Prior to the publication of *De revolutionibus*, Osiander corresponded with Copernicus in an effort to persuade him of the wisdom of presenting the work in an instrumentalist light in order to secure a sympathetic and unprejudiced reading:

> I have always felt about hypotheses that they are not articles of faith but the basis of computation. Thus, even if they are false, it does not matter, provided that they reproduce exactly the phenomena of the motions. For if we follow Ptolemy's hypotheses, who will inform us whether the sun's nonuniform motion occurs on account of an epicycle or on account of the eccentricity? For, either arrangement can explain the phenomena. It would therefore appear to be desirable for you to touch upon this matter somewhat in an introduction. For in this way you will mollify the peripatetics and theologians, whose opposition you fear.[49]

Talk of mollifying the peripatetics and theologians indicates that Copernicus must have communicated his fears in this regard to Osiander.[50] And of course, both peripatetics and theologians had powerful reasons for objecting to Copernicus' conclusions. Osiander made a point of addressing Rheticus along the same lines: 'The peripatetics and theologians will be readily placated if they hear that there can be different hypotheses for the same apparent motion: that the present hypotheses are brought forward, not because they are in reality true, but because they regulate the computation of the apparent and combined motion as conveniently as may be'.[51] In the preserved fragment of his letter to Rheticus, Osiander reveals a more subtle grasp of the rhetoric at play behind the commendation of instrumentalism. He suggests that by allowing for different hypotheses to save the phenomena, potential critics are not only disarmed but invited to conceive more convenient hypotheses; and failing to do so, will be converted to the Copernican hypothesis:

> ... each and every man is at liberty to devise more convenient hypotheses: and ... if he succeeds, he is to be congratulated. In this way they will be diverted from stern defense and attracted by the charm of inquiry; first their antagonism will disappear, then they will seek the truth in vain by their own devices, and go over to the opinion of the author.[52]

This shows a more devious mind at work. It also reveals an element of hypocrisy. If Osiander was genuinely committed to instrumentalism, he would hardly be talking about 'truth-seeking', let alone collapsing truth with opinion. Was he expecting that the Copernican system would come, by this subterfuge, to be accepted as true? This seems unlikely; but its acceptance would amount to much the same thing.[53] On this approach Osiander could be said to be recommending a way of shielding *De revolutionibus* from certain kinds of critical response, in the hope that conversion would follow from the successful use of the system.[54]

Neither Copernicus nor Rheticus, however, was persuaded to adopt the instrumental stance. As it happened, Rheticus, who was editing *De revolutionibus* and overseeing its publication by the Nuremberg printer Johannes Petreius, was offered a professorship in Leibzig, and could not complete the task. Osiander took over, adding an epistle recommending an instrumentalist interpretation of the hypotheses in the book to readers. Unacknowledged as it was, the epistle seemed to many at the time like an introduction by Copernicus; but Copernicus had written his own preface, which suggested a different reading altogether. Had Osiander been seriously intent on presenting the work instrumentally, he had the opportunity not only to add the epistle, but to modify Copernicus' own preface. As it is, the epistle and preface contradict one another, and speak to the reader with different voices. Confronted with a work whose truth is both affirmed and denied, the reader is strangely cut loose from authorial intent.

This heightens the interest of the book from a printing perspective. Despite Copernicus' hopes and intentions, the book is not a private letter to close friends, speaking with a sincere and univocal tongue. Printing disrupts the culture of private conversation by dissipating speakers and listeners. In the case of Copernicus and *De revolutionibus*, there is the author, too old and sick, and too far away from the centres of printing, to oversee the production of the book. There is Rheticus, a reliable (but unacknowledged) friend who set things in train only to abandon the work to the epistemic traitor, Osiander. And there is the printer, Petreius, who was after all engaged in a commercial venture, and was far from being a go-between for Copernicus and 'Pythagorean' friends. Although Osiander's deceit was unmasked, Petreius 'refused to add a correction to the text'.[55] Further, the book was eventually condemned by Copernicus' intended Catholic audience;[56] and his appeal to the authority and protection of the Pope would hardly have impressed a Protestant audience. The immediacy of dialogical interaction thus falls apart, giving way to interference, failed intentions, unreliable mediation, bad faith, unpredictability and lack of control. On top of all this, the commercialisation of the book introduces a fundamental uncertainty about who might buy it, and the uses to which it might be put.

Tycho Brahe provides an interesting example of the kind of reader Copernicus could not have envisaged. A precocious teenager, Tycho made a point of teaching himself astronomy by 'sneaking books past his tutor and pouring over them alone at night'.[57] He had access to printed books by both Ptolemy and Copernicus; and 'copies of the (Ptolemaic) *Alphonsine Tables,* the (Copernican) *Prutenic Tables* and Stadius' *Ephemerides* (which were based on the latter)'.[58] So equipped, he could see the inadequacies of both the old data and the inherited systems. Printing, in effect, enabled Tycho to step outside the master/disciple relationship. And as a self-made astronomer, he was not beholden to any particular intellectual tradition.

Printing is not just a technology. It alters networks of possible relations—economic, social, political and intellectual. Printing affected Copernicus' lifeworld in ways which he could not grasp. He yearned for Pythagorean secrecy and elitism, but was caught up in an expanding and increasingly open intellectual forum, beyond the control not only of authors like himself, but of Popes and established epistemic order. Even when attempts were made to control the presses, the authorities (State, Catholic and Protestant) were reduced to using those same presses to distribute printed lists of banned books;[59] and such lists only served to make these books more attractive. Printing permitted the growth of private libraries, which effectively decentred learning and fed both intellectual eclecticism and epistemic iconoclasm. Orthodoxy was thus undermined.

Such disruption, however, also paved the way for a new epistemic order and a new intellectual ethos. This change can be characterised as a move from a Pythagorean 'closed-shop' esoteric ethos, in which knowledge is contained within, and controlled by, an elite local community, to an 'open-shop' ethos, in which knowledge is public, and the communication of ideas, via print, cuts across different communities. Copernicus attempted to remain within the former ethos, and addressed his book to such a circumstance. *De revolutionibus*, however, fell into the public domain into which print put it, and went beyond the control of particular closed shops. The book may well have been esoteric in some technical-mathematical sense, but it was certainly not esoteric in the sense of being confined to a circle of chosen initiates, a 'whispered word' in the ears of the worthy. Little did Copernicus realise that in exclaiming that 'Mathematics is written for mathematicians',[60] he was giving voice to the same exclusive mentality which would see the book cast out of the Catholic circles to which it was addressed.

In so far as print distanced readers from authors, is also distanced them from established authority. They could thus read, and criticise, outside the networks of Church and State (though of course, there were limits—as the unfortunate fate of Bruno so graphically illustrates). This does not, however, entitle us to that quintessentially modernist illusion of the autonomous rational subject. Printing altered networks, and helped establish new ones; but it did not place people outside all networks. It did, however, decentre subjects. Who speaks in *De revolutionibus*? Copernicus? Osiander? Reformist[61] Catholicism? Horatian aesthetics? Mathematics? and who listens? The Pope? Rheticus? Osiander? Tycho? Kepler? No one voice and one one listener; many voices and many listeners—what is the difference? Though mute, the materiality of print speaks and moves through networks in ways in which individuals cannot. Thus, even the notion of 'public' is transformed. Instead of co-presence and dialogue, the new 'public' of print is non-localised and non-dialogical. The older formations of knowledge/power are disrupted. In vain, then, did Copernicus seek the protection of the old order.

Astronomy, of course, had its place in the old order, and subserved a greater network of both power and 'higher' knowledge. In helping to loosen the bonds which kept it in place, print contributed significantly to its public reconstitution.

More particularly, commentary and the enforcement of orthodoxy gave way to discovery and heterodoxy. Neither established authorities nor authors could hope to control the presses (as a whole); and the presses were not vying for epistemic power. They were, after all, commercial ventures. Print did not, therefore, simply facilitate communication in established networks and leave social relations in place. It made possible new forms of interaction and helped changed social relations.

This consideration should not be taken to imply a strong determinist view of print. Nor should it hide the fact that print was after all only '*an* agent of change',[62] as Eisenstein so rightly puts it. Thus, printing acted within the social and political order. Had those orders been different, printing would have had different effects. Print must after all be used by people, produced, bought and consumed by them. Who uses print, the uses to which it is put, the control or lack of control of print—these are the sorts of social and political factors which must be taken into account in the assessment of the impact of printing.

Well may Copernicus have desired to pass on the mysteries of this astronomy to friends 'not in writing but by personal contact'.[63] Among such friends he could perhaps have been taken at his word, defended that word, articulated it further. He was happy to play at such intimate politics. Of course, he was obliged to play on a bigger stage, and to consider 'outsiders'. Even so, he sought protection. Copernicus understood enough about the politics of knowledge at the time to attempt to control the reception of his work; but he failed to grasp the ways in which that politics was being affected by printing. The communications shift did transform the network of relations, and did so in ways which none of the players fully comprehended. Fortunately for the development of astronomy, it was out of their hands.

School of Science and Technology Studies, University of New South Wales

NOTES

[1] N. Copernicus, *De revolutionibus orbium coelestium,* trans. by A.M. Duncan, *Copernicus: The Revolutions of the Heavenly Spheres* (Newton Abbot, 1976) p. 24. I am using the translation by Duncan in preference to that of E. Rosen, *On the Revolutions* (Baltimore, 1978), because it is based on the Nuremberg edition (that is, the first printed edition—Rosen eclectially combines both the Nuremberg edition and Copernicus' autograph manuscript), and also because it is a more

literal translation which, according to N.M. Swerdlow & O. Neugebauer in their *Mathematical Astronomy in Copernicus's* De Revolutionibus (New York, 1984), brings 'the reader quite close to the original Latin' (p. 91).

[2] Copernicus, *De revolutionibus*, p. 24.

[3] An attitude aided by postmodern incredulity towards metanarratives (see J-F Lyotard, *The Postmodern Condition,* Minneapolis, 1984), though there seem to be some murmurings about the need for 'big pictures' (see A. Cunningham & P. Williams, 'De-centring the 'big picture': *The Origins of Modern Science* and the modern origins of science', *British Journal for the History of Science* 26 (1993) 407-32).

[4] The case is argued by Cunningham & Ross, 'De-centring the 'big picture''. See also D.C. Lindberg & R.S. Westman (eds), *Reappraisals of the Scientific Revolution* (Cambridge, 1990), for further critiques of grand narrative 'revolutionary' historiography.

[5] While the more radical thesis is stated by Marshall McLuhan, *The Gutenberg Galaxy: The Making of Typographic Man* (Toronto, 1962), the major historical case for the 'Printing Revolution' is argued by Elizabeth Eisenstein, *The Printing Press as an Agent of Change,* 2 vols (Cambridge, 1979) and *The Printing Revolution in Early Modern Europe* (Cambridge, 1983), an abridgment of the two-volume version. For critiques of Eisenstein's work, see M. Hunter, 'The impact of print', *The Book Collector* 28 (1979) 335-52; A Grafton, 'The importance of being printed', *Journal of Interdisciplinary History* 11 (1980) 265-86; and S. Hindman (ed.), *Printing and Written Word: The Social History of Books, circa 1450-1520* (Ithaca, 1991). For a general critique of the thesis that printing marks a radical break with the past, see M.T. Clanchy, 'Looking back from the invention of printing', in D.P. Resnick (ed.), *Literacy in Historical Perspective* (Washington, 1983) pp. 7-22.

[6] G. Sarton, *Six Wings* (Bloomington, 1957) and *Appreciation of Ancient and Medieval Science During the Renaissance 1450-1600,* 2nd edn (New York, 1958); D. de Solla Price, *Science since Babylon* (New Haven, 1975); S. Drake, 'Early science and the printed book: The spread of science beyond the university', *Renaissance and Reformation* 6 (1970) 38-52; and H.E. Lowood & R.E. Rider, 'Literary technology and typographic culture: the instrument of print in early modern science', *Perspectives on Science* 21 (1994) 1-37. The most extensive treatment of the impact of printing on science, however, is to be found in Eisenstein, *The Printing Press.* That this should be so is particularly noteworthy, since it is clear that Eisenstein's history is not focussed on science but on printing.

[7] This does not mean that historians more generally have not shown a keen interest in printing, but this interest does not usually carry over to an interest in the history of science; and why should it, given that they are writing histories of printing (or, in more inclusive efforts, histories of writing; see, for example, H-J. Martin, *The History and Power of Writing,* trans. L.G. Cochrane (Chicago & London, 1994). Interestingly, there are only two page references given to 'science' in the index, and none at all to Copernicus).

[8] It may be objected that this is not the case; that although it might be true of so-called 'internalist' historians, it is not true of 'externalists'. This seems like a plausible objection at first sight, but the internalist-externalist divide is only of interest in so far as both sides make *conflicting* claims about how the content of science comes about; that is, in so far as there is some question about the sorts of factors which explain the content and direction of science.

[9] Archer Taylor puts the point nicely: 'The powers which shape men's lives may be expressed in books and type, but by and of itself printing ... is only a tool, an instrument, and the multiplication of tools and instruments does not of itself affect intellectual and spiritual life' (cited in Eisenstein, *The Printing Press,* p. 703).

[10] This response can, of course, be expressed in terms of Marshall McLuhan's adage, 'the medium is the message'. I will not, however, be discussing McLuhan's statement of the thesis (though see McLuhan's *The Gutenberg Galaxy*).

[11] Eisenstein, *The Printing Press,* p. xv.

[12] *Ibid.*

[13] *Ibid.,* xvi.

[14] See Note 5 for critical sources.

[15] Eisenstein, *The Printing Press,* p. 36.

[16] *Ibid.,* p. 703.

[17] *Ibid.,* p. 689.

[18] *Ibid.,* p. 578.

[19] *Ibid.,* p. 579.

[20] *Ibid.,* p. 603-5; though Eisenstein does not make the point as sharply as this, and puts it in terms which constitute an attack on the Kuhnian notion that Copernicus was reacting to a breakdown of normal problem-solving activity (thus precipitating a 'revolution' in astronomy).

[21] Eisenstein, *The Printing Press,* p. 629.

[22] Of course, despite the severity of these criticisms, I would not have taken printing seriously were it not for the inspiration provided by Eisenstein's work.

[23] Copernicus, *De revolutionibus*, p. 24.

[24] *Ibid.*

[25] See E. Rosen, *Copernicus and the Scientific Revolution* (Malabar, Florida, 1984) p. 113; and *Three Copernican Treatises,* 3rd ed. (New York, 1971) pp. 6-7, 343-4 for details on the *Commentariolus.*

[26] *Ibid.*

[27] Cited in R.S. Westman, 'Proof, poetics, and patronage: Copernicus's preface to *De revolutionibus*', in Lindberg & Westman, *Reappraisals of the Scientific Revolution,* pp. 167-205, at p. 182.

[28] See Swerdlow & Neugebauer, *Mathematical Astronomy,* for a plausible estimate of the time it would have taken Copernicus to complete the work—without wasting any time: 'Considering the magnitude of his undertaking and the time that it necessarily required, it can be seen that he carried out his work as rapidly as could be expected' (p. 10).

[29] Copernicus, *De revolutionibus*, p. 24. Interestingly, the 'friends' who published the work are not acknowledged by name—unlike the Catholic dignitaries to whom Copernicus refers. But then, Rheticus and Osiander were Protestants; and the preface was addressed to the Pope Paul III.

[30] Plato, *Phaedrus* 275a-b. Trans. by R. Hackforth, *Plato's Phaedrus* (New York, 1952) p. 157.
[31] Plato, *Phaedrus* 275d.

[32] See Swerdlow & Neugebauer, *Mathematical Astronomy,* pp. 23-89 for an assessment of the role which Rheticus played in the publication of *De revolutionibus*. See also Owen Gingerich, '*De revolutionibus*: an example of Renaissance scientific printing', in *The Eye of Heaven: Ptolemy, Copernicus, Kepler* (New York, 1993) pp. 252-68 (reprinted from G.P. Tyson & S.S. Wagonheim (eds), *Print and Culture in the Renaissance* (Newark, Delaware, 1986) pp. 55-73). Gingerich, who probably knows more about the printing of *De revolutionibus* than anyone else, is much more direct in his assessment, and claims that Copernicus 'would never have seen his work printed

except for the intervention of a young professor of astronomy from Lutheran Wittenberg, Georg Joachim Rheticus' (p. 252).

[33] Interestingly, Rheticus referred to Copernicus as his *Dominus Praeceptor* in the *Narratio prima.* Despite the possible ulterior motives which Rheticus may have had in approaching Copernicus, he did not fail to observe the traditional sign of respect in the disciple's approach to the master—the giving of a gift as a gesture of good faith. In this case, the gift is more revealing—a number of recently published books on astronomical and mathematical matters (see Swerdlow & Neugebauer, *Mathematical Astronomy*, p. 24, for details).

[34] Plato, *Phaedrus* 275e.

[35] *Ibid.*

[36] For an interesting account of such uses, see Westman, 'Proof, poetics, and patronage', pp. 169-75.

[37] Westman, 'Proof, poetics, and patronage', p. 168.

[38] Copernicus, *De revolutionibus,* p. 24.

[39] *Ibid.,* p. 23.

[40] *Ibid.,* p. 26.

[41] *Ibid.,* p. 27.

[42] *Ibid.*

[43] Westman, 'Proof, poetics, and patronage', p. 168.

[44] *Ibid.,* p. 192.

[45] Copernicus, *De revolutionibus*, p. 22.

[46] *Ibid.*

[47] Copernicus, *De revolutionibus*, p. 23.

[48] *Ibid.*

[49] Rosen, *Copernicus and the Scientific Revolution,* p. 193.

[50] *Ibid.,* pp. 193-4.

[51] *Ibid.,* p. 194.

[52] *Ibid.*

[53] Guy Freeland has suggested to me that Osiander may have held a concept of instrumental truth. This is a wonderfully subtle possibility. 'Convenience' would then become a mark of probable truth. In so far as the peripatetics could be classed among philosophers who 'will perhaps look more for probability', as Osiander characterises philosophers in his preface to *De revolutionibus*, then this would make some sense. But being already committed to the truth of geocentrism, it is unlikely that peripatetics would have been inclined to 'convert' to heliocentrism along such lines, since presumably they did not believe in the truth of geocentrism for instrumental reasons.

[54] See B. Wrightsman, 'Andreas Osiander's contribution to the Copernican achievement', in R.S. Westman (ed.), *The Copernican Achievement* (Berkeley, 1975) pp. 213-43, for an extended defence of Osiander's strategy: 'for over a century', "Ad Lectorum" [Osiander's unacknowledged epistle to the reader in *De revolutionibus*] *protected* the work from this kind of scrutiny during an extremely tense period of ideological and political conflict and thus, actually *permitted* the work to be used and pondered during that period by those with such scruples, by advocating the way it was, in fact, being regarded and used' (p. 240).

[55] J.D. Moss, *Novelties in the Heavens: Rhetoric and Science in the Copernican Controversy* (Chicago, 1993) p. 39. See also Rosen, *Three Copernican Treatises,* pp. 404-6.

[56] Decree XIV of the Holy Congregation of the Index (March 1616) declared heliocentrism not only 'false and altogether opposed to Holy Scripture' (cited in O. Gingerich, 'The censorship of *De revolutionibus*', *The Eye of Heaven,* pp. 269-85, at p. 274), but also prejudicial to 'Catholic truth'. The relative slowness of the official condemnation, however, should not blind us to the fact that, as Westman puts it, '*De revolutionibus* was immediately perceived as a resource of disciplinary disruption' (Westman, 'Proof, poetics, and patronage', p. 187).

[57] Eisenstein, *The Printing Press,* p. 596.

[58] *Ibid.*

[59] See, for example, P.F. Grendler, 'Printing and censorship', in C.B. Schmitt (ed.), *The Cambridge History of Renaissance Philosophy* (Cambridge, 1988) pp. 25-53.

[60] Copernicus, *De revolutionibus*, p. 27.

[61] The 'reformists' indicated here are the humanist curial reformers identified by Westman in 'Proof, poetics, and patronage'—a very different crowd to the reactionary reformists who prevailed at the Council of Trent.

[62] Eisenstein, *The Printing Press,* p. xv.

[63] Copernicus, *De revolutionibus*, p. 24.

NEIL THOMASON

1543—THE YEAR THAT COPERNICUS DIDN'T PREDICT THE PHASES OF VENUS*

'There is a beautiful tradition respecting the phases of Venus, which appears, however, to partake rather of the nature of an allegory or illustration, than of a historical anecdote.'
—The Hon. Mrs. Ward
Telescope Teachings (1859)

For many good (and some bad) reasons, philosophers of science are enjoined to study the history of science. Some even propose to use that history as a touchstone to evaluate philosophies of science. Clearly, the value of this advice depends, *inter alia*, on the historical accuracy of the proffered history. Of course, history is complex and, as with other areas of scholarship, we should expect many histories of science to contain errors or to mislead in some way. Most such errors will be trivial or limited to one author or, if they are of some interest, they will be corrected by other historians. But if there is a systematic bias among historians dealing with many different areas of the history of science, the resulting general understanding of science will be distorted. This is so especially if that bias among historians has the effect of reinforcing a widespread tendency among philosophers.

In the long run, the best way to demonstrate such a bias would be to show that there is a distinct pattern of distortions in the standard accounts of many episodes in the history of science. I won't do this here. Rather I will take one of the best known episodes in the history of science, Galileo's discovery of the moon-like phases of Venus, and trace certain distortions in standard accounts over the last two centuries. This is not a case where the true history is

* This is a much developed version of an earlier paper 'Sherlock Holmes, Galileo, and the Missing History of Science' to be published in Arthur Fine (ed.) *PSA* 1994, vol. 1.

G. Freeland and A. Corones (eds.), 1543 and All That, 291 – 331

unknown—the actual history is too simple and the sources are too well-known and too unequivocal to have been universally overlooked. In fact, there are two traditions of the recounting of Galileo's discovery, one which gets the story right and one which distorts it. My interest here is in the distorted tradition, why it is so strong when it is so clearly inconsistent with the historical record, and what its effect might be on the philosophy of science.

1. THAT COPERNICUS DIDN'T PREDICT THE PHASES OF VENUS

As we shall see below, there is a widespread but not universal view among historians and philosophers that the moon-like phases of Venus are predicted Copernican theory and that Copernicans did indeed predict them. This view is often supported by diagrams looking like those in fig. 1. It appears, then, that Ptolemaic theory predicts Venus will only be seen as a crescent, but that Copernican theory predicts that Venus will go through an entire cycle of phases as the moon does—from full to half to crescent and back again. (Hereafter, for brevity, I will refer to the prediction that Venus goes through an entire set of phases like the moon as the prediction that Venus has phases.)

However persuasive such diagrams and claims may be, Copernicus, Galileo, Kepler, and their contemporaries knew and emphasised that an additional problematic auxiliary hypothesis must be assumed for the Copernican theory to predict the phases of Venus: *Venus is a dark, opaque body*. If Venus is self-illuminated like the sun or translucent like crystal or amber, Venus would always appear fully illuminated—from the earth, one would never see Venus as crescent-shaped or gibbous. These diagrams implicitly presuppose that Venus is a dark, opaque body that shines by reflected sunlight.

The source of planetary and stellar light had been an open question since the Greeks, with Plato in *Timaeus* 39B appearing to hold that the sun was the sole source of planetary and stellar light. As al-Biruni wrote:

> Opinions of intelligent people differ ... as to whether the planets are self-luminous like the sun, or merely illuminated by the rays of the sun falling on them. Many assert that light is exclusively the property of the sun, that the stars [and planets] are destitute of it. ... But others believe that all the planets are luminous by nature with the exception of the moon.[1]

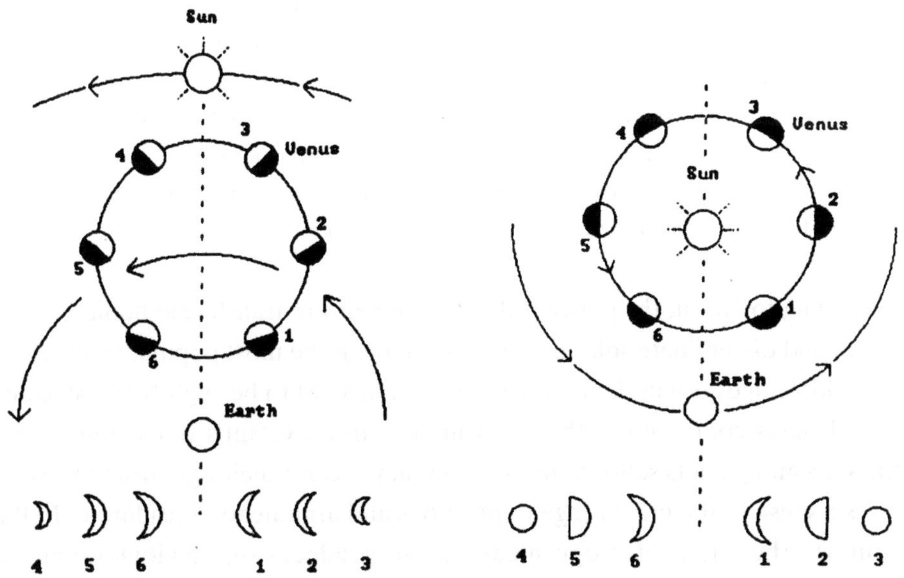

Figure 1. *Comparison of the Ptolemaic (left) and Copernican (right) systems with respect to the appearance of Venus.*

Among others, Ibn al-Haytham and Levi Ben Gerson (1288-1344) held that the planets were self-luminous, because they (in particular Venus) never appear crescent-shaped.[2] By the middle of the fourteenth century, Albert of Saxony clearly was aware that the nature of Venus' substance could not be resolved by the available evidence. In his widely studied *Questiones super quattuor libros de coelo et mundo*, Albert wrote:

> The question, do the stellar bodies other than the sun and moon receive their ight from the sun can be thought as neutral; *the reasons one gives for one side can be as easily refuted as those one gives for the other side*. Therefore, for the love of Aristotle, the prince of philosophers, I will refute the six opinions formulated against Aristotle's opinion, in favour of Avicenna's opinion, and I will assert that all stellar bodies other than the sun and moon, whether they are planets or fixed stars, receive their light from the sun.[3]

Ariew comments:

> Of course, neither Avicenna nor Albert think that Venus and Mercury actually have phases. So Albert has to resort to an *ad hoc* explanation to defend Aristotle against Avicenna's attack: If Venus and Mercury received their light from the sun, we would see their phases; and we don't. In defense of Aristotle, Albert replies that 'Venus and Mercury are so transparent that the ight of the sun becomes incorporated with these stars and gets soaked up in all their parts, which does not happen for the moon.'[4]

According to Grant, the position that the planets are translucent bodies able to retain and disseminate solar light may have been the most popular view during the middle ages. Even the moon was generally held to be slightly translucent.[5] The debates continued in the following centuries. Grant's discussion of 'Are the stars and planets self-luminous, or do they receive their light from the Sun?' gives a sense of the great range of positions and arguments considered.[6] In this context, the view of Copernicus' great predecessor Regiomontanus is straightforward:

> The bodies of the planets other than the moon absorb sunlight into themselves. They do so to no greater extent than the moon. Yet, perhaps on account of the different variations in the planets and stars, the planets other than the moon receive the sun's rays into their very depths. On the other hand, on account of its greater density, the moon is not illuminated down to its centre. Hence it looks to us like a crescent. But Venus, even though it is quite close to the sun, never appears in this way as a crescent, because its body is penetrated throughout by sunlight.[7]

In Book I, Chapter 10, of *De revolutionibus*, Copernicus shows he was clearly aware of this long-standing controversy. In discussing the various hypotheses about the order of the heavenly spheres, he writes:

> According to those who follow Plato, since they consider that all stars, being otherwise dark bodies, shine by the solar light which they receive, if they were below the Sun, on account of their short separation from it, they would be seen only as halves, or at most as not completely round. For they would generally reflect upwards, that is towards the Sun, the light which they have received, as we see in the new or waning Moon. ... On the other hand, those who place Venus and Mercury below the Sun ... do not admit that these

> heavenly bodies have any opacity like the moon's. On the contrary, these shine either with their own light or with the sunlight absorbed throughout their bodies.[8]

There is no evidence that Copernicus ever accepted the opacity of the planets or believed that his hypothesis was committed to Venus having phases like the moon. As far as our historical sources show, Copernicus left the issue completely open, just listing alternative accounts of planetary light.

Galileo was well-acquainted with Scholastic natural sciences and the possibility that the planets are self-illuminated or transparent.[9] His understanding of the situation comes through clearly in his 1 January 1611 letter to Giuliano de' Medici publicly announcing the discovery of Venus' phases. After describing them in detail, he continued:

> From this marvellous experience we have a sensible and sure proof of two great suppositions which have been doubted until now by the greatest minds of the world. One is that all planets are dark by nature (the same for Mercury as for Venus). The other is that Venus must necessarily revolve around the sun, just like Mercury and all the other planets...[10]

In his polemical response to "Apelles", the *Letters on Sunspots* of 1613, Galileo was again unequivocal about there being two issues, although incorrectly claiming that Copernicus declared Venus to be either self-illuminating or translucent. After describing the phases, Galileo continued:

> These things leave no room for doubt about the orbit of Venus. With absolute necessity we shall conclude, in agreement with the theories of the Pythagoreans and of Copernicus, that Venus revolves about the sun just as do all the other planets. ... No longer need we employ arguments that allow any answer, however feeble, from persons whose philosophy is badly upset by this new arrangement of the universe. For these opponents, unless constrained by some stronger argument, would say that Venus either shines with its own light or is of a substance that may be penetrated by the sun's rays, so that it may be lighted not only on its surface but also throughout its depth. They take heart to shield themselves with this argument because there have not been wanting philosophers and mathematicians who have actually believed this—meaning no offence to Apelles, who says otherwise. Indeed, Copernicus

> himself was forced to admit the possibility and even the necessity of these two ideas, as otherwise he could give no reason for Venus failing to appear horned when beneath the sun. As a matter of fact nothing else could be said before the telescope came along to show us that Venus is naturally and actually dark like the moon, and like the moon has phases.[11]

Twenty years later, in his *Dialogue Concerning the Two Chief World Systems*, Galileo's incorrect account of Copernicus' view on the matter remained the same. Salviati is speaking:

> Add to these another difficulty [for the Copernican system]; for *if* the body of Venus is intrinsically dark, and like the moon it shines only by illumination from the sun, which seems reasonable, then it ought to appear horned when it is beneath the sun ...—a phenomenon which does not make itself evident in Venus. For that reason, Copernicus declared that Venus was either luminous in itself or that its substance was such that it could drink in the solar light and transmit this through its entire thickness that it might look resplendent to us. In this manner Copernicus pardoned Venus its unchanging shape ...[12]

As we shall see, in reporting that Copernicus declared that Venus was either self-luminous or translucent, Galileo erred in the direction opposite to that of many modern historians.[13] Whatever the explanation for Galileo's error, at a minimum these well-known passages show that Galileo did not hold that Copernicus had predicted the phases of Venus or that Copernican theory was committed to them.

Kepler's enthusiastic 28 March 1611 response to Galileo's discovery further emphasises how open was the issue of the source of Venus' light:

> *Unexpected by me in any way was your observation*, for on account of the unusual brightness of Venus I believed light of its own to be inherent in it. Astonishing, unless Venus is all gold; or, as I said in my *Foundations of Astrology*, amber.[14]

Kepler had previously argued in *Ad vitellionem paralipomena* of 1605 that both planets and stars are self-luminous—indeed one of his arguments was that Venus did not have phases.[15] Even allowing for some Keplerian hyperbole, this highly sophisticated Copernican clearly did not regard the phases of Venus as an inevitable consequence of Copernican theory. He certainly appears not to have regarded the apparent non-existence of the phases as a difficulty.

Clearly Copernicus, Galileo and Kepler all understood that Venus would have a complete set of phases like the moon only if two controversial hypotheses were true: (1) *Venus revolves around the Sun* and (2) *Venus is an opaque, intrinsically dark body*. They clearly saw that heliocentric theory (plus unproblematic auxiliary hypotheses) by itself does not predict that Venus has phases like the moon. In fact, as far as I have been able to discover, only one contemporary held that Copernican theory predicted the phases of Venus. Galileo's student, the committed Copernican Benedetto Castelli, wrote to Galileo on 5 December 1610 proposing that Galileo closely examine Venus with his new telescope:

> From that notice you [Galileo] gave me, *after various thoughts passed through my mind, I finally hit on this*: that being true, as I hold most true, the Copernican arrangement of the world, Venus would have to have, at the elongations from the Sun, sometimes a hornéd appearance and sometimes not hornéd, according as it is found beneath or beyond the Sun ...[16]

With the exception of this private letter written almost a year after Galileo's initial telescopic discoveries and just before the public announcement of the phases, I have found no evidence in the primary sources that the apparent absence of the (moon-like) phases of Venus was a difficulty for Copernican theory.[17] In the midst of all the difficulties Copernicanism faced, the existence of the phases was not predicted and their absence was not raised as a difficulty for Copernican theory. This is not surprising, since if the absence of phases was a difficulty for Copernican theory, it would also be a difficulty for Ptolemaic and Tychonic theories. For an astronomical realist, the apparent absence of phases could not count as evidence against one of the three major theories without counting as evidence against the other two.[18] It could not be used as evidence against just one of them, say Copernicanism.

For reasons that will soon be all too clear, it is worth emphasising that the correct story is not complex. Clear and correct versions sometimes appear in that favourite target of historians of science, the introductory science textbook. This is from the fourth edition of Snow's *The Dynamic Universe*:

> Galileo quickly realized that this was analogous to the phases of the Moon—we see from the Earth varying portions of the sunlit side of Venus ... This had two important implications: It showed that Venus shines by reflected sunlight rather than by its own power; and it demonstrates that Venus orbits the Sun instead of the Earth ...[19]

In general, however, the debate over Venus' opacity is not mentioned in the textbooks and Copernican theory is often reported as predicting the phases. (The prize for the most-technically-correct-while-missing-the-point account goes to Jastrow and Thompson: '... Galileo first turned the telescope on Venus in 1610 and discovered that the planet was a round body like the earth and the moon...'.[20])

Many historians (although, as we shall see, few philosophers) get the history right: Caspar[21] and Cohen[22] who provide especially clear explanations; Ariew, in his 'The Phases of Venus before 1610', to which I am indebted; Abetti, who quotes Galileo's letter to Guiliano de' Medici quoted above, which succinctly sets out the facts;[23] and Van Helden who cites Ariew's article in his new translation of *Siderius nuncius*. Here is Cohen's account in his popular *Birth of a New Physics*. It is accompanied by two particularly useful figures:

> The first [discovery after the *Sidereus nuncius*] was the discovery that Venus exhibits phases ... In the first place, it proved that Venus shines by reflected light, and not by a light of its own; this meant that Venus is like the moon in this regard, and also like the earth (which Galileo had previously shown to shine by reflected light of the sun). ...[And] if Venus moves in an orbit around the sun, not only will Venus go through a complete cycle of phases, but under constant magnification the different phases will appear to be of different sizes because of the change in the distance of Venus from the earth.[24]

Although Drake had earlier written that the phases 'were required by [Copernicus'] theory', later he was quite clear about the facts. Here is his reply to Westfall's charge that Galileo had not actually observed Venus for as long as Galileo claimed he had:

> Now, there is another reason that no phases of Venus might be observable even if Copernicus were right, and that reason had so strongly gripped the mind of Kepler that he was entirely in the dark about Galileo's anagram. ...
>
> If Venus possessed light of its own, phases occasioned by reflection of sunlight could hardly be detected telescopically or in any other way. Castelli's deduction was formally invalid without explicit assumption that Venus was as dark as the moon. Galileo's critics may suppose that he would overlook that and risk his reputation with Kepler, but I should like to know from them what evidence Galileo had that Venus was dark, before he began to observe its phases. Since even the possibility was unthinkable to Kepler, it is hardly objective to attribute to Galileo its automatic assumption.[25]

If a historian does not want so much detail, or chooses not to discuss the opacity issue, there are perfectly adequate, widely adopted alternatives, such as 'Galileo used the moon-like phases of Venus as strong evidence against Ptolemy' or 'Copernican and Tychonic theories explained the moon-like phases of Venus—which are inconsistent with the Ptolemaic theory' and the like.[26]

2. THAT NOT WITHSTANDING THE PRIMARY TEXTS, MANY HISTORIANS SAY THAT COPERNICAN THEORY DID PREDICT THE PHASES OF VENUS AND THAT THIS FAILED 'PREDICTION' WAS A FAVOURITE ARGUMENT OF ANTI-COPERNICANS

Despite the clarity of these well-known texts, the simplicity of the issues, and the fact that many historians get the history right, there is a surprisingly long and popular tradition of seriously mis-reporting the episode. It is very often presented as a straightforward example of a single hypothesis (heliocentricity) or as an example of some astronomers *predicting* a novel fact (that Venus has moon-like phases). We will examine the historians first, starting with John Keill's Oxford lecture notes.[27] Although I know of no earlier examples, my search has been far from systematic and I would be surprised if Keill actually started this tradition:

(1) John Keill, 1721:

Before the Invention of this Noble Instrument, when Copernicus first revived the ancient *Pythagorean System*, and proposed it to the Learned in *Astronomy*, ... it was objected to him, That if the Motions of the *Planets* were such as he supposed them to be, that then *Venus* ought to undergo the same Changes and Phases as the Moon does. *Copernicus* answered, That perhaps the astronomers in after-ages would find, that *Venus* does really undergo all these Changes. The Prophecy of *Copernicus* was first fulfilled by that great *Italian* Philosopher Galileus, who directing his telescope to *Venus* observed her Appearances to emulate the moon, as *Copernicus* had foretold: And these Observations did surprisingly confirm the old system revived by Copernicus.[28]

(2) Robert Smith, 1738:

When Copernicus revived the ancient Pythagoric system, asserting that the earth and planets moved round the sun in the centre of their orbits, the Ptolemaics objected, if this were true, the phases of Venus should resemble hose of the moon. Copernicus replied, that some time or other, that resemblance would be found out.[29]

(3) d'Alembert, 1772/1774:

When Copernicus proposed his system, at a time when telescopes were not invented, *the non-existence of these phases was brought against him. He predicted that they would be discovered one day*, and telescopes have verified his prediction.[30]
Copernicus predicted that the centuries to come would discover that Venus underwent the same changes as the moon.[31]

(4) Bailly, 1785:

In the system of Copernicus on the other hand, where they revolve around that star, *they must show* sometimes a full disc, sometimes an obscure disc, and all the intermediate phases that one observes in the changing of the moon from its weak crescent to its full bright light. *Copernicus had dared to announce that if our organ had the ability to see these two inferior planets as we see our satellite, we would see that they were subject to the same variations.*[32]

(5) Adam Smith, 1795:

It had been objected to Copernicus, that, if Venus and Mercury revolved round the Sun, in an orbit comprehended within the orbit of the Earth, they would show all the same phases with the Moon, present, sometimes their darkened, and sometimes their enlightened sides to the Earth, and sometimes part of the one, and part of the other. *He answered, that they undoubtedly did all this*; but that their smallness and distance hindered us from perceiving it. This very bold assertion of Copernicus was confirmed by Galileo. ...

Succeeding telescopical observations, discovered, in each of the Five Planets, spots not unlike those which Galileo had observed in the Moon, and thereby seemed to demonstrate what Copernicus had only conjectured, that the Planets were naturally opaque, enlighted only by the rays of the Sun, habitable, diversified by seas and mountains, and, in every respect, bodies of the same kind with the Earth; and thus added one other probability to this system.[33]

(6) Robert Small, 1804:

... and, when *in answer to another equally powerful objection*, that no varieties of phase were seen in the planets, *Copernicus could only express his hopes that such varieties would be discovered in future times*, his reply, though it now raises admiration, could not in his own times make the least impression on those who opposed his system.[34]

(7) Thomas Morell, 1827:

The phases of Venus, resembling those exhibited by the moon, had been *conjectured by Copernicus as highly probable phaenomena*, but were fully demonstrated by the telescopic observations of Galileo.[35]

(8) J.P. Nichol, 1847:

... and the demonstrable fact, that she ought to have phases, was objected to Copernican theory at the time of its discovery. *The mode in which Copernicus dealt with that objection, is another and most emphatic proof of the greatness of his mind. An inferior person would at once have 'denied the fact,' and bought forward metaphysical reasons of a kind then very much in vogue, why Venus should not be subjected to such laws; but after some wavering, our astronomer boldly acknowledged the accuracy of the deduction, and in the finest spirit of prophecy, added without hesitation, that should men ever see Venus better, they would discern her phases.* And singularly enough, the verification of this confident prediction was one of the earliest achievements of the telescope![36]

(9) Oliver Lodge, 1893:

The discovery of the phases of Venus] was a dreadful blow to the anti-Copernicans, for it removed the last lingering difficulty to the reception of Copernican doctrine.

Copernicus had predicted, indeed, a hundred years before, that, if ever our powers of sight were sufficiently enhanced, Venus and Mercury would be seen to have phases like the moon. And now Galileo with his telescope verifies the prediction to the letter.[37]

(10) Andrew White, 1895:

But the new truth could not be concealed; it could neither be laughed down nor frowned down. Many minds had received it, but within the hearing of the papacy only one tongue appears to have dared to utter it clearly. This new warrior was that strange mortal, Giordano Bruno. He was hunted from land to land, until at last he turned on his pursuers with fearful invective. For this he was entrapped in Venice, imprisoned during six years in the dungeons of the Inquisition at Rome, then burned alive, and his ashes scattered to the winds. Still, the new truth lived on. Ten years after the martyrdom of Bruno the truth of Copernicus's doctrine was established by the telescope of Galileo.

Herein was fulfilled one of the most touching prophecies. Years before, the opponents of Copernicus had said to him, 'If your doctrines were true, Venus would show phases like the moon'. Copernicus answered: 'You are right; I know not what to say; but God is good, and will in time find an answer to this objection'. The God-given answer came when, in 1611, the rude telescope of Galileo showed the phases of Venus.[38]

(11) John Fahie, 1903:

It had always been a formidable objection to the Copernican theory that Venus and Mercury did not exhibit the same phases as the Moon, which they should if they revolved round the sun. Copernicus himself had endeavored to account for this, by supposing that the sun's rays passed freely through the body of the planets, and Galileo took occasion to praise him for not being deterred from adopting the system ...[39]

(12) J. Dreyer, 1906:

Before the end of 1610 ... the discovery of the phases of Venus *deprived the opponents of Copernicus of a favourite weapon.*[40]

(13) Stillman Drake, 1957:

The other discovery ... was that Venus passes through a regular series of changes in shape precisely like those of the moon. *Copernicus had been puzzled* at the apparent absence of such changes, *which were required by his theory.* [Drake's Footnote #12]

[Here is Drake's Footnote #12:]

De Revolutionibus, i, 10: 'Neither do they grant that any darkness similar to that of the moon is found in the planets, but they assume that these are either self luminous or are lighted by sunlight throughout their whole bodies'. Copernicus refrained from giving his own opinion on the problem. Galileo was much impressed by the fact that this much apparent contradiction of the senses had not deterred Copernicus from adhering to the heliocentric system; cf. *Dialogue*, pp. 334-335.[41]

(14) Thomas Kuhn, 1957:

Many other arguments were derived from telescopic observation, but only the observations of Venus provided sufficiently direct evidence for Copernicus' proposal to concern us here. Copernicus himself had noted in Chapter 10 of the First Book of *De Revolutionibus* that the appearance of Venus could, if observable in detail, provide direct information about the shape of Venus' orbit. If Venus is attached to an epicycle moving on an earth-centered deferent, and if the center of the epicycle is always aligned with the sun, then ... an observer on the earth should never be able to see more than a crescent edge of the planet. *But if Venus's orbit encircle the sun ..., then an earthbound observer should be able to see an almost complete cycle of phases, like the moon's; only phases near 'new' and 'full' would be imperceptible, because Venus would then be too close to the sun.*

... *Copernicans, or at least the cosmologically more radical ones*, had anticipated the sort of universe the telescope was disclosing. *They had predicted a detail, the phases of Venus, with precision* ... There are few phrases more annoying or more effective than 'I told you so'.[42]

(15) A. Pannekoek, 1961:

In the same way [Galileo] announced in December that Venus imitates the igures of the moon. *Some followers of Copernicus had predicted it; others, Kepler among them, believed that the planets partly radiated their own light or were saturated by absorbed sun light.*[43]

(16) R. Westfall, 1971:

In the geocentric system, Venus is always more or less between the sun and the earth, and *must always appear as a crescent*. In the heliocentric system, it travels behind the sun and can appear nearly full ...[44]

(16) Stillman Drake, 1972:

The phases of Venus removed *a serious objection* to the Copernican system ...[45]

(17) A. Van Helden, 1989:

The appearance of Venus predicted by the Ptolemaic and Copernican systems ...[46]

(18) M. Segre, 1991:

Galileo also succeeded in seeing the phases of Venus for the first time; *the existence of planetary phases was predicted by the Copernican theory*, and those of Venus are more easily seen than others.[47]

(19) O. Gingerich, 1992:

> There was, however, another key role for the Cytherian planet, Venus. … With the telescope, however, it is a simple matter to check out the phases. *In the Ptolemaic system, Venus could never show a fully illuminated face*, because it is never on the far side of the sun from the earth. Hence in that scheme it shows only crescent phases. *Quite the contrary is true in the Copernican system, where Venus will show a complete range of phases from crescent to full.*[48]

(20) J. North, 1994:

> The upshot of this is that *all* models produce a set of phases of Venus, but the Ptolemaic model *does not have the full set*. For traditional astronomers, Venus should at best show a crescent shape. The opponents of Copernicus had pointed out that the variation in Venus' appearance was not enough to support the idea of a full set of phases.[49]

There are more examples in the writings of other, less well-known historians of astronomy. Not surprisingly, the claim that Copernican theory and/or Copernicus himself predicted the phases of Venus also appears in general histories of science. The following is from Crombie's *Augustine to Galileo*:

> [Galileo] also confirmed *Copernicus' deduction* that Venus, because of the position he held it to have inside the earth's orbit, would have phases like the moon …[50]

By and large, these are judicious historians who know the key texts very well. Yet, some say that Copernican theory predicts that Venus has phases and others that Copernicus or unnamed 'more radical Copernicans' predicted them; while still others state that the absence of the phases of Venus was a major objection to Copernican theory. In this tradition of Copernican scholarship, the texts discussing the possibility that Venus is translucent or self-illuminating usually are not mentioned.

Let me expand a point I made above. There are two major traditions among historians, one accurate and one strikingly inconsistent with the key passages in the very well known textual sources. Many of the historians and philosophers of science I have talked with have known that the source of planetary light was an open question and were surprised to hear that much of the scholarly literature indicated the contrary; others, particularly philosophers, have been surprised to hear that Copernican theory did not predict the phases of Venus.

Just as the inaccurate tradition has a long history, so the accurate tradition has a long history. A nice example of how these two traditions have so peacefully co-existed over the centuries comes from a recent edition of Descartes' *Principles of Philosophy* of 1644. Descartes accurately, if briefly, reports Galileo's telescopic discoveries:

> 10. That the moon and other Planets derive their light from the Sun.
>
> ... from the fact that the Moon shines only on the side facing the Sun, we must conclude that it has no light of its own and merely reflects toward our eyes the rays which it has received from the Sun. The use of the telescope {recently} revealed the same thing to be true of Venus. ...
>
> 16. That Ptolemy's hypothesis is not in conformity with appearances.
>
> Ptolemy devised the first of these [hypotheses about the planets]; but, as it is already commonly rejected by all Philosophers, because it is contrary to several {recent} observations (especially to the change in light, similar to hose which occur on the Moon which we observe in Venus), I shall not speak further of it here.

Here is the recent editorial footnote to Descartes' point 16:

> ... In the Ptolemaic system, ... an observer on the Earth should never see more than a small crescent of Venus illuminated. Through the telescope, however, Venus sometimes appears as a large crescent and sometimes as a much smaller almost circular disk. *The effect was predicted by Copernicus* and shows that Venus, at least, must orbit the Sun.[51]

In regard to one aspect of the faulty tradition, such mutually oblivious co-existence has been broken down at least twice. In 1847, the logician August De

Morgan wrote a rather scathing attack on the claim that Copernicus himself had predicted the phases:

> The common story is, that Copernicus, on being opposed by the argument that Mercury and Venus did not show phases, answered that the phases would be discovered some day. The first place in which I find this story is in Keill's Lectures. It is also given by Dr. Smith, in his well-known Treatise on Optics, by Bailly, and by others. But I cannot find it mentioned either by Melchior Adam or Gassendi, in their biographies of Copernicus; nor by Rheticus, in his celebrated *Narratio*, descriptive of the system of Copernicus; nor by Kepler, nor by Riccioli, in their collections of arguments for and against the heliocentric theory; nor by Galileo, when announcing and commenting on the discovery of the phases; and, what is most to the purpose, Müller, in his excellent edition of the great work of Copernicus, when referring to the discovery of the phases of Venus, as made, since, and unknown to Copernicus, does not say a word on any prediction or opinion of the latter.
>
> This story may then be rejected, as the gossip of a time posterior to Copernicus.[52]

De Morgan's attack had some effect for a few years. No less a personage than Alexander von Humbolt said that, as a result of De Morgan's 'strict examination', the story of Copernicus' prediction 'has become altogether doubtful'.[53] In her popularisation of astronomy a decade later, Ward continues the critique citing De Morgan's article in the *Penny Cyclopedia*:

> There is a beautiful tradition respecting the phases of Venus, which appears, however, to partake rather of the nature of an allegory or illustration, than of a historical anecdote. It is said that when Copernicus announced his theory of the solar system, it was objected that were his theory true, Venus ought at certain positions of its orbit, to exhibit the various forms of the Moon. The invention of the telescope had not then been dreamt of; but it is said that Copernicus, in a fine spirit of prophecy answered, that should men ever see Venus better, they would discern these phases. No mention is made of this story by Galileo, who discovred the varying forms of Venus in the year 1611, with the aid of the telescope; or by Gassendi, the biographer of Copernicus. Copernicus, indeed, was not spared to answer any objections to his system, as he barely lived to lay his hand upon a copy of his own work, and never opened it.

> In that work he is not altogether silent on the subject, and in fact, proposes a different theory to account for the circumstances of Mercury and Venus always appearing circular; namely, that these near neighbors of the Sun are possibly self-luminous, or completely saturated with the solar rays.[54]

But, as the quotations above show De Morgan failed and Ward failed. Within a few years, we find Chambers' reporting De Morgan's view, only to cast doubt on it:

> It was one of the objections urged to Copernicus against his theory of the solar system that if it were true then the inferior planets ought to exhibit phases. He is said to have answered that if ever men obtained the power of seeing them more distinctly, they would be found to do so. Prof. De Morgan believes the anecdote to be apocryphal. ... But 'se non é vero, é ben trovato'.[55]

'*Se non é vero, é ben trovato*' means 'If it is not true, it should be'; or, more literally, 'If it is not true, it is a good find/invention'. I think Chambers is on to something psychologically deep, that many feel that Copernicus *should have* predicted the phases. This feeling may partially explain the accounts given by philosophers, to whom we shall soon turn.

Over a century later, in 1965, Edward Rosen published 'Copernicus on the Phases and the Light of the Planets' in the Polish journal, *Organon*. Dealing with the biographical question of Copernicus' 'prediction', it was rarely cited by anyone but Rosen, and has had little apparent impact.

3. THAT A LOT OF PHILOSOPHERS HAVE SAID THAT COPERNICAN THEORY PREDICTED THE PHASES OF VENUS

On the whole, philosophers have done worse than historians. The claim that the Copernican theory per *se predicted* Venus' phases is a popular philosophical example. It is often cited as evidence for the claim of the historical and evidential importance of predicting 'novel facts'. Only a few philosophers mention that the origin of planetary light was an open question or that Galileo's observations of the phases resolved two issues, simple though the basic facts are. Clavelin[56] is one of that elite group and Andersson nicely states the basic facts: 'Both theories predicted that the planets would show phases *if* they radiated reflected light. But they predicted different phases'.[57]

Here are some philosophers on the phases of Venus:

(1) W. Whewell, 1857:

> *It had always been a formidable objection* to the Copernican theory that this appearance of the planets had not been observed. The author of that theory had endeavoured to account for this, by supposing that the rays of the sun passed freely through the body of the planet; and Galileo takes occasion to praise him for not being deterred from adopting the system which on the whole appeared to agree best with the phenomena, by meeting with some appearances which it did not enable him to explain. Yet while the fate of the theory was yet undecided, this could not but be looked upon as a weak point in its defences.[58]

(2) B. Russell, 1961:

> [Galileo] observed the phases of Venus, which *Copernicus knew to be implied by his theory*, but which the naked eye was unable to perceive.[59]

(3) K. Popper, 1963:

> We have no reason to regard the new theory as better than the old theory—to believe that it is nearer to the truth—until we have derived from the new theory *new predictions* which were unobtainable from the old theory (the phases of Venus, the perturbations, the mass-energy equation) and until we have found that these new predictions were successful.[60]

(4) K. Popper, 1963:

> ... *if Copernicus was right the inner planets (and they alone) should, when observed from the earth, show phases like the moon*; and Galileo had seen in his telescope the phases of Venus.[61]

(5) T. Kuhn, 1969:

> Sometimes the looser practice that characterizes extraordinary research will produce a candidate for paradigm that initially helps not at all with the problems that have evoked crisis. When that occurs, evidence must be drawn from other parts of the field as it often is anyway. In those other areas particularly persuasive arguments can be developed if the new paradigm permits *the prediction of phenomena that had been entirely unsuspected while the old one prevailed.*
>
> *Copernican theory, for example, suggested that planets should be like the earth, that Venus should show phases* and that the universe must be vastly larger than had previously been supposed. As a result, when sixty years after his death the telescope suddenly displayed mountains on the moon, the phases of Venus, and an immense number of previously unsuspected stars, those observations brought the new theory a great many converts, particularly among non-astronomers.[62]

(6) W. Shea, 1972:

> We can understand why philosophers, who grappled with the difficult problem of motion, and practical astronomers, who were interested in computing accurate tables, became impatient with the adolescent outcries of the young rebels. All they could see in their protest was the immature and self-conscious revolt of a rising generation against the authority of their elders. They were convinced that in time these young men would come to see the light, especially after they had read Tycho Brahe's *Astronomical Letters* in which the latest Copernican claims had been put to rest.
>
> Tycho Brahe took care to summarise his arguments in non-technical language, and his work rapidly became a convenient handbook for the non-specialist who wished to quote an expert when attacking an opponent who believed in the motion of the earth. To *the difficulties already recognized by Copernicus, that the heliocentric theory would require Venus and Mercury to show phases like the moon,* ... Tycho added further objections.[63]

(7) Lakatos, 1978:

> For the inductivist and the falsificationist it does not really matter whether the discovery of a fact preceded or followed a theory: only their logical relation is decisive. The 'irrational' impact of the historical coincidence, that a theory

happened to have *anticipated* a factual discovery, has no internal significance. Such anticipations constitute 'not proof but [mere] propaganda'.[Lakatos' Footnote #1]

[Here is Lakatos' Footnote #1:]

This is Kuhn's comment on Galileo's successful *pre*diction of the phases of Venus. Like Mill and Keynes before him, Kuhn cannot understand why the historic order of theory and evidence should count, and he cannot see the importance of the fact that Copernicans *pre*dicted the phases of Venus, while the Tychonians only explained them by *post hoc* adjustments. Indeed, since he does not see the importance of the fact, he does not even care to mention it.[64]

[Later, Lakatos' text continues:]

A favourite hunting ground of externalists has been the related problem of why so much importance is attached to—and energy expended on—*priority disputes*. This can be explained only *externally* by the inductivist, the naive falsificationist, or the conventionalist; but in the light of the methodology of research programmes some priority disputes are vital *internal* problems, since in this methodology *it becomes all-important for rational appraisal which programme was first in anticipating a novel fact and which fitted in the by now old fact only later.* Some priority disputes can be explained by rational interest and not simply by vanity and greed for fame. It then becomes important that Tychonic theory, for instance, succeeded in explaining—only *post hoc*—the observed phases of, and the distance to, Venus which were originally precisely anticipated by Copernicans.[65]

(8) I. Lakatos and E. Zahar 1978:

Copernicus's programme was certainly theoretically progressive. It anticipated novel facts never observed before. It anticipated the phases of Venus. It also predicted stellar parallax, though this was very much a qualitative prediction, because Copernicus had no idea of the size of the planetary system. ...

But the phases of Venus prediction was not corroborated until 1616. ... The Copernican system may have constituted heuristic progress within the Platonic tradition, it may have been theoretically progressive but it had no novel *facts* to its credit until 1616.[66]

(9) A. Chalmers, 1982:

Later, Galileo was to confirm that Venus had phases like the moon, *as Copernicus had predicted* but which clashed with Ptolemy's system.[67]

(10) M. Finocchiaro, 1989:

... one has to look at the counter-arguments and there were plenty of them.
...
The appearance of the planet Venus was the basis of another objection. For if the Copernican system were correct, then this planet should exhibit phases similar to those of the moon but with a different period; however, none were visible (before the telescope). The reason why Venus would have to show such phases stems from the fact that in the Copernican system it is the second planet ...[68]

(11) G. Andersson, 1991:

In order to falsify the Ptolemaic theory, it is sufficient to show that Venus near superior conjunction is almost 'full', *as predicted by Copernicus. According to the Ptolemaic theory Venus should only show a crescent at that time.*[69]

(12) R. Giere, 1991:

In the Ptolemaic picture, ... when viewed from the earth, Venus is mostly dark since it is always illuminated from behind by the sun.
Step 4. There are likewise two predictions: (1) the Ptolemaic prediction that Venus can never be seen fully illuminated ...; and (2) the Copernican prediction that Venus can be seen going through a complete set of phases, including being fully illuminated ...[70]

Wallace's *Galileo's Logic of Discovery and Proof*[71] is a slightly different case. It spends about four pages analysing Galileo's logic in the phases of Venus argument without mentioning that the origin of Venus' light was an open question before Galileo's telescope, or that this very argument resolved it.

4. THAT THE ABSENCE OF VISIBLE PHASES OF VENUS WAS NOT A 'FORMIDABLE OBJECTION' TO COPERNICAN THEORY

Above, I wrote that 'As far as I have been able to discover' Galileo's one-time student Castelli was the only contemporary who held that Copernican theory required/predicted that Venus has phases. But my argument above requires a much stronger claim than just the autobiographical one that *I know of* no one other than Castelli. My argument that the tradition is wrong on this point requires that (virtually) no one else, critic or supporter, claimed that Copernican theory predicted the phases; and further, that the absence of phases was not seen as an objection to Copernican theory and thus as evidence for Ptolemaic or Tychonic theory.

As we've seen, my claim runs contrary to what many major historians of astronomy have claimed. So how can I justify it? I haven't done a thorough review of the archives—a review that I do not have the temperament, the archives, the time, the linguistic sophistication, or the historical knowledge to do. I am not a Westman or a Swerdlow. Nor do I know of anyone who has done such a review. I will discuss the more general problem below, but for the moment, here are my reasons for holding firstly that Castelli in 1610 was (virtually) the only one who held that Copernicanism predicted a full set of phases for Venus, and secondly that the absence of visible phases was not an objection raised against Copernican theory.

1. Not only are Copernicus' words on this topic unambiguous, but they clearly explain why his theory does not predict phases.

2. The absence of visible phases could not be used as an argument for the superiority of Ptolemaic over Copernican theory. Ptolemaic theory placed Venus between the earth and the sun. So situated, if Venus were dark and opaque, it would always have a crescent shape facing the earth. This was a (maybe *the*) major source of the tradition that the planets were not dark and opaque. It would not have been easy to use the apparent absence of the phases to attack Copernican theory without thereby attacking Ptolemaic and Tychonic theory as well. An astronomical instrumentalist such as Osiander in his anonymous preface to *De revolutionibus* could have used their absence for an attack on realist interpretations of *all* astronomical theories. Osiander did not do this—in

fact, the argument in the preface actually presupposes that Venus is always fully illuminated when viewed from the earth.

3. The weight of authority of so many historians of astronomy is considerably weakened by the fact that so many also say that Copernican theory or Copernicus himself 'predicted', 'required' or 'deduced' the moon-like phases. There is no evidence for this claim and it is inconsistent with Copernicus' comments in *De revolutionibus*.

4. Although many knowledgeable historians of astronomy wrote that the absence of visible phases was 'a major (or formidable or powerful) objection' to Copernicanism, none of these historians names anyone who actually used this 'favourite weapon' against Copernicans.

5. Both the story of Copernicus predicting Venus' phases and of this being a major objection to Copernican theory is over two centuries old. And yet it is only recently that historians of astronomy have started to examine views other than those of such major figures as Copernicus, Brahe, etc., carefully.

6. If in Galileo's day it had been widely objected that Copernican theory incorrectly predicted the phases, then it would not have been necessary for Castelli to 'finally hit' on the phases, 'after various thoughts passed through my mind'. Castelli appears to be presenting the phases to Galileo as a novel idea, not as an opportunity to reply to a well-known difficulty for Copernicanism.

7. The highly sophisticated and knowledgeable Kepler found the phases surprising: 'Unexpected by me in any way was your observation'. These are not the words of a person who thought either that Copernican theory did predict the phases or that many had objected to Copernican theory because the phases were not observed.

8. As we shall see, there is a plausible account of how the story of the 'powerful objection' was introduced into the history of astronomy and why it continues, even if the absence of phases was not raised as an objection to Copernican theory.

9. Crucially, no historian I have corresponded or talked with named anyone who raised the absent phases as an objection to Copernican theory. Most have thought there were no such people, although few were certain. No historian has named anyone except Castelli who held that Copernican theory predicted the phases. This consensus has been very reassuring to me.[72]

5. WHAT IS GOING ON HERE?

Having misinterpreted more than my fair share of arguments, I certainly do not want to claim that it is easy to get the history of science right. Still, when the well-known primary texts are so clear and when so many first-rate scholars misinterpret them so strikingly, with some chutzpah one can propose that scholars are falling prey to some cognitive bias. Or, more precisely, biases. For just as it would be surprising if there was only one way scientists go astray, so it would be surprising if there were only one way that historians of science go astray. It certainly would be surprising if the same source produced 'Copernican theory predicts that Venus has phases like the moon' as well as White's Copernicus replying to his critics '*You are right; I know not what to say; but God is good, and will in time find an answer to this objection*' and the claim that the objection was widespread.

I suspect that, in addition to the all too self-explanatory human tendency to err, the full source of this particular history has at least three mutually reinforcing components which we might call: 'Copernicanism Does Predict the 'Phases of Venus'; 'Galileo Galilei, Master Rhetorician'; and 'If It's Good Enough for Dreyer, It's Good Enough for me'.

6. 'COPERNICAN THEORY *DOES* PREDICT THE PHASES OF VENUS'

One can argue that, despite all I have said above, Copernican theory really does predict the phases. Consider this proposal:

> 'Predict' is used in two ways: biographical and logical. The biographical sense is when some individual says that a certain phenomenon will be observed. Thus Castelli biographically predicted the phases of Venus. A theory logically

> predicts when it (plus unproblematic background assumptions) entails the reality of a phenomenon. Thus a theory can logically predict a phenomenon even though no one biographically predicted that phenomenon. Logical prediction is what people mean when they say, for example, 'General Relativity predicts the precession of the perihelion of Mercury'—even though the precession of Mercury's perihelion was well known before Einstein's theory was developed, Einstein's theory plus unproblematic background assumptions entails the precession. In the same way, Copernican theory (logically) predicts the phases of Venus.

There is something to this proposal. From the beginning, Copernican theory (plus unproblematic background assumptions) logically predicted the conditional that *if* Venus is dark and opaque, its phases would run a full cycle like the moon's. And after Galileo's telescope showed that there were phases, Copernican theory unconditionally logically predicted that there was a full cycle of phases.

At best, however, this proposal can only explain a few of the quotations above. It does not explain why scholars so often said that Copernicus predicted the phases. It does not explain why so many historians claimed that the absence of phases was an objection to Copernicanism. It doesn't explain Kuhn's cosmologically more radical Copernicans predicting a detail 'with precision' or the relevance of the fact that there are 'few phrases more annoying or more effective than 'I told you so'.' And so on …

Further, this proposal ignores that in some situations, the locution 'Copernican theory predicts the phases of Venus' is seriously misleading, if not false. A theory logically predicts a phenomenon if it *plus* more-or-less unproblematic background facts entails the phenomenon. The critical element here is the phrase 'unproblematic background facts'. These change over time as scientists learn things. Before 1610, the opacity of Venus was not an unproblematic background fact, after 1610 it was. Copernican theory plus unproblematic background facts did not entail the phases of Venus before 1610 and did entail them after 1610. So, one might say that Copernican theory did not logically predict the phases before 1610 and did logically predict them after 1610. Certainly Copernican theory now logically predicts them.

In some cases, marking the distinction before 1610 and after 1610 is the difference between truth and falsity. Consider Gingerich's quotation from and paraphrase of a 1615 letter of Cardinal Bellarmine:

> 'I say that if there were a true demonstration that the sun was in the center of the universe, then it would be necessary to be careful in explaining the Scriptures that seemed contrary, and we should rather have to say that we do not understand them than to say that something is false. But I do not think there is any such demonstration, since none has been shown to me. *To demonstrate that the appearances are saved by assuming the sun at the center and the earth in the heavens is not the same thing as to demonstrate that in fact the sun is in the center and the earth in the heavens*. I believe that the first demonstration may exist, but I have very grave doubts about the second'.
>
> In other words, Cardinal Bellarmine is saying that *although the Copernican system predicted the phases of Venus*, this did not necessarily imply the converse. ...
>
> ... *[Galileo] knew that the Copernican system not only predicted the phases of Venus*, but that as a model it explained many other things.[73]

There are two things to note here. First, note the naturalness of Gingerich's move from Bellarmine's words 'to demonstrate' to Gingerich's word 'predicted'. We shall return to this when discussing Psychlogical Predictivism below.

Second, consider the truth of such statements as 'the Copernican system ... predicted the phases of Venus'. My initial attempt to assess such statements was a rather silly legalistic one. I attempted to determine whether the statement was arguably true by distinguishing whether the author was discussing pre- or post-1610, whether the verb 'predict' was present or past tense, etc. Gingerich is clearly discussing post-1610 when the source of Venus' light was known. So, Gingerich is literally correct in saying that Galileo knew that Copernican theory (post-1610) predicted the phases.

But, I soon realised that literal truth is not the key issue. After all, a sentence can be literally true on one possible reading and still seriously misleading. This paper is concerned with whether, in evaluating theories of science, philosophers of science can rely on historians. For philosophers of science, the literally true but seriously misleading is little better than the literally false. And statements to the effect that Galileo in 1615 'knew that Copernican theory predicted the phases of Venus' can seriously mislead unless accompanied by some qualifying statements such as: 'However, he also knew that before 1610, it only predicted 'that *if* Venus was dark ...' or 'Such prediction of the theory, however, depends on the opacity of Venus, which was not established until ...'. Without such a qualifying statement, the bald statement 'Copernican theory predicted ...', especially when combined with the standard diagrams, will almost inevitably lead even the most conscientious reader to believe that it always predicted the phases. In lecturing to academics and students, I have repeatedly put the standard diagram on the blackboard, given the standard explanation of how Copernican 'theory predicts the phases, announced that my explanation contained an assumption that was highly problematic in Copernicus' lifetime, and asked my audience to guess what the problematic assumption was. No one has ever guessed it.

'Predicts' when used to mean 'entails' is, so to speak, a tenseless verb. And read tenselessly, 'Copernican theory predicts the phases of Venus' is bound to mislead, for it will generally be read as saying that Copernican theory now predicts the phases and always has done so. Having read it this way, the conscientious reader will often come to believe that someone, possibly Copernicus, biographically predicted them. And once one believes that Copernicus predicted the phases, it is a natural if not inevitable step to infer that opponents of Copernicus must have raised the apparent absence of the phases as an objection to the theory. The literal truth that Copernican theory (now) predicts the phases of Venus can, unless qualified, naturally lead to the history's being seriously misconceived.

And, as philosophers are susceptible to being misled in this way, the literal truth of one interpretation of 'Copernican theory predicts the phases of Venus' should not reassure philosophers desiring to use historians' accounts to test their philosophical proposals. But the semantic ambiguity of 'predicts' is not the only possible source of the faulty tradition.

7. GALILEO GALILEI, MASTER RHETORICIAN

Because of the considerable impact of his writings, I initially expected that Galileo's exuberances could largely explain the faulty tradition. Basically, I was wrong, although this can explain one aspect of it.

In the *Dialogue*, the absence of phases appears in a list of three 'difficulties' for Copernican theory that were resolved by the telescope. And this list is presented in dramatic enough language. It begins with the suitably theatrical, 'A while ago I sketched for you an outline of the Copernican system, against the truth of which the planet Mars launches a ferocious attack', passes through 'O Nicholas Copernicus, what a pleasure it would have been for you to see this part of your system confirmed by so clear an experiment!', and more-or-less ends with 'The illnesses are in Ptolemy, and the cures for them in Copernicus'. Galileo's presentation is not, by any stretch of the imagination, a disinterested rational evaluation of the epistemic and historical situations.

Consider, for example, the first item in Galileo's list—that if Copernicus is right, the size of Mars' disc should vary sixty-fold from minimum to maximum but that to the naked eye, Mars' disc varies only four or five fold. Galileo then explains that the telescope shows that Mars' disc has Copernicus' predicted 60-fold variation. The telescopic observations are presented as a triumph for Copernicus and a devastating blow to Ptolemy. Brahe is not mentioned at all. However, as is (and was) well known, the problem of the variation of the naked-eye size of Mars' disc was equally a difficulty (or non-difficulty) for Ptolemaic and Brahean theory as for Copernican.[74] In fact, Kepler had already publicly taken Galileo to task for mis-stating these issues:

> In the same place [in Galileo's *Assayer*], Galileo denies that the Ptolemaic hypothesis could be refuted by Tycho, Copernicus, or others, and says that it was refuted only by Galileo through the use of the telescope for observation of the variation of the discs of Mars and Venus, the latter being forty times and the former sixty times larger at perigee than at apogee; for it is in this way that the arrangement of their orbits around the sun is proved.
>
> Nothing is more valuable than that observation of yours, Galileo; nothing is more advantageous for the advancement of astronomy. Yet, with your indulgence, if I may state what I believe, it seems to me that you would be well advised to collect those thoughts of yours that go wandering from the course of reason and memory in that vastness of many interrelated things. ...Your own observation of the discs confirms the proportion for the eccentric to the epicycle in Ptolemy, as it does the orbit of the sun in Tycho or of the *orbus magnus* in Copernicus...[75]

It appears likely that in his *Dialogue* list of contributions the telescope made to Copernican theory, Galileo knowingly mis-stated the facts. Still, deliberate or not, Galileo's mis-statements cannot explain the tradition of reporting that Copernicus (and/or Copernican theory) predicts the phases. In fact, this tradition is *inconsistent* with what Galileo clearly says:

> Add to these another difficulty [for the Copernican system]; for *if* the body of Venus is intrinsically dark, and like the moon it shines only by illumination from the sun, which seems reasonable, then it ought to appear horned when it is beneath the sun ... —a phenomenon which does not make itself evident in Venus. For that reason, *Copernicus declared that Venus was either luminous in itself or that its substance was such that it could drink in the solar light and transmit this through its entire thickness that it might look resplendent to us*. In this manner Copernicus pardoned Venus its unchanging shape; but he said nothing about its small variation in size; much less of the requirements of Mars.[76]

Galileo's faulty history here can explain the claim that the absence of phases was raised as a considerable difficulty for Copernican theory—although Galileo only says it was a difficulty and doesn't directly state that anyone raised it.

But, Galileo's history cannot explain the claim that Copernicus and/or Copernican theory predicted the phases of Venus. Galileo here says that Copernicus declared that Venus was self-luminous or translucent. Thus, Copernicus predicted that there should not be phases. Galileo clearly says that Copernicus got it wrong and Galileo got it right. The tradition says that both Copernicus and Galileo got it right.

So, we must leave the historical record behind us and turn to the speculative psychology of historians.

8. WHIG HISTORY ABIDETH FOREVER OR THE-KNEW-IT-ALL-ALONG-EFFECT

Our quest is for a psychological explanation of why so many highly competent historians have implicitly presupposed that Venus was known to be dark and opaque before the telescope, when the historical record is so clear. While we will never know for sure, I strongly suspect that the fact that Venus subsequently was shown to be dark and opaque is crucial. Had Venus proven to be self-luminous, I find it hard to imagine that many historians or philosophers would have reported that Copernicus and the Copernican theory *incorrectly* predicted that Venus had phases. In fact, I suspect that had Venus proved to be self-luminous, Galileo's 'testimony' about Copernicus' views often would have been taken as definitive and Copernicus often would have been given credit for a successful prediction—that Venus does not have phases.

I suspect that we have here an example of what psychologists call the 'Knew-It-All-Along-Effect' and historians call 'Whig History.' It is not limited to historians of astronomy. Consider Elkana and Goodfield's analysis of 'the current view' among historians that Harvey predicted the existence of capillaries:

> When a problem in science is clearly formulated by a scientist but not solved, historians become subject to an understandable temptation: to attribute to the scientist his anticipation, in some sense or other, of the solution that finally turned up. The impression is that once the problem has been specified the form of the subsequent solution must *necessarily* have followed.
>
> This has been the case with Harvey and the problem of the 'capillaries.' Realising that he could not see the anatomical connections between the arteries and veins, historians nevertheless write as though he 'postulated,' 'guessed,' or 'posited' the existence of capillaries and was in fact *bound to do so*.[77]

What, then, is the psychological mechanism that tempts many historians and philosophers to find such non-existent necessities in the history of science? I suspect that it is not directly a desire to attribute to Copernicus and other Heroes all the great achievements—the classical type of Whig history. Instead I suspect it is inadvertent Whig history, the by-product of that constant nemesis of the historian—anachronism.

Actually, we often are remarkably good at avoiding anachronisms: If you tell me that in January 1610 Galileo wanted desperately to contact Kepler, I don't ask, 'Why didn't he just telephone?' But in subtler matters, we sometimes fail. We naturally tend to attribute to others what we now know to be true—unless we are vividly aware that it was not known. We find it hard to imagine possible alternatives that were viable then but aren't now. When I first heard that Harvey might not have predicted the capillaries, I could not imagine any alternative way of getting blood from the arteries to the veins—even though Harvey had several alternative proposals. This tendency to fail to imagine long-dead possible alternatives will produce 'predictions' and thus, however inadvertently, Whig History. But this is not the only psychological source of such anachronistic reports of predictions.

9. PSYCHOLOGICAL PREDICTIVISM

I also suspect that many of our historians were implicitly thinking along these lines: 'Copernican theory *must predict* the phases of Venus—why else would the phases have been such excellent evidence for it?' I've elsewhere called this form of argument 'Psychological Predictivism'—the visceral sense that if a discovered phenomenon is good evidence for a theory, the theory must predict that phenomenon.

I think that most of us have strong tendencies to Psychological Predictivism. Here is a test for your tendency for this dread cognitive disfunction:

> Suppose that you have a friend Mary who has a non-probabilistic hypothesis that you only partially understand. You ask Mary 'Does your hypothesis predict phenomenon P?' and Mary says 'No; on my hypothesis, P is unlikely.' The next day, Mary tells you, 'Wonderful news—P has been observed! I'm really surprised! Still, the observation of P means that my hypothesis is almost certainly right.'

Even as I write Mary's story, my immediate gut reaction is the MARY CAN'T DO THAT! She can't both predict that something probably won't happen and then, when it does, conclude that it strongly supports her theory. Insofar as you share my reaction, you are a Psychological Predictivist, regardless of your sophisticated philosophical views on evidential support. In this reaction you are in good philosophical company, for in more sophisticated forms Logical Predictivism has permeated the philosophy of science. The historical influence of the view of Hempel, Popper, and Lakatos is a tribute to the power of Psychological Predictivism.

But, however psychologically attractive it is, Psychological Predictivism leads to serious misunderstandings of the history of science. It does not distinguish predictions from, among other things, conditional predictions. In our case, it does not distinguish between Copernican theory predicting Venus has phases and Copernican theory predicting that *if* Venus is dark and opaque it will have phases.

10. 'IF IT'S GOOD ENOUGH FOR DREYER, IT'S GOOD ENOUGH FOR ME'

There is an almost universal tendency for people (and this includes even scholars of the highest calibre) to sometimes just plain not rethink what they are writing. It would be mad to write a history of astronomy without consulting previous histories. And when one reads that '[Galileo] also confirmed Copernicus' deduction that Venus, because of the position he held it to have inside the earth's orbit, would have phases like the moon' (Crombie) and '... the discovery of the phases of Venus deprived the opponents of Copernicus of a favourite weapon' (Dreyer), it is natural enough to conclude that Copernicus did deduce that Venus had phases like the moon and that their apparent absence was for many years a favourite weapon against Copernicanism. Unless there was a specific reason to think through the issue, why wouldn't one semi-automatically assume that Crombie and Dreyer were right, especially when Kuhn, Drake, and so many others apparently say roughly the same thing?

That Crombie's statement is overtly inconsistent with what Copernicus and Galileo said is obvious, once it is brought to one's attention. But it is very easy to not notice such inconsistencies, unless they are brought to one's attention. This truism was recently brought home to me when I much belatedly noticed that the way I spell my first name (Neil) is inconsistent with the childhood ditty I had relied on for decades—'*I* before *E* except after *C* or when it sounds like 'A' as in 'neighbour' or 'weigh'. For well over a third of a century, I had failed to notice that the very first word I could spell contradicted my spelling rule.

Dreyer's claim about the absence of the phases being a 'favourite weapon' of Copernicus' opponents poses subtler problems. Since Dreyer does not tell the names of those opponents of Copernicanism for whom the absent phases was a favourite weapon, a historian simply can't double-check the cited texts without considerable effort. Without some reason to suspect otherwise, why would a sane historian go to that effort? Why shouldn't a historian believe that Dreyer (and North and Finocchiaro and ...) had specific opponents and incidents in mind—people and incidents that they just didn't bother to name? Especially since Galileo might appear to have said the same thing? And so why shouldn't the later historian, in good faith, paraphrase their claims? Why should a historian doubt Dreyer on this issue, given the manifest high scholarly standards and critical intelligence of Dreyer's book—especially if, as in this case, there does not seem to have been any critical literature actually showing that it was not a weapon, and it seems plausible that it was.

Obviously, the problem of scholars relying on plausible but incorrect sources is not limited to the history of astronomy. For a similar history, see Jeffrey Russell's *Inventing the Flat Earth: Columbus and Modern Historians*, which recounts the origin and development of the nineteenth-century canard that in the Middle Ages most educated people believed that the earth was flat, whereas, in fact:

> In the first fifteen centuries of the Christian era, five writers seem to have denied the globe, and a few others were ambiguous and uninterested in the question. But nearly unanimous scholarly opinion pronounced the earth spherical, and by the fifteenth century all doubt had disappeared.[78]

Russell plausibly traces the nineteenth-century invention of the Flat Earth to a few highly creative historians pursuing financial (Washington Irving) or political (Andrew White) goals, and to subsequent historians passing on the story, in no small part because it contained the potential for a fine bit of melodrama—Columbus facing the Inquisition of Bible-besotted fundamentalists who were hell bent on denying the scientific evidence because of a fanatical Christianity.

But, the platitude that 'historians should always double-check sources,' while undoubtedly sensible, is often practically impossible to apply. Consider the slightly embarrassing example of my quoting Russell's claim about the rarity of medieval flat-earthers. I believe him. But I have not checked the original sources to see if he is right. Should I attempt to check him by a thorough review of texts from the Patristic to fifteenth-century geography texts? Even I were so inclined, I have neither the linguistic expertise (Hebrew, Greek, Latin for starters) nor access to the archives. My feeling is 'if it is good enough for Russell, it is good enough for me'. And, for all practical purposes, I have no choice in the matter. Virtually all historians are in this situation from time to time. Certainly those writing more general histories have no choice.

And philosophers relying on historians of science have little choice. Consider the amount of research Franssen had to do before he could answer 'No' to the question 'Did King Alfonso of Castile Really Want to Advise God Against the Ptolemaic System?' The story is that King Alfonso X of Castile (1252-1284) had reproached God for making the system of heavenly orbs unnecessarily complex and had claimed that the celestial system would have been better had he, Alfonso the Learned, been present at the Creation to give God the benefit of his advice. Kuhn, apparently relying on Dreyer, used this story as evidence for the claim that Ptolemaic theory was in a state of crisis before Copernicus. Was this unreasonable? Should Kuhn have done the research required to conclude that the story about Alfonso's astronomical advice to the Lord apparently first appeared in 1676, 'only when the controversy over the system of the heavens had largely subsided...'?[79]

11. THE USE OF THE HISTORY OF SCIENCE FOR THE PHILOSOPHY OF SCIENCE

One final, rather pessimistic reflection: Galileo's discovery of the moon-like phases of Venus is a remarkably simple episode in the history of science. There are no complex conceptual issues involved; there are no difficult problems of seriously ambiguous, defective, or rare primary sources. The issues and historical evidence are about as clear as one will ever find in the history of science. Here, if ever, we philosophers should find ourselves in the happy situation described by Gilbert and Sullivan's Don Alhambra: 'In the entire annals of our history there is absolutely no circumstances so entirely free from all manner of doubt of any kind whatever!'

But we find ourselves far from being in that happy situation. If much of the historiographical tradition for this simple case is so confused and downright false, what is happening when the history is complex? Consider the history of concepts of space and time, of physical laws, of DNA, of phlogiston, of the calculus, of ...? Of what value to philosophers can history be, if we can't be reasonably confident that even the best historians won't mislead? The proposal that philosophers should do the history themselves is a non-starter, if only because philosophers so rarely have either the temperament or the training to spend years working in archives. Further, the sad record of philosophers' accounts of the phases of Venus gives us little reason to think that philosophers would do nearly as well as historians. Their record only reinforces one's fear that many philosophers are strongly inclined to read history to verify their philosophical views.

Of what value, then, is the history of science for philosophy of science? We philosophers should use historians' accounts as heuristics, as a rich source of subtle and interesting hypothetical cases worthy of analysis, whether these cases are historically accurate or not. They are a source far richer, far subtler, than those we can imagine in our armchairs of an afternoon—the favourite alternative. To paraphrase The Hon. Mrs. Ward, philosophers should view the accounts of historians as partaking rather of the nature of allegories or illustrations, than of historical anecdotes.[80]

Department of History and Philosophy of Science, University of Melbourne

NOTES

[1] Cited by E. Rosen, 'Cosmology from Antiquity to 1850', in P. Weiner (ed.), *Dictionary of the History of Ideas: Studies of Selected Pivotal Ideas* (New York, 1973) 535-54, p. 546. The quotation from al-Biruni is from *Book of Instruction in the Elements of the Art of Astrology*, p. 67, para. 156.

[2] W. 'Arafat & H.J.J. Winter, 'The light of the stars—a short discourse by Ibn al-Haytham', *British Journal for the History of Science* 5 (1971) 282-8; and B.R. Goldstein, *The Astronomy of Levi ben Gerson (1288-1344)—A Critical Edition of Chapters 1-20 with Translation and Commentary* (New York, 1985) p. 8.

[3] R. Ariew, 'The phases of Venus before 1610', *Studies in History and Philosophy of Science* 18 (1987) 81-92, pp. 84-5, my italics.

[4] *Loc.cit.*

[5] E. Grant, *Planets, Stars and Orbits: The Medieval Cosmos, 1200-1687* (Cambridge, 1994) pp. 396, 398, and 404.

[6] Grant, *Planets, Star and Orbits*, pp. 392-418.

[7] E. Rosen, 'Copernicus on the Phases and the Light of the Planets', *Organon* 2 (1965) 61-78, p. 78.

[8] Nicholas Copernicus, *De revolutionibus orbium coelestium*, trans. E. Rosen, *On the Revolutions* (Baltimore, 1978) pp. 18-19.
[9] Ariew, 'The phases of Venus', pp. 83-4.
[10] G. Abetti, *The History of Astronomy*, trans. by B. Abetti from *Storia dell' astronomia* (London, 1954) p. 106.
[11] S. Drake, *Discoveries and Opinions of Galileo* (New York, 1957) pp. 93-4, my italics.
[12] Galileo Galilei, *Dialogue Concerning the Two Chief World Systems—Ptolemaic & Copernican*, trans. by Stillman Drake, 2nd ed. (Berkeley, 1967) p. 334, my italics.
[13] There are several explanations of Galileo's claim that Copernicus rejected the Venus is dark and opaque theory. First, Galileo was not a historian of science and he might have given a 'rational reconstruction' of Copernicus' views. The words I italicised in the excerpts from his letter on sunspots supports this. Indeed, Copernicus himself was forced to admit the possibility and even the necessity of one of these two ideas, as otherwise he could give no reason for Venus failing to appear horned when beneath the sun. As a matter of fact nothing else could be said before the telescope came along to show us that Venus is naturally and actually dark like the moon, and like the moon has phases. Rosen has proposed that Galileo relied on the faulty first edition of *De revolutionibus*, which at the key point has *'fatemur'* ['We say'] instead of the holograph manuscript's *'fatentur'* ['They say']. That is, instead of 'They [Plato's followers] say that in the planets there is no opacity like the moon's', the key passage reads 'We say that in the planets there is no opacity like the moon's' (Rosen, 'Copernicus on the Phases', pp. 73-4). In addition, Galileo might have also noted that Osiander's preface to *De revolutionibus* clearly presupposes that Venus is not dark and opaque. He then might reasonably enough have inferred that, whether the Preface was written by Copernicus or by someone close to him, this view of Venus must have been accepted by Copernicus.
[14] In S. Drake, 'Galileo, Kepler and phases of Venus', *Journal of the History of Astronomy* 15 (1984) 198-208, p. 204, my italics. *The prima facie* plausibility of Kepler's earlier view that Venus is self-illuminating can be seen by Ball's reflections well over two centuries later: 'The beautiful evening star is often such a very conspicuous object that it may seem difficult at first to realise that the body is not self-luminous' [R.S. Ball, *The Story of the Heavens* (London, 1891) pp. 140-141].
[15] Cf. *Op*. II p. 293 also Vol. I, p. 424.
[16] Drake, 'Galileo, Kepler and the phases of Venus', p. 206, my italics.
[17] Below, I will discuss how to evaluate such negative evidence.
[18] An astronomical instrumentalist (such as Osiander in his anonymous preface to *De revolutionibus*) *could* count their absence as evidence that no astronomical theory was literally true and so astronomical theories should only be viewed as calculating devices. But the absence of phases was not Osiander's argument. The situation is actually somewhat more complex than this, since there was one theory at the time that could reconcile the non-existence of the phases with a dark, opaque Venus—the theory that Copernicus attributed to the 'followers of Plato.' The theory is that Venus remained always on the far side of the sun from the earth, and so we always see Venus' illuminated side. Since this theory was rarely held in the 16th century, I will ignore it in what follows.
[19] T.P. Snow, *The Dynamic Universe*, 4th ed (St Paul, 1991) p. 71.
[20] R. Jastrow and M. Thompson, *Astronomy: Fundamentals and Frontiers*, 2nd ed. (New York, 1974) p. 412.
[21] M. Caspar, *Kepler*, trans. C. Doris Hellman (London, 1959) p. 200.
[22] I.B. Cohen, *The Birth of a New Physics* (London, 1961) pp. 81-3.
[23] Abetti, *The History of Astronomy*, p. 106.
[24] Cohen, *Birth of a New Physics*, pp. 82-3.
[25] Drake, 'Galileo, Kepler and the phases of Venus', pp. 203-4.
[26] For ease of presentation, I will write as though the truth of Ptolemaic theory requires the assumption that Mercury and Venus circle between the earth and the sun. One can, as Kepler did,

hold that this is a minor, easily changed aspect of Ptolemaic theory—that Ptolemaic theory was only committed to holding that the centres of Mercury's and Venus' epicycles are somewhere near the line originating at the stationery earth and going through the sun, in which case, the discovery of Venus phases would only resolve a minor technical issue within Ptolemaic astronomy and would not count as direct (indeed almost conclusive) evidence against it. Here is Kepler's admonition of Galileo in the Appendix to the *Hyperaspiste* (1625):

> This observation of yours that those planets circle around the sun does not refute the very distinguished system of Ptolemy nor add to it. Indeed, this observation of yours refutes not the Ptolemaic system but rather, I say, it refutes the traditions of the Ptolemaics regarding the least differences of planetary diameters, traditions resulting from the observation with the naked eye ... [S. Drake & C.D. O'Malley (trans. & eds), *The Controversy on the Comets of 1618* (Philadelphia, 1960) pp. 344-5]

While this makes the history more complex and interesting, it does not affect what I report below. Interestingly, the Keplerian interpretation of Ptolemy was rarely made by opponents of Copernicanism, who generally opted for Brahe's theory. It is not obvious why this should be so.

[27] For additional examples, see Rosen, 'Copernicus on the Phases'.

[28] J. Keill, *An Introduction to the True Astronomy or Astronomical Lectures read in the school of the University of Oxford* (London, 1721) Lecture XV, p. 163.

[29] R. Smith, *A Compleat System of Opticks* (Cambridge, 1738) p. 415, section 1050. Cited in Rosen, 'Copernicus on the Phases'.

[30] D. Diderot & J. d'Alembert (eds), *Encyclopédie ou dictionnaire raisonneé des sciences, des artes et des metiers* (Paris, 1772) 'Copernic' v. IV p. 160, my italics.

[31] Diderot & d'Alembert, *Encyclopédie* (Paris, 1774) 'Phases' v. XII p. 453. Unsigned, presumably d'Alembert, my italics.

[32] Bailly, *Histoire de l'astronomie moderne—de l'ecole d'Alexandrie jusqu'a l'epoque de MDCCXXX*, 3 vols. (Paris, 1785) v. 3 p. 93. My italics. Bailly's 3 volume *Histoire de l'astronomie moderne* was for many years a (maybe *the*) standard scholarly work.

[33] Adam Smith, *The Principles Which Lead and Direct Philosophical Enquiries; Illustrated by the History of Astronomy* (1795) IV.47 and IV.49, reprinted in W.P.D. Wightman, J. C. Bryce and I.S. Ross (eds), *Adam Smith: Essays on Philosophical Subjects* (Oxford, 1980), my italics. This is Volume III of 'The Glasgow Edition of the Works and Correspondence of Adam Smith'. Possibly, Adam Smith's claim that Copernicus had conjectured that all the planets were habitable comes from Kepler's dramatic inference from Galileo's discovery of the moons of Jupiter in his 1610 *Conversation with the Sidereal Messenger*:

> The conclusion is quite clear. Our moon exists for us on the earth, not for the other globes. Those four little moons exist for Jupiter, not for us. Each planet in turn, together with its occupants, is served by its own satellites. From this line of reason we deduce with the highest degree of probability that Jupiter is inhabited. [Galileo, *Siderius nuncius or the Sidereal Messenger*, trans. A.V. Helden (Chicago, 1610/1989) p. 99]

[34] R. Small, *An Account of the Astronomical Discoveries of Kepler: a reprinting of the 1804 text with a forward by Wiliam D. Stahlman* (Madison, 1963) p. 125, my italics.

[35] T. Morell, *Morell's Elements—Elements of the History of Philosophy of Science* (London, 1827) p. 369, my italics.

[36] J.P. Nichol, *Contemplations on the Solar System*, 3rd ed. (London, 1847) pp. 208-9.

[37] O. Lodge, *Pioneers of Science* (London, 1893) pp. 110-11.

[38] A.D. White, *A History of the Warfare of Science with Theology in Christendom* (London, 1895) vol. 1, pp. 129-30, my italics.

[39] J.J. Fahie, *Galileo, His Life and Work* (London, 1903) p. 125, my italics. As I'll discuss in some detail below, the tradition of holding that the apparent absence of phases was an 'objection' or 'difficulty' apparently began with Galileo. Fahie is somewhat unusual in reporting, albeit in an

oversimplified form, both parts of Galileo's account, saying that it was a 'formidable objection' to his system *and* giving Copernicus' response.

[40] J.L.E. Dreyer, (1906/1953) *History of the Planetary Systems from Thales to Kepler* (Cambridge, 1906) reprinted as *A History of Astronomy from Thales to Kepler* (New York, 1953) p. 414, my italics.

[41] Drake, *Discoveries and Opinions of Galileo*, pp. 74-5, my italics. This is from Drake's 1957 introduction to his translation of Galileo's 1613 'Letters on Sunspots'. It is rather unusual in that the apparent contradiction between the passage from *De revolutionibus* that Drake quotes in footnote 12 shows why such changes were not 'required' by Copernicus' theory. Further, on the cited pages of Drake's translation of the *Dialogue* we find Galileo writing, 'In this manner Copernicus pardoned Venus its unchanging shape ...' and on the sidebar: 'Venus, according to Copernicus, is either luminous by itself or is of transparent material'.

[42] T. Kuhn, *The Copernican Revolution: Planetary Astronomy in the Development of Western Thought* (Cambridge, Mass., 1957) pp. 222-4, my italics. Kuhn's translation skips over the following key words from I:10 of *De revolutionibus*: 'On the other hand, those who place Venus and Mercury below the Sun ... do not admit that these heavenly bodies have any opacity like the moon's. On the contrary, these shine either with their own light or with the sunlight absorbed throughout their bodies' (Copernicus, *De Revolutionibus*, p. 19. Omitted from Kuhn, *The Copernican Revolution*, p. 177).

[43] A. Panekoek, *A History of Astronomy* (New York, 1961/1989) p. 230, my italics.

[44] R. Westfall, 'Science and patronage: Galileo and the telescope', *Isis* 76 (1971) pp. 11-30, my italics. The legend to the accompanying diagram states: '... In the Ptolemaic system, Venus must always appear more or less crescent shaped'.

[45] S. Drake, 'Galileo, Kepler and the phases of Venus', in Gillispie, *Dictionary of Scientific Biography*, p. 241, my italics.

[46] Galileo, *Siderius nuncius*, trans. A. Van Helden, p. 108. This is the caption of a diagram showing the full set of phases for the Copernican system and a set of crescent phases for the Ptolemaic system. Immediately beneath this figure, Van Helden rightly reports (pp. 108-9):

> Now this progression of phases proved several things. First, Venus shines with light borrowed from the Sun, just as our Moon does. Second, Venus (and by implication Mercury) goes around the Sun.

[47] M. Segre, *In the Wake of Galileo* (New Brunswick, N.J., 1991) p. 19, my italics.

[48] O. Gingerich, *The Great Copernicus Chase and Other Adventures in Astronomical History* (Cambridge, Mass., 1992) pp. 100-1, my italics. Also, '... the Tychonic system also predicted them' on p. 111. On pp. 102-3, Gingerich clearly and accurately reports the situation facing Galileo:

> By September 1610, Venus, the brightest celestial object apart from the sun or moon, was well placed for observation in the western evening sky. ... As the figure shows however, hardly enough could have been said to warrant even a sketch. A natural conclusion might have been that tiny Venus was shining with its own light.
>
> ... [Galileo] knew that the round disklike appearance was incompatible with the Ptolemaic arrangement if Venus shone by reflected light, but until the phases began to appear he could not rule out the possibility that Venus shone by its own light or lay always beyond the sun.

Below, I will discuss the implications of scholars such as Van Helden and Gingerich sometimes writing that Copernican theory 'predicted' the phases even though they fully understand the debate over Venus' luminosity.

[49] J. North, *The Fontana History of Astronomy and Cosmology* (London, 1994) p. 336.

[50] A.C. Crombie, *Augustine to Galileo* (London, 1952) p. 318; see also p. 48 and the revised edition of this book (Harmondsworth, 1959) p. 210.

[51] R. Descartes, *Principles of Philosophy*, trans., with explanatory notes, by V.R. Miller and R.P. Miller. *Synthese Historical Library*, Volume 24 (Dordrecht, 1983) pp. 87-90, my italics.

[52] A. De Morgan, 'On the opinion of Copernicus with respect to the light of the planets', *Proceedings of Learned Societies*, Royal Astronomical Society, LXXVII, June 11 (1847) 528-31, p. 528.
[53] A. von Humbolt, *Kosmos* (Berlin, 1850) Vol. 3, p. 538. Cited in Rosen, 'Copernicus on the Phases', p. 62.
[54] The Hon. Mrs.' Ward, *Telescope Teachings* (London, 1859) pp. 89-90.
[55] G.F. Chambers, *A Handbook of Descriptive and Practical Astronomy* (Oxford, 1889) p. 105, my italics.
[56] M. Clavelin, *The Natural Philosophy of Galileo: Essay on the Origins and Formation of Classical Mechanics*, trans. by A.J. Pomerans (Cambridge, Mass., 1974) p. 200.
[57] G. Andersson, *Criticism and the History of Science: Kuhn's, Lakatos's and Feyerabend's Criticisms of Critical Rationalism* (Dordrecht,.1994) p. 121, my italics.
[58] W. Whewell, *History of the Inductive Sciences from the Earliest to the Present Time*, 3rd ed., in 3 vols (London, 1857) v.1 p. 302, my italics. Whewell is unusual among both philosophers and historians in at least roughly paraphrasing Galileo's incorrect report of Copernicus' position. Few philosophers do so well.
[59] B. Russell, *History of Western Philosophy and Its Connection with Political and Social Circumstances from the Earliest Times to the Present Day*, new edition (London, 1961) p. 520, my italics.
[60] K. Popper, *Conjectures and Refutations* (New York, 1963) p. 246.
[61] Popper, *Conjectures and Refutations*, p. 98, my italics.
[62] T. Kuhn, *The Structure of Scientific Revolutions*, 2nd ed. (Chicago, 1969) pp. 154-5, my italics.
[63] W. Shea, *Galileo's Intellectual Revolution* (London, 1972) p. 110, my italics.
[64] I. Lakatos, 'History of science and its rational reconstructions', in J. Worrall and G. Currie (eds), *The Methodology of Scientific Research Programmes: Philosophical Papers*, vol.1 (Cambridge, 1978) p. 115. Lakatos is here using the 'prediction' as an illustration of (and evidence for?) the superiority of his philosophy of science. God only knows where Lakatos got this 'fact' that Kuhn did 'not even care to mention', but certainly not from the historical record. It is not easy to imagine what the Tychonian '*post hoc* adjustments' could have been. (Or to imagine why the phases are better evidence for Copernicus than Brahe, even if Copernicus had predicted them and Brahe hadn't!)
[65] I. Lakatos, 'History of science and its rational reconstructions', p. 116.
[66] I. Lakatos and E. Zahar, 'Why Copernicus's programme superseded Ptolemy's', in J. Worrall and G. Currie (eds), *The Methodology of Scientific Research Programmes: Philosophical Papers*, vol.1 (Cambridge, 1978) pp. 183-4. I suspect that Lakatos and Zahar's '1616', instead of 1610, is a transcriptional error. The history in Lakatos and Zahar's article, while less historically creative than Lakatos' prior effort, apparently comes from Kuhn. In his illuminating account of the origin and presuppositions of Kuhn's *The Copernican Revolution*, Westman briefly reports that Lakatos and Zahar's 'conception of the problem was based entirely on evidence to be found in [*The Copernican Revolution*]' [R. Westman, 'Two cultures or one: A second look at Kuhn's *The Copernican Revolution*', *Isis* 85 (1994) 79-115, p. 9]. Kuhn's omitting the luminosity issue and his proclaiming that the phases had been predicted by 'cosmologically more radical' Copernicans seem to have directly affected the philosophy of science in this and other cases.
[67] A. Chalmers, *What Is This Thing Called Science*? 2nd ed. (St. Lucia, 1982) p. 71.
[68] M. Finocchiaro (ed. & trans.), *The Galileo Affair: A Documentary History* (Berkeley, 1989) pp. 17-8.
[69] G. Andersson, 'Falsifications, Galileo, and Lady Reason', in G. Munévar (ed.), (1991) *Beyond Reason: Essays in the Philosophy of Paul K. Feyerabend* (Dordrecht, 1991) 281-96, p 287. Andersson here cites Kuhn, *The Copernican Revolution*, p. 223.
[70] R. Giere, *Understanding Scientific Reasoning*, 3rd ed. (New York, 1991) pp. 65-7. Giere's

version of the history fits well into his predictivist view of science as presented in *Understanding Scientific Reasoning*. He interprets 'prediction' in the atemporal way that I will discuss below: '... often, predicting the data simply means using the model to determine what the data should look like, even though the experiment has already been done' (p. 31).

[71] W. Wallace, *Galileo's Logic of Discovery and Proof*, Vol. 137 in the series, *Boston Studies in the Philosophy of Science* (Dordrecht, 1992) pp. 203-7.

[72] For assistance on this key point, I would like to thank Roger Ariew, Robert Westman, Albert Van Helden, Owen Gingerich, Maurice Finocchiaro, Jim Lattis, and Keith Hutchison.

[73] O. Gingerich, 'Galileo's Astronomy', in W. Wallace (ed.), *Reinterpreting Galileo* (Washington DC, 1986) p. 120, my italics apart from italics within the quotation from Bellarmine, which are those of Gingerich.

[74] In fact, the inconsistency of both Ptolemaic theory and Copernican theory with the observed variation in planetary size had already played a role in the history of Copernicanism. In Osiander's anonymous preface to *De revolutionibus*, the inconsistency between all astronomical theories and the observed changes in the diameter and size of Venus was the only empirical consideration mentioned for holding that astronomy cannot tell us the real planetary motions.

[75] Kepler, *Appendix to the Hyperaspistes* (1625) in Drake and O'Malley, *The Contoversy on the Comets*, pp. 344-5.

[76] Galileo, *Dialogue Concerning the Two Chief World Systems*, p. 334, my italics.

[77] Y. Elkana & J. Goodfield, 'Harvey and the Problem of the 'Capillaries'', *Isis* 59 (1968) 61-73, p. 61; italics in original.

[78] J. Russell, *Inventing the Flat Earth: Columbus and Modern Historians* (New York, 1991) p. 26.

[79] M. Franssen, 'Did King Alfonso of Castile Really Want to Advise God Against the Ptolemaic System? The Legend in History', *Studies in History & Philosophy of Science* 24 (1993) 313-25, p. 323.

[80] I am particularly grateful to Michael Ellis. I am also grateful to Keith Hutchison, Roger Ariew, Robert Westman, Ross Phillips, Brian Ellis, Len O'Neill, Rod Home, Maurice Finocchiaro, Howard Sankey, Martin Tamny, and the Victorian Centre for the History and Philosophy of Science.

KEITH HUTCHISON

THE NATURAL, THE SUPERNATURAL, AND THE OCCULT IN THE SCHOLASTIC UNIVERSE

1. INTRODUCTION: A PROBLEM AND ITS CONTEXT

To understand historical *change*, one must obviously compare something old with something new. So analysis of scholasticism is a vital part of any sustained discussion of the events which this volume symbolically attaches to the year 1543. The reason for this is not simply the bland fact that medieval philosophy is part of the 'background', but something far sharper: a rejection of scholastic attitudes is central to the Scientific Revolution itself.[1] One of the things here rejected was medieval matter theory; another (perhaps) was the scholastic view of supernatural causes. The present paper explores the connections between these two doctrines, with a view to clarifying the fortunes of belief about the supernatural in the course of the Scientific Revolution. What does the abandonment of the medieval view of matter by the mechanists of the seventeenth century tell us about changing attitudes to the function of supernatural actions in the operation of the world? This is the question which links my discussion to 1543, but it is not a question that I will be directly answering here.[2] Instead, I shall concentrate on the rejected views themselves, one part only of the larger topic.

Behind my chosen focus, furthermore, lies the traditional notion that the couple of centuries that began around 1543, was a time in which philosophy and science—and 'thought' in general—became increasingly secular. An important part of this secularisation, so continues the tradition, was the concurrent abandonment of confidence in the occult sciences—astrology, alchemy, magic, etc. In thinking about the precise meaning of this claim, however, I have much difficulty with the word 'secular', and accordingly have chosen to discuss a slightly different problem, phrased in terms of the supernatural. In what senses were the occult sciences deemed supernatural in the scholastic worldview? That is the primary question attended to here.

G. Freeland and A. Corones (eds.), 1543 and All That, 333 – 355

2. HYLEMORPHISM AND NATURALISM

Matter theory gets involved in answering this question, because the occult sciences were routinely articulated in terms of the scholastic doctrine of 'hylemorphism'.[3] Adapted from the writings of Aristotle, with an indirect input from Plato, hylemorphism became a standard scholastic notion, though there was much uncertainty, variation, and dispute about its details. It viewed the universe as made up of objects composed of two very different ingredients: a passive and unchanging material substratum (*hyle*); and internal 'natures', 'essences', 'qualities', 'powers' or '[substantial] forms' (morphism), real entities, separate from, and more noble than mere matter. These forms were liable to alteration; and (far more significantly for our discussion here) were responsible for the activity displayed by objects. When snow melts in the spring, it does so by responding to warming powers in the sun and air (etc.), to exchange a quality of coldness for a quality of warmth, and a quality of solidity for a quality of fluidity. When an object falls, it does so because it contains some internal power, 'gravitie', which actively generates the motion which conveys its matter groundwards; just as holy water removes sin (according to some theologies) as a result of some facility, a *virtus justificanda*, it acquired during its manufacture, when it was blessed by a priest.[4]

In the seventeenth century, these beliefs were widely attacked by proponents of a new philosophy of nature, mechanists, who totally abandoned the hylemorphic interpretation of change, to portray it instead in purely geometrical terms, as nothing but the spatial reorganisation of portions of matter. And they drastically retreated from the scholastic practice of hylemorphism, in general eschewing the notion of implanted power. Their explanations did not abandon this notion totally, but presumed only scant innate activity in material objects, though (as with the scholastics) they differed amongst themselves as to the details. Descartes implicitly allowed matter an ability to rebound after collision, and an associated ability to resist penetration; others allotted it an ability to move; some an ability to gravitate-at-a-distance. Yet all sought to reduce the diverse behaviours of material objects to a few simple causes, most of which were conceived of as invariable properties of matter itself, rather than as some attached, yet separate and mutable, activating power.[5]

To many historians, this abandonment of hylemorphism has seemed an unequivocal move in the direction of naturalism. Such an interpretation probably evolved out of the rhetoric of seventeenth century philosophical debate itself, where innate powers were often castigated for being 'mysterious' and 'unintelligible'.[6] It must also have been supported by earlier Protestant complaints about hylemorphic interpretation of the sacraments, where seemingly supernatural powers were attributed to material objects by Catholic theorists.[7] The obvious affinities between hylemorphism and animism (the belief that objects are activated by resident spirits or souls) would strengthen the notion, and so would ongoing confusion in the literature between hylemorphism in general, and the specific belief that objects are sometimes activated by *occult* powers. Assembling all these ingredients, it was easy to make it seem that scholastic matter theory was profoundly supernaturalistic.[8]

Historians familiar with the details of scholasticism, however, would commonly reject this view as seriously anachronistic: it unreasonably substitutes the opinions of later observers for those of the participants. Aristotle's philosophy is very self-conscious of its focus on natural causes, and overtly uses its hylemorphism to characterise natural actions—as those governed by an internal principle. Furthermore, when Aristotle's philosophy was later taken up in thirteenth-century Europe, it was part of a naturalistic reaction against an earlier supernaturalistic vision of the universe, which had tended to locate all causality in God and to deny the potency of secondary causes. The notion of a natural cause was thus vital to the scholastic programme, and hylemorphism provided one of the main vehicles for its articulation. The powers and virtues that it attached to matter were real entities, created by God to act as *bona fide* substitutes for His own activity, and the scholastic universe normally ran itself via these causes—with some form of Divine sustenance certainly, but with something else as well.[9] On relatively rare occasions, events happened that were not caused by implanted powers, and it was these events which were deemed miraculous and supernatural. Hylemorphic actions were thus diametrically opposed to supernatural actions.

A fleeting remark by Bodin illustrates this contrast particularly well, for Bodin interprets God's biblical promise not to allow the Great Flood to recur, as a promise of supernatural intervention to prevent the natural astrological powers of the planets from producing another flood whenever the configuration of the sky returns to that it had in the time of Noah. 'One will then find much the same effects and changes', he observes, 'unless God restrains the effects of the celestial causes by his power'. This passing remark is taken from a discussion where Bodin is asking himself about the causes of political change—the rise and fall of 'republics'. Are such changes caused supernaturally, or naturally, or through human free-will, he asks? And as 'natural causes' here, he envisages the 'secret virtue' (as he calls it) of the stars.[10]

3. OCCULT AND MANIFEST QUALITIES

In calling the natural activity of the stars 'secret' here, Bodin is almost certainly agreeing with a view common among his contemporaries (and to be elaborated upon below), that this power was 'mysterious' and 'unintelligible'—but this is not a consequence of the hylemorphism implicit in his discussion. For those who accepted this doctrine routinely divided the powers attached to matter into two classes, those which were mysterious and those which were not—'occult' qualities and 'manifest' ones, to use a standard piece of scholastic terminology.[11] Occult qualities (as we also see further below) were controversial, and not universally accepted by hylemorphists—while manifest qualities were seen as *prëeminently* intelligible. For these qualities were often defined as those directly detected by the senses, and this meant (in scholastic psychology) that they were the only features of the external world which could be accurately reproduced in the intellect. Hotness, coldness, wetness, and dryness (and their various combinations) were the primary agents of all our soul's understandings of anything. As the familiar slogan put it, in a slightly distorted paraphrase, 'Nothing at all can enter the intellect which has not come from the qualities of objects'.[12]

This quintessential intelligibility of manifest qualities in scholastic philosophy is clearly attested by its later critics, for to them this was one of the grand mistakes of the earlier view, and they were to insist over and over again that our senses do not give us an accurate image of the world—that what we perceive as redness, for instance, is radically different from that which causes redness in the external world. 'All philosophy' observes Fontenelle, in a nice statement of this new view, 'is based on two things only: curiosity and poor eyesight ... We want to know more than we can see [and] we see things other than as they are. So true philosophers spend a lifetime not believing what they do see …'. The scholastics, by contrast, had placed extreme confidence in their ability to sense the manifest qualities of objects. To claim that they were wrong here, is to insist that they did not hold the view that qualities and powers were in themselves mysterious.[13]

Yet many scholastics certainly did hold a view that *some* qualities were unintelligible. For it was accepted that objects sometimes had actions associated with them which did not directly reveal themselves to the senses—like the attraction of a magnet for iron (which could not in itself be inspected, but which displayed itself only through its effect on the iron). Such actions, however, became an intellectual puzzle—for (as we have just noted) scholastic psychology entailed that entities which could not be sensed could not be fully understood—and there was great uncertainty about how to account for this sort of behaviour.[14] This gave multiple opportunities for connections to emerge between hylemorphism and the supernatural. The word 'connections' is a very vague one however, and it should be emphasised immediately that the *primary* connection between occult qualities and supernatural actions is the one already glimpsed in Bodin—opposition and negation. To attribute dry weather to a miracle was to deny that it was caused by powers in the stars; and vice versa, to attribute a flood to innate powers was to deny that it was supernatural. For the idea that an effect like magnetic attraction was caused by a special unintelligible power in the magnet—an occult quality—was only one of a range of options to account for the behaviour of iron near a magnet, and it was the other options—to be canvassed further below—that came closer to being supernatural. So once again scholastic hylemorphism can be seen to support a naturalistic conception of the functioning of the universe, for it emphasises the locating of causality in the powers that belong to natural objects.

This incidentally is one important sense in which the common tendency to declare occult qualities (or manifest ones for that matter) vacuous is quite seriously wrong. This claim is very wide indeed in the secondary literature, and common in the writings of early modern opponents of scholasticism, but it misrepresents the scholastic doctrine. Thus when Aquinas answers the question 'Why is wood heated in the presence of fire?' by saying 'Because heating is the natural action of fire … because heat is its proper accident', he might seem to be saying very little—until we look a little closer at the context, and note that he is not citing this analysis to inform us about fire, but to make the point that the answer 'Because God willed it', though correct, is inappropriate.[15] So to declare that an action was caused by a quality was to assert *much* about its character, and in particular to declare that it was not miraculous, that its immediate cause was relatively remote from God. Indeed hylemorphism was sometimes attacked by early modern opponents of scholasticism for this reason. Thus Calvin objects to astrology because of the way it *reduces* supernatural activity: 'When unbelievers transfer the government of the universe from God to the stars', he says, 'they fancy that their bliss or misery depends upon ... the stars, not upon God's will ...'. Yet there is 'no erratic power, or action, or motion in creatures ...'. For they 'are governed by God's secret plan in such a way that nothing happens except what is knowingly and willingly decreed by him'. Luther similarly disapproves of the scholastic opinions about causality. 'Aristotle makes the Prime Mover the cause', he says, in a commentary on *Genesis*, 'while Averroes declares that forms ... are the causes of the [celestial] motions, [but] we follow Moses and declare that all these phenomena occur and are governed simply by the Word of God'.[16]

Similar denials of the existence of occult powers were also common among the scholastics for, as suggested above, such entities were invoked only in situations where the effects and their causes were very unclear. Despite the unquestionable gullibility of medieval commentators in this domain, there was also ongoing and severe doubt as to whether some mysterious effect—like the alleged ability of a magnet to engender nightmares in adulterous women, or the alleged curing of an illness by saintly relics, or the alleged transformation of a piece of bread into the body of Christ—was real, illusory or fabulous; and if real, whether the cause was internal or external. So it was common for occult qualities to be denied outright; while other scholastics did so partly, insisting on principle that there are no genuinely occult powers, and that all apparently occult qualities, stellar influences for example, acted through such manifest powers as heat, light and motion. 'Many indeed', says Albertus Magnus:

> seem to doubt whether there are in stones any of the [non-manifest] powers which are regarded as belonging to them ... and they assert that there is nothing in a composite substance except [what is due to] its constituents and the way they are combined .... But the opposite is proved most convincingly by experience ... stones do have powers of wonderful effect ... .[17]

4. THE CONTRAST BETWEEN POWERS AND SPIRITS

Furthermore, even when it was accepted that an effect was real, there was doubt as to whether it was caused by an implanted power, or by some spiritual alternative, such as an angel or a demon, or by God Himself. So another of the important 'connections' between hylemorphism and supernaturalism is that occult qualities came up for consideration in contexts where a supernatural cause was under consideration, but again the hylemorphic explanation is the naturalistic choice amongst the range of potential answers. The supernaturalism resides in the context where the question of causation gets asked, not in the answer. The supernatural answer is a rejection of the hylemorphic one.

So when Aquinas considers the transformation of bread into flesh, he concludes that the effect is genuinely miraculous, an intervention of God's 'infinite power' in the everyday world. There is, he insists, a power in the words used in the ceremony, but this power is merely instrumental, like that in a craftsman's hammer: control and guidance in each case come from a vastly superior agent. The power in the words is not the main cause: 'No creature, acting as a principal cause, can bring about miraculous effects'.[18] Similarly, when he asks about the action of saintly relics, he concludes that the cures experienced by pilgrims are not caused by some 'implanted form' or 'intrinsic principle' or 'from some power residing and permanent in the [relics]', for if that were the case, everyone visiting the relics would be cured. The fact that the cures are selective indicates that a will is involved, so Aquinas concludes that it is angels who generate the cures, not the relics.[19]

Since, however, the precise causalities involved in such obscure actions were always unclear, there was an ongoing tension between explanations that invoked occult qualities and those that denied them by invoking immaterial spirits. 'Because of the hidden nature of [such causes]', notes Oresme, 'certain stupid necromancers have said that these powers in precious stones are there as the result of the presence of certain incorporeal spirits'. Albertus Magnus had taken a similar view: 'stones ... have no souls', he insists, 'but they do have substantial forms'. Some radical Aristotelians (like Pomponazzi) went much further than Oresme and Albertus here, and eliminated virtually all spiritual intervention in the everyday world by presuming a very large supply of occult powers to be attached to material objects. But, as Oresme and Albertus indicate, others took the opposite view, and insisted that many of the strange powers which seemed to be attached to matter were not so attached in reality—for the effects that seemed to be produced by material objects were really produced by spirits. This (as is well-known) became a standard interpretation of magic. 'There is', insists the mid-seventeenth-century Aristotelian, Alexander Ross,

> a naturall magick, by which you may do strange things, and anticipate the times prefixed by nature, in producing of divers effects, by applying [actives to passives]: so you may produce a rose in winter, and raise parsly out of the ground within a few hours after the seed is sowne. There is also a mathematicall magick, by which strange things are done; as was that woodden Pigeon, which Architas caused to flie; and that brasen head, which Albertus Magnus made to speak ... . Such things, and many more, may be done without witchcraft.[20]

But there is also, he continues,

> a diabolicall magick in working strange things by the power of Sathan, by a contract which witches make with them, God permitting, in his secret judgement, the affectors of such evill things to be deluded and abused by the evill Angels.

In analogous fashion, Aquinas had rejected many of the claims of astrologers and magicians: what they take to be powers in the stars, he says, are often angelic, or demonic actions. Magicians claim to be able to make statues move by using stellar powers, he says. But 'it is impossible', he asserts in rebuttal, 'for something inanimate to be made able to move itself by the power of the celestial bodies'. 'So it is not possible for [this] effect of the arts of magic to be done by celestial power'. The statues must be moved by a spirit.[21] Note that Aquinas does not take this view because he rejected the idea that there are powers in the stars: on the contrary, he attributed many terrestrial actions to the stellar influences, and used the latter as a standard way of accounting for the occult operations of nature, specifically citing magnetism and the occult virtues of herbs and stones as examples. 'It is obvious', he says, 'that inanimate bodies ... obtain certain powers and abilities from the celestial bodies'. But, he adds, 'what is done by the power of celestial bodies is a natural effect, for the forms that are caused in lower bodies by the power of celestial bodies are natural'.[22]

Implicit in this common demonic interpretation of magic is the strong contrast that we have met before between hylemorphic actions and those that are supernatural. In some cases, however, the language of hylemorphism could in fact be used to refer to spiritual actions. In one of Luther's discussions of the sacraments, for example, the possibility is accepted that it might be reasonable to talk of holy water as if it had a power to wash away sins. Yet in Luther's view this action is a supernatural one, and it was one of his principal complaints against the Catholic theory of justification that it had allowed some causal efficacy to the water of baptism:

> It is not baptism that justifies or benefits anyone, but it is the faith in that word of promise to which baptism is added ... . It cannot be true, therefore that there is contained in the sacraments a power efficacious for justification ... .

Yet 'if anyone should want to call this Word or this promise a power that has been given to the water of Baptism, I shall not object', he says elsewhere. 'But the idea of the sophists was something different. They do not attribute this power to the Word, but concerning the element they state that it has a special power given to it'.[23] So although Luther seems happy to talk of 'supernatural powers' attached to matter, he is not endorsing a hylemorphic theory, but a form of nominalism: the power he speaks of is not a real entity attached to the matter of the water, it is just a *façon de parler*, a name for a phenomenal effect, not an identification of the cause. Real qualities—scholastic qualities—remain opposed to supernatural causation.[24]

5. ARE DEMONS AND ANGELS SUPERNATURAL?

But let us return to the theory that some magical actions are performed by spirits rather than occult powers, for in citing this standard view above I invited you to interpret it as further evidence of the contrast between hylemorphic actions and supernatural ones. Though I do not wish to withdraw this interpretation, I must now make some modification to it, because of an ongoing scholastic tradition that angelic and demonic actions were not strictly supernatural—but rather natural. Stuart Clark has demonstrated that this belief was very widespread in early modern Europe, and Aquinas (as we have seen) is quite explicit about it: true miracles can only be done by God. Though angels and demons seem to work miracles by performing effects that are beyond the natures of the tools they use (etc.), their actions are no more outside nature *in toto*, than when a human throws a stone into the air, and overrules its innate tendency to fall. Indeed, like humans who throw stones, angels and demons are creatures, with their own internal natures, and their actions in the world are a consequence of the powers natural to them:

> An angel's power is said to be infinite over the lower world, inasmuch as his power is not enclosed in matter .... But it is not infinite in respect of things above it ... , because the angel receives a finite nature from God, so that his substance is confined to a particular genus, and consequently his power is confined to a particular mode of action .... Although an angel does wonderful things as the result of art, they are not miracles ... . God alone can work miracles by acting independently of the natural course.[25]

Furthermore (as this passage hints) the natures of angels and demons were sometimes articulated in terms of the hylemorphic theory. Thus, for Aquinas, such spirits were pure form unattached to matter, 'separated substances' as he calls them, and contrasted with the human soul, a form attached (in this life at least) to matter. Others, more faithful to authentic Aristotelian doctrine perhaps, insisted that form could never be separated from matter, and posited a special 'spiritual matter' for the spiritual forms to reside in.[26] In both cases, however, the forms continued to function as the locus of natural powers—powers delegated by God to act as alternatives to His own activity, powers which enabled objects to act, yet which also placed limits on the actions they could perform. They are naturalistic ingredients in a world-view which consistently portrays unconstrained supernatural activity as rare and unusual.

Yet part of the reason we have been able to endorse this strongly naturalistic interpretation of scholastic forms, has been the way the initial problem was posed, for the question implicit in the preceding discussion has been an either-or question: were the actions exerted by qualities natural, or supernatural? Our ongoing answer to this question has accurately captured an important feature of scholastic thought, the sharp contrast between an infinite God and His finite creation—but it has obscured another feature of scholastic thought, a tendency to think in terms of hierarchy and gradings. So we should also ask ourselves some slightly different questions. Were angelic actions more supernatural than everyday actions? Were some qualities and forms more supernatural than others?

It seems that we must accept some sort of 'Yes' answer here. Aquinas himself agrees that many demonic actions are *like* miracles, in that they cause wonder and puzzlement, and are outside the course of nature as known to mankind. He also accepts that some apparently angelic actions are really miraculous, in the same way as transubstantiation is supernatural: the angel is not the principal cause, but merely an instrument. Furthermore the fact that demons and angels are immaterial entities, means that they are very different from the objects of the everyday world. They may not be strictly supernatural, but they have vastly superior abilities, so a strong contrast between the normal and the exceptional is certainly retained in the scholastic view of spirits. Indeed the actions of magic and demons are sometimes accommodated by drawing a distinction between *preter*natural actions and supernatural ones: this verbal move clearly indicates recognition of an affinity between the two, while retaining the characteristically scholastic contrast between Divine actions and those of lower entities.[27]

Then again, in an earlier example, we glimpsed Bodin contrasting supernatural causation, not with natural activity alone, but with a duality: natural actions and voluntary ones, and we saw the same contrast in Aquinas' discussion of the action of relics. This distinction between natural actions and those caused by a will—either human, angelic or demonic—was very common in medieval times, and had been used in antiquity to argue for the divinity of the planets.[28] Christian monotheism could not accept this last conclusion, but in allowing that the planets were moved by angels it was giving some recognition to the affinity between angels and God. Indeed, it had to draw the ongoing contrast between the two in order to maintain its monotheism, but the contrast had to be drawn only because there was a *prima facie* similarity between the two.

Certainly, humans had wills as part of their nature, and while these wills enabled them to perform actions which were classified by Aristotelians as artificial or even violent, and though contrary to the natures of the objects they acted on, their voluntary actions were part of nature as a whole. Yet in separating wilful actions into a special category, and distinguishing them from natural actions, our sources are revealing a sense that rational creatures have something about them that is worth contrasting with unreasoning nature. Their proximity to divinity in the scale of nature is itself a separation from the remainder of nature. The human soul, the entity that gives us our ability to make voluntary choices, is *like* God, Aquinas admits, made in his image indeed—but is not Divine, for the whole of nature is like God in some sense. Yet, this same soul was widely believed to be created supernaturally, and its capacity to survive the death of the body distinguished it from that of the brutes.[29]

6. A SUPERNATURAL EPISTEMOLOGY

So, despite the unequivocal counter-evidence, it remains reasonable to think of angelic and demonic actions as, in a number of ways, non-natural. It has been important for me to make this view plausible, because of a significant change in direction I now wish to make. For in most of the preceding discussion, we having been approaching our problem—the connection between qualities and the supernatural—*ontologically*, by exploring scholastic beliefs about reality, and the types of agencies envisaged as operating in the world. I now wish to return to a alternative approach, and look again at our problem *epistemologically*, exploring scholastic beliefs about how we find out about occult qualities and the like. For here a quite different picture emerges, and this is why it is important to distinguish these two approaches to our question. Though occult qualities were preëminently natural as causes, they were (as briefly noted above) preëminently unintelligible by natural means—by the unaided exercise of our natural faculties; and though angels and demons were similarly classified among natural objects, there was wide agreement our knowledge of such things was also severely limited, and equally dependent on external supplementation. 'In the present life', observes Aquinas, 'the intellect depends on the sense for the origin of knowledge; and so those things that do not fall under the senses cannot be grasped by the human intellect except in so far as the knowledge of them is gathered from sensible things'.[30]

So in scholastic philosophy a strong contrast is drawn between knowledge acquired naturally—by the exercise of the innate processes of the human mind acting on sensations received from the external world—and that acquired artificially, from an external source.[31] The latter includes that acquired from other humans—testimony—and that acquired from superior intelligences—angels, demons, or God. This distinction is of course fundamental to scholasticism, for it is the basis of that philosophy's explication of the traditional Christian notion of revelation, and its accommodation of Christian mysteries like the Trinity. Implicit in it was the notion that there were many truths which humans could not attain to of their own accord (i.e., naturally)—so supernatural supplementation was vital if these additional truths were to be reached. Aristotle had used natural reason to claim that the world was eternal, for instance, and different commentators had different views about the precise status of his arguments, but virtually all agreed that *Genesis*—non-natural testimony—tells us that he was wrong here. 'This is why the Ancients', explained John of Jaudan, 'who used to draw their knowledge from rational arguments verified by sensible experience, never succeeded in conceiving such a mode of production [of the universe]'.[32]

The pagan philosophers of late antiquity had attacked Christianity here for its apparent irrationality, while it, in defence, adopted a 'cult of human frailty' (to use Brown's especially apt phrase), a partial scepticism which urged the profound incompetence of human reason. Thus Augustine cites the characteristics of such natural objects as quicklime or the magnet to indicate the limitations of the human faculties—and uses this to make the enigmata of Christianity seem more acceptable. It is, he implies, unreasonable for the pagans to deny something simply because they cannot understand it. 'Many things whose reason cannot be discovered are still undoubtedly true', he observes. The belief was also adopted that there were certain facts about the universe which were not merely beyond the reach of human inquiry, but which were also forbidden to humanity by Divine fiat. So Augustine quotes Christ's reply to his disciples, 'It is not yours to know the times that the Father has fixed by his power', to indicate that the time of the second coming is deliberately hidden from us.[33]

The scholastics, as we have already noted, marked the boundary of natural knowledge with the human senses, and pessimistically restricted the scope of human reason. But in doing this, they also endorsed an optimistic view: within its field of action, human reason was believed to work extremely well, so well indeed that it was sometimes capable of giving us understandings of things as they really were—essences etc. It was additionally agreed that our faculties could give us some grasp of insensible reality—for God in particular was not totally unknowable to the unaided intellect, but the natural knowledge thus acquired of God was severely limited.[34] The same was true of material objects: their insensible properties could perhaps be recognised by us, but we could not grasp their essences and causalities, so we could not have a scientific knowledge of them. Oresme, for example, does allows us a limited knowledge of these things, but sees too much interest here as dangerous:

> It is expedient to know those aspects of the occult powers of ... natural things which are suitable to ... good living. But we ought to be content with the understandings of such things as are known by physicians, surgeons, goldsmiths, and others. The other more secret things nature herself, like a modest mother so to speak, does not wish divulged .... In those books which have been written concerning such things there is little truth .... Those who ... seek ... to violate the secrets of nature ... have been fittingly cursed by the Author of nature.[35]

Ross is perhaps more pessimistic:

> In aiming at such abstruse causes [as the cause of magnetism] ... all men must confesse, that our science here is but ignorance: and we see the natures of things, as that blind man, who saw men walk like trees. Who can tell how Rhubarb purgeth choler; Agarick phlegme? How the Torpedo stupefieth the hand thorow the cane, and the Remora stayes the ship? ...
>
> 'Tis modesty and ingenuity to confesse our ignorance in those secrets which God hath purposely concealed from us, to teach us humility, for the pride of our first Parents, in affecting the forbidden fruit of knowledge; and that we should account all knowledge here but ignorance, in respect of the excellent knowledge of Christ, in whom are hid all the treasures of wisdom and knowledge. This we know, there are divers contrary, and also sympathising principles in nature, which are the causes not only of occult, but likewise of manifest qualities: but to demand the reason of these, is to search into those secrets of God, the knowledge of which is reserved for us in a happier life ... .[36]

Aquinas had similarly cited the occult action of the magnet, as an illustration of the limitations of the human mind when dealing with insensible causes, things 'which man is not capable of explaining' [*quarum ratio ab homine assignari non potest*], and he took a similar view of our knowledge of God. This too is limited by our nature as humans, and in particular our dependence on sense-perceptions.[37] In both cases we experience the effects, and can tell that the cause exists, but we are unable to appreciate the nature of the cause.

Yet further knowledge of God could, of course, be obtained from revelation, and by implication the same was true of magnetism. Christ presumably understood magnets much better than we do, and He could tell us about them if that seemed to matter to Him. Adam indeed was widely reputed to have had a thorough understandings of the inner workings of nature, and this had been supplied to him by God in some supernatural fashion, but after the fall this knowledge had been taken away from mankind. Hence the common magical quest to recover the ancient wisdom of Adam.[38]

Yet demons and angels also understood natural things better than we did, for they were not encumbered with a material body, so were not forced to rely on sense-perceptions to obtain knowledge. 'Being of a fine substance', explains the late sixteenth-century physician, Weyer:

> they far surpass the awkwardness of earthly bodies and the sluggishness of earthly senses [and] their lifetimes extend down from the very beginning of time [so] they have acquired a remarkable familiarity with far greater things than men can know, given the moment-like brevity of man's life.[39]

Demons, as we saw above, may perform magic by carrying out the actions that some were inclined to attribute to material actions. But since they have this superior knowledge, they may also teach us about the properties of material objects which we are not able to discover on our own. The occult sciences could be *taught* by demons, as well as *performed* by them.[40]

This of course was part of the standard view of magic, and one of the grounds for its frequent condemnation. Indeed, anyone with a special knowledge of a difficult field was suspected of dealings with demons, simply because of their having unusual knowledge: 'Let a man know more than a common student', complained the mid-seventeenth-century astrologer Harflete, 'then he is accounted a conjurer: he deals with the devil'. Osiander's preface to *De revolutionibus* claims that 'neither [astronomer nor philosopher] will grasp or convey anything certain, unless it has been Divinely revealed to him', and Keith

Thomas cites the example of an unfortunate in England who was accused of dealings with the devil simply because he possessed a copy of Sacrobosco's *Sphere*! Aquinas similarly speaks of a belief that people can learn a foreign language quickly by this method, and Grosseteste had earlier condemned astrology for being written by the devil.[41] So a good case can be made for a positive connection between occult qualities and the supernatural, if we interpret the category 'supernatural' broadly enough to include angels and demons, and focus on epistemology. But they remain natural causes in their action.

7. THE CONTRAST WITH MECHANISM

As hinted at in my opening remarks, these questions are important for understanding the Scientific Revolution, because of the contrast with the mechanical philosophies of the seventeenth century.[42] I shall not explore these contrasts in any detail here, but let me, in concluding, at least sketch the sort of thing I have in mind.

In both the *Meditations* and the *Principles*, Descartes puts forward the idea that God has to recreate the universe afresh every instant. This is not just a casual opinion of Descartes, for he claims that he has proven it to be true, and even uses it to argue the certainty of God's existence. He does this by reversing the argument against astrology noted in Calvin above. Calvin said that God causes everything, therefore there are no powers in the stars. Descartes says they are no powers in objects ('there is no strength in us whereby we may conserve ourselves'); therefore substances cannot of themselves persist in time; therefore they must be recreated, and this requires Divine power.[43] So everything that occurs in Descartes' universe is—by scholastic standards—supernatural: he has returned to a world-view like that we saw Aquinas objecting to earlier—one which attributes all causality to God.

Indeed when Gassendi (whose own version of the mechanical philosophy does not claim matter to be stripped of all qualities) objects to the Cartesian position here, in his reply to the *Meditations*, he insists on something very like the innate powers of the scholastic alternative: 'natural constitution[s]' are real, he says, and these give things 'a power ... which suffices to enable [them] to continue to exist, unless some destructive cause supervenes'. Descartes is right to acknowledge dependence on a higher power, he agrees, but only for initial production, not for continued existence. Descartes' world-view is too supernatural for Gassendi, and he recommends a retreat in the hylemorphic direction.[44] This suggests that the decline in hylemorphism is associated with an increased emphasis on Divine activity in the universe.

This is not however a particularly convincing example, and that is *paradoxically* one reason I have chosen to use it here. Descartes did not make a big issue of his doctrine of constant creation, and it is hard to feel confident he was fully committed to it. Furthermore, we know that some of his near-contemporaries, Henry More for example, came to fear his philosophy might lead to atheism, and this could hardly be so if continuous divine creation were recognised as a truly unavoidable component of the Cartesian system. Indeed, one can easily imagine a version of Descartes' philosophy, similar to its original in virtually all respects, except that it lacks ongoing Divine causation. Descartes' philosophy then is perhaps better described as indifferent to the question whether the world runs by natural causes—whatever Descartes personally thought. But at a minimum this means that it is not committed to natural causes, and can readily allow a supernatural interpretation—unlike its scholastic competitor, where a supernatural interpretation of qualities is impossible. For that leads to the nominalism we have seen in Luther—and that is not hylemorphism: implanted powers then become vacuous.

Something very similar is also true of Boyle. He uses his abandonment of innate powers to argue against the traditional Aristotelian view that the world does not have a creator, he opposes traditional notions of nature, and comes very close to insisting that God must also be involved in all mundane causations. Yet it is very difficult for us to tell precisely how much Boyle thinks the world to operate in a purely supernatural manner. It is not vitally important for Boyle to decide this question, for as with Descartes', his philosophy is indifferent here, and equally capable of accepting a natural or supernatural interpretation. It is easy to give a similar reading of the Newtonian account of gravitational motions. Newton and Clarke both seem to have viewed gravity as a 'continuous miracle', to have denied that gravity could be an implanted power, and to have seen their philosophy as providing a guarantee of Divine involvement in the universe, but it was not essential to agree with them here to practice Newtonian science. Other Newtonians did not share these views.[45]

So while scholastic hylemorphism unequivocally distanced God from His creation, the mechanical philosophies could accommodate supernatural causation—and, to some participants, even seemed to demand it. Furthermore the new science appeared to do this without appeal to revelation. Descartes, it is true, makes appeal to theological premises about the character of God to justify his methodology, and Glanvill (at least) does much the same thing in England: 'Reason is an exercise of faith', he says.[46] But these appeals are defences of human reason, and though they continue that blurring of the natural-supernatural distinction we have just been examining, they involve no transmission of specific facts or techniques from God or angels to humanity. A key ingredient here is the new attitude to occult actions like magnetism. Though the idea that these are sometimes caused by a special implanted power is widely rejected, so too is the idea that such actions are in principle suspect, and either unreal or unintelligible because of the belief they are beyond the range of human investigation.[47] For natural philosophy was now deemed capable of dealing with the insensible properties of objects, at least as capable as it is in dealing with anything else. So a specific necessity for supernatural assistance in the study of the material world was now removed. Though an occult action like gravity may well be purely supernatural in Newton's conception of the world, that does not prevent us from investigating it naturally. The scholastic conception of a world run by natural causes, whose investigation requires supernatural assistance has been turned on its head: the new science investigates supernatural activity by natural means.

Department of History and Philosophy of Science, University of Melbourne

NOTES

[1] The sort of scholasticism studied for such purposes should, of course, emphasise that set out and taught within early modern Europe, for that is most likely to be the prime target of contemporary hostility to the older outlook. But there are severe obstacles to doing this, the most notable of which is the general lack of a sympathetic secondary literature to guide one's researches. As a consequence, and as a definite compromise, I am forced to look at scholasticism over a much broader time-scale, and to use an eclectic mixture of sources from far earlier periods, thus blurring changes in the scholastic outlook. Such a technique is obviously dangerous, but the risks are undoubtedly worth taking—by comparison with the certainty that the many accounts of early modern science which effectively ignore scholasticism are taking far greater liberties, and making

very real errors—witness the various discussions that presume the notion of natural law to be a novelty of seventeenth-century philosophy! Such claims are easily refuted by the most cursory glance at earlier traditions.

[2] I have sketched an answer to this question elsewhere: see K. Hutchison, 'Supernaturalism and the mechanical philosophy', *History of Science* 21 (1983) 297-333 and 'Dormitive virtues, scholastic qualities, and the new philosophy', *History of Science* 29 (1991) 245-78, esp. pp. 262ff. The present discussion differs from those earlier ones in (a) its emphasis on the occult sciences, (b) its contrasting the naturalistic aetiology of such sciences with their supernaturalistic epistemology, and (c) its portrayal of the mechanical philosophy as neutral on the classification of fundamental cause.

[3] E.g., B. Hansen, 'Science and magic', in D. Lindberg (ed.), *Science in the Middle Ages* (Chicago, 1978) 483-506, pp. 489-91; B. Copenhaver, 'Astrology and magic', in C. Schmitt, Q. Skinner, E. Kessler & J. Kraye (eds), *The Cambridge History of Renaissance Philosophy* (Cambridge, 1988) 264-300, pp. 282-4; B. Copenhaver, 'Scholastic philosophy and renaissance magic in the De vita of Marsilio Ficino', *Renaissance Quarterly* 37 (1984) 523-54, *passim*, but esp. pp. 524, 531-2; Thomas Aquinas, 'On the occult works of nature', in J.B. McAllister, *The letter of Saint Thomas Aquinas De occultis operibus naturae* (Washington, 1939) 20-30. There is of course a great deal of uncertainty in the literature as to exactly what the word 'scholastic' should be taken to mean, but that problem seems of only minor importance in the present discussion, where the focus is on the specific doctrine of hylemorphism. Though this doctrine was central to scholasticism, it was also widely accepted in the various deviant philosophies of the Renaissance, and nothing in the discussion below hinges significantly on how broadly we conceive scholasticism. Yet Copenhaver does note ('Astrology and magic', pp. 292-3; B. Copenhaver and C. Schmitt, *Renaissance Philosophy* (New York, 1992) pp. 318-9) significant departure from Peripatetic hylemorphism in the Renaissance, though he also observes a retention of the terminology (on which see Note 24 below).

[4] I know of no sustained exposition or analysis of medieval (as opposed to ancient) hylemorphism in the secondary literature, but fragmentary accounts, especially of the disputes centred on it, are common: see e.g.: Copenhaver, 'Scholastic philosophy and renaissance magic', pp. 539-49; Copenhaver and Schmitt, *Renaissance Philosophy*, pp. 303-5; E. Dijksterhuis, *The Mechanization of the World Picture*, trans. C. Dikshoorn (Oxford, 1961) pp. 200-4; A. Maurer, *Medieval Philosophy* (New York, 1962) pp. 213-6; A. Crombie, *Augustine to Galileo: The History of Science A.D. 400-1650* (London, 1952) pp. 44-52. For my illustration via the melting of snow, compare the brief remarks in Dijksterhuis, *Mechanization*, pp. 19-21 (on Aristotle); Albertus Magnus, *Book of Minerals*, trans. Dorothy Wyckoff (Oxford, 1967) I.i.9 = p. 33; Aquinas, 'Occult works', p. 20; Nicole Oresme *Le livre du ciel et du monde*, [Middle French translation of, and commentary on, Aristotle, *On the heavens*, together with English trans.], eds A. Menut and A. Denomy, trans. A. Menut (Madison, WI, 1968) p. 683. For the power of 'gravitie', see: Aquinas, 'Occult works', p. 20; A. Ross, *The Philosophicall Touch-Stone* (London, 1645) pp. 13-4, 16-7; for the *virtus justificanda*, see: Hutchison, 'Dormitive virtues', pp. 268-70, 277-8nn.62-3, citing e.g. Thomas Aquinas, *Summa theologica*, [*Summa theologiae*, Latin and English, various translators] Dominican edn. 61 vols (London, 1963-80) 3a.66,2 = v.57, pp. 10-11.

[5] See e.g.: Dijksterhuis, *Mechanization*, pp. 431-3; B. Easlea, *Witch-Hunting and the New Philosophy*, (Brighton, Sussex, 1980) pp. 111-43; B. Van Fraassen, *The Scientific Image* (Oxford, 1980) pp. 1-2; Hutchison, 'Dormitive virtues', *passim*; J. Henry, 'Occult qualities and the experimental philosophy: Active principles in pre-Newtonian matter theory' *History of Science* 24 (1986) 335-81, esp. pp. 337-51.

[6] E.g.: René Descartes, *The Philosophical Works of Descartes*, 2 vols. trans. E. Haldane and G. Ross (Cambridge, 1931) pp. 120-1, 254-5; Robert Boyle, *Selected Philosophical Papers of Robert Boyle*, ed. and annot. M. Stewart (Manchester, 1979) pp. 57, 63; Nicolas Malebranche, *The*

Search After Truth, [*De la recherche de la verité*, 6th French edn, 1712], ed. G. Rodis-Lewis, trans. T. Lennon and P. Olscamp (Columbus, Ohio, 1980) with commentary by Thomas Lennon, p. 446; Leibniz, Letter to Hartsoeker of 10 Feb., 1771, as trans. by Cajori in Issac Newton, *Mathematical Principles of Natural Philosophy*, 2 vols, trans. A. Motte, rev. F. Cajori (Berkeley, 1966) v.2, pp. 668-9; *idem*, fifth letter to Clarke, G. W. Leibniz, and S. Clarke. *The Leibniz-Clarke Correspondence*, ed. H. G. Alexander (Manchester, 1956; 1717), Leibniz' letters trans. by Clarke, p. 94.

[7] K. Thomas, *Religion and the Decline of Magic* (London, 1971) pp. 33-40; Hutchison, 'Dormitive virtues', pp. 270-1.

[8] For various examples of these misleading presumptions, see: N. Steneck, *Science and Creation in the Middle Ages: Henry of Langenstein (d.1397) on Genesis* (Notre Dame, IND., 1976) pp. 108 (celestial and supernatural influences identified), 114 (celestial influences then paralleled to other occult powers), 119 (occult and supernatural causes identified), 129 (contrast between occult and natural causes); D. Walker, *Spiritual and Demonic Magic: From Ficino to Campanella* (London, 1958) pp. 109-10 (contrast between occult and natural causes); S. Talmor, *Glanvill: The Uses and Abuses of Scepticism* (Oxford, 1981) p. 25; A. Guerrini, 'Ether madness: Newtonianism, religion and insanity in eighteenth-century England' in P. Theerman and A. Seeff (eds), *Action and Reaction* (Newark, NJ., 1993) 232-54, esp. pp. 232-4 (contrast between supernatural and mechanical).

[9] For the naturalism associated with the revival of Aristotle, see the discussion, and sources cited, in Hutchison, 'Supernaturalism and the mechanical philosophy', pp. 304-11; *idem*, 'Dormitive virtues', pp. 265-8; *idem*, 'Individualism, causal location, and the eclipse of scholastic philosophy', *Social Studies of Science* 21 (1991) 329-35.

[10] Jean Bodin, *La République* (Paris, 1583; facs. reprint, Aalen, 1961) pp. 542-3, 550. For other examples of this sharp contrast between causation by implanted powers and supernatural causation, see: Aquinas, *Summa theologica*, 3a.78,4 (= v.58, pp. 182-5); Buridan, *Questions on the Physics*, in M. Clagett, *The Science of Mechanics in the Middle Ages* (Madison, 1959) p. 536; Oresme, *Nicole Oresme and the Marvels of Nature: A Study of his* De causis mirabilium, ed. and trans. etc. Bert Hansen (Leiden, 1986) pp. 136-7; M. Luther, *Luther's Works*, (v.1: Lectures on Genesis, ed. J. Pelikan, trans. G. Schick, (Saint Louis: Concordia, 1958); v.35, ed. E. Bachmann, various translators, (Philadelphia: Fortress, 1960); v.36, ed. A. Wentz, various translators, (Philadelphia: Fortress, 1959)) v.1, pp. 29-30, v.36, pp. 64-7; E. Grant, 'Medieval and Renaissance scholastic conceptions of the influence of the celestial region on the terrestrial' *Journal of Medieval and Renaissance Studies* 17 (1987) 1-23, p. 22; Ross, *Philosophicall Touch-Stone*, pp. 16-7; Malebranche, *Search after Truth*, pp. 446-9, 466, 658.

[11] For the distinction between occult and manifest qualities, see: Hutchison, 'Occult qualities', esp. pp. 233-5; Dijksterhuis, *Mechanization*, pp. 157-8. According to Copenhaver ('Scholastic philosophy and renaissance magic', pp. 525-6), the distinction comes from Galen.

[12] For these limitations on human understanding, see: Aquinas, *Commentary on the* De anima *of Aristotle*, trans. K. Foster and S. Humphries (London, 1951) p. 456 (on Aristotle, *On the soul*, 432a1-10); Aquinas, *Summa theologica*, 1a.84,7, 1a.2ae.91,4 (= v.12, pp. 38-43, v.28, pp. 30-1).

[13] B. Fontenelle, *Conversations on the Plurality of Worlds*, trans. H. Hargreaves (Berkeley, 1990; 1686) p. 11; cf. Galileo, *Discoveries and Opinions of Galileo*, trans. and annot. edn Stillman Drake (New York, 1957) p. 274; Boyle, *Selected Papers*, pp. 31-2; J. Glanvill, *Vanity of Dogmatizing: The Three Versions*, ed. S. Medcalf (Hove, Sussex, 1970) pp. 171-2; Hutchison, 'Occult qualities', pp. 242-9; G.M. Ross, 'Occultism and philosophy in the seventeenth century', in A. Holland (ed.), *Philosophy, its History and Historiography* (Dordrecht, 1985) 95-115, p. 102.

[14] See, for example: Augustine, *De civitate dei contra paganos*, xxi.4-5 (= Loeb Classical Library edition (London, 1957-72), trans. G. McCracken [and other(s)] v.7, pp. 14-33); Aquinas, *Summa theologica*, 2a.2ae.96,2 (= v.40, pp. 74-5). See also Notes 30, 33, 35, 36, 37, 40 below.

[15] Aquinas, *Summa contra gentiles*, III.97.17 (= bk.3, pt.2, p. 72 of University of Notre Dame Press ed., trans. A. Pegis [and other(s)], Notre Dame, 1975). For an extended discussion of this issue, see Hutchison, 'Dormitive virtues', *passim*.

[16] J. Calvin, *Institutes of the Christian Religion*, 2 vols, ed. J. McNeill, trans. F. Battles (London, 1961) v.1, pp. 181, 201; Luther, *Works*, v.1, pp. 29-30.

[17] For various examples of these claims, doubts, and uncertainties, see n.1, plus: Pseudo-Galen, *Diagnosis and Cure of Kidney Diseases* (K¸hn XIX, 677-8) as trans. Copenhaver, 'Scholastic philosophy and Renaissance magic', p. 526; ongoing discussion of power of talismans, Copenhaver, *op.cit., passim*; Aquinas, *Summa theologica*, 1a.110 (= v.15, pp. 2-18), 3a.77,1 (= v.58, pp. 124-31), 3a.75,2 (= v.58, pp. 58-63); Steneck, *Science and Creation*, pp. 97, 184n.38 (Henry of Langenstein on magnetism); Oresme, *Tractatus de configurationibus qualitatum et motuum*, II.xxvi-II.xxx (trans. in M. Clagett, *Nicole Oresme and the Medieval Geometry of Qualities and Motions*, Madison, 1968; 1351-6, pp. 336-55); Grant, 'Scholastic conceptions', p. 17 (Richard of Middleton on Maimonides); Hutchison, 'Occult qualities', pp. 241-2; S. Clark, 'The scientific status of demonology', in B. Vickers (ed.), *Occult and Scientific Mentalities in the Renaissance* (Cambridge, 1984) 351-74, pp. 353-4, 358-60; L. Thorndike, *A History of Magic and Experimental Science*, 8 vols (New York, 1923-53) v.2, pp. 342-3, 346, 358, v.4, pp. 208, 227, 287, 499; Albertus Magnus, *Minerals*, II.i.1 = pp. 55-57, from where the quoted passage is taken. It is clear from its context that Albertus is not discussing manifest properties in this passage, for the opening paragraph of his second Book refers the reader to Book I for these simpler properties. His reference to constituents (in the quoted words) confirms this: see Hutchison, 'Occult qualities', p. 240.

[18] Aquinas, *Summa theologica*, 3a.64,1 (= v.56, pp. 100-5); 3a.78,4 (= v.58, pp. 182-5).

[19] Aquinas, *Contra gentiles*, III.105.10 (= bk.3, pt.2, pp. 96-7); *idem, Occult Works*, pp. 21-3. Cf. Steneck, *Science and Creation*, pp. 102-3.

[20] Ross, *Philosophicall Touch-Stone*, p. 29 (with adjustments to capitalizations, fonts etc.); Oresme, *De configurationibus*, I.xxv, II.xxvi-xxxv (= pp. 236-9, 336-75); Albertus Magnus, *Minerals*, I.i.6, II.i.1 (= pp. 24-5, 56-7). For other examples, see: Aquinas, *Summa theologica*, 1a.110,4 (= v.15, pp. 14-7); *idem, Contra gentiles*, III.103-107 (= bk.3, pt.2, pp. 86-99); Thomas, *Decline of Magic*, pp. 203, 255-7, 362, 368; Walker, *Spiritual Magic*, pp. 107-11; Clark, 'Scientific status', pp. 364-5; Ross, 'Occultism and philosophy', p. 102.

[21] Aquinas, *Contra gentiles*, III.104 (= bk.iii, pt.2, pp. 89-93).

[22] Aquinas, *Summa theologica*, Ia.110.1 (= v.15, pp. 4-5: planets exert some power), 1a.110.3 (= v.15, pp. 12-13: tides caused by power of the moon); Aquinas, *Contra gentiles*, III.92-3, 104 (= bk.3, pt.2, pp. 42-50, 89-93).

[23] Luther, *Works*, v.1, pp. 227-8, v.36, pp. 64-7. Cf. v.1, pp. 95-6; v.35, pp. 33-44.

[24] Whether such nominalism was ever applied to hylemorphism by philosophers much closer to the scholastic camp than Luther, I do not know (but see above, Note 3). The existence of such philosophies would help make sense of those frequent seventeenth-century claims noted above that the scholastic doctrine was vacuous—for (as we have already seen, Note 15) this is patently false of the standard realist version of the theory. For the views of the English Reformers see Kirsten Birkett's paper in this volume.

[25] Aquinas, *Summa theologica*, 1a.110,1-1a.111,1; 1a.114,4 (= v.15, pp. 2-23, 80-5); *idem, Contra gentiles*, III.101-3 (= bk.iii, pt.2, pp. 81-9); *idem, On the power of God* [*Quaestiones disputatae de potentia Dei*], 3 vols, trans. English Dominican fathers (London, 1933; reprinted on demand, Xerox) VI.6.3-5 (= v.2, pp. 167-88, quoting from pp. 177, 180); Oresme, *Livre du ciel*, p. 293; Clark, 'Scientific status', pp. 360-3.

[26] See, e.g.: Maurer, *Medieval Philosophy*, pp. 177-8; J. Wippel, 'Essence and existence', in N. Kretzmann (ed.), *The Cambridge History of Later Medieval Philosophy* (Cambridge, 1982) 385-

410, pp. 394, 407-10; Aquinas, *Summa theologica*, 1a.50,1-2; 1a.76,4 (= v.9, pp. 2-15; v.15, pp. 64-71); *idem, Contra gentiles*, I.17, II.49-56, 68-71 (= bk.1, pp. 101-3, bk.2, pp. 146-68, 203-13).

[27] Clark, 'Scientific status', pp. 362-5; J. North, 'Celestial influence—the major premiss of astrology', in P. Zambelli (ed.), *'Astrologi Hallucinati': Stars and the End of the World in Luther's Time* (de Gruyter, 1986) 45-100, pp. 76-8; Aquinas, *Summa theologica*, 1a.110, 4 (= pp. 14-7); idem, *Power of God*, VI.6.3 (= v.2, pp. 167-78); *idem, Contra gentiles*, III.103.9 (= bk.iii, pt.2, p. 89).

[28] North, 'Celestial influence', p. 47 (citing Cicero, *De natura deorum*, 44 = ii.16 = pp. 242-3); B. Ward, *Miracles and the Medieval Mind: Theory, Record and Event 1000-1215* (London, 1982) pp. 4-5; Copenhaver, 'Scholastic philosophy and renaissance magic', p. 539; Aquinas, *loc. cit.* Note 19, plus: *Contra gentiles*, III.86.10 (= bk.3, pt.2, p. 28).

[29] Aquinas, *Contra gentiles*, II.79-89 (= bk.2, pp. 254-308), III.19-20 (= bk.3, pt.1, pp. 75-81); *idem, Power of God*, III.3.9 (= v.1, pp. 151-3); J. Randall, 'Introduction to: Pomponazzi, *On the Immortality of the Soul.*' in E. Cassirer et al. (eds), *The Renaissance Philosophy of Man* [selections, in trans. with commentary etc.], (Chicago, 1967) 257-79.

[30] Aquinas, *Contra gentiles,* I.3 (= bk.1, pp. 63-6); W. Hine, 'Marin Mersenne: Renaissance naturalism and Renaissance magic', in B. Vickers (ed), *Occult and Scientific Mentalities in the Renaissance*, (Cambridge, 1984) 165-76, p. 166.

[31] E. Gilson, *Reason and Revelation in the Middle Ages* (New York, 1938) *passim*; J. Randall, *The Career of Philosophy*, (New York, 1961) v.1, pp. 31-6; D. Knowles, *The Evolution of Medieval Thought* (London, 1962) pp. 261-4; M. Wilks, *The Problem of Sovereignty in the Later Middle Ages* (Cambridge, 1963) pp. 122-3; Aquinas, *Summa theologica*, 1a.117,1 (= v.15, pp. 131-5); idem, *Contra gentiles*, III.52-3 (= bk.3, pt.1, pp. 177-82).

[32] Quoting John of Jaudan, from Gilson, *Reason and Revelation*, pp. 56-63. Cf. E. Grant, 'Cosmology', in D. Lindberg (ed.), *Science in the Middle Ages* (Chicago, 1978) 265-302, p. 269; Maurer, *Medieval Philosophy*, pp. 195-6; Copenhaver and Schmitt, *Renaissance Philosophy*, pp. 108-9; Aquinas, *Contra gentiles*, II.18-21, III.41-6 (= bk.2, pp. 55-64, bk.3, pt.1, pp. 133-8).

[33] Augustine, *De civ. Dei*, xi.5 (= Loeb v.3, pp. 440-7); xviii.53 (= Loeb v.6, pp. 78-9, citing Acts 1.7 on time of second coming); xxi.4-5 (= Loeb v.7, pp. 14-33); P. Brown, *Augustine of Hyppo: A Biography* (London, 1967) pp. 152. For later resonances of these notions, see: Oresme, *De configurationibus*, II.xxxi = pp. 358-61; C. Webster, *The Great Instauration: Science, Medicine and Reform 1626-1660* (London, 1975) pp. 514; Thorndike, *History of Magic*, v.3, pp. 156-9; Clark, 'Scientific status', p. 367. For the common idea that human ignorance was a punishment for original sin, see Note 36 below. The notion of a sacred mystery is not, of course, intrinsically Christian, and can be found in classical authors, e.g., Lucan, *Civil War*, trans. J. Duff (London, 1928, Loeb Classical Library) I.419 (= pp. 34-5: on the cause of the tides).

[34] Aquinas, *Contra gentiles*, III.38-57 (= bk.3, pt.1, pp. 125-92.)

[35] Oresme, *De configurationibus*, II.xxxi = pp. 358-61.

[36] Ross, *Philosophicall Touch-Stone*, pp. 34, 56-7 (with a large gap in the quotation). For other examples of Ross' notion here that human ignorance results from original sin, see Van Leeuwen, *Problem of Certainty*, pp. 73-4; : Cf. Webster, *Great Instauration*, pp. 324-5, 482, 508; J. Webster, *Academiarum examen*, [1653, pp. 26-30]. Facs. reprint on pp. 67-192 of: *Science and Education in the Seventeenth Century: The Webster-Ward Debate*, ed. annot. etc. A. Debus (London, 1970) p. 27 (=†Debus edn, p. 109).

[37] Aquinas, *Summa theologica*, 1a.84.7 (v.12, pp. 38-43: limited knowledge of God), 2a.2ae.96,2 (= v.40, pp. 74-5: with *non* mistranslated here as 'not yet'); *idem, Contra gentiles*, I.3 (= bk.1, pp. 63-6).

[38] See, e.g.: Giambattista Andreini, *L'Adamo*, [1613]. Trans. on pp. 227-67 of: W. Kirkconnell, *The Celestial Cycle. The theme of Paradise Lost in world literature with translations of the major analogues* (New York, 1967) pp. 245-8; Webster, *Academiarum examen*, pp. 108-12; Thorndike,

History of Magic, v.7, pp. 333, 434, 484, 490-1, v.8, p. 267; G. Molland, 'Roger Bacon and the Hermetic tradition in medieval science', *Vivarium* 31 (1993) 140-60, pp. 141ff.; Thomas, *Decline of Magic*, pp. 271-2.

[39] J. Weyer, *Witches, Devils, and Doctors in the Renaissance*, [*De praestigiis daemonum*, 1583], trans. J. Shea (Binghamton, NY, 1991) p. 26; cf. Aquinas, *Contra gentiles*, II.96-101 (= bk.2, pp. 325-42); Thorndike, *History of Magic*, v.2, pp. 407-8, v.4, pp. 271-2, 284, 324-5.

[40] Thomas, *Decline of Magic*, pp. 270, 271-2, 634; Webster, *Great Instauration*, p. 86; Clark, 'Scientific status', pp. 364-5; Thorndike, *History of Magic*, v.2, pp. 281, 321-2, 865, v.4, pp. 116, 170-1, 224-5, 499.

[41] Osiander, 'To the reader', p. 22 of Copernicus, *On the revolutions*; Thomas, *Decline of magic*, pp. 362-3.

[42] For a selection of sources, see above, Note 5.

[43] Descartes, *Philosophical Works*, v.1, pp. 168-9, 227-8, v.2, pp. 219-20.

[44] Gassendi, in Descartes, *Philosophical Works*, v.2, pp. 168-170.

[45] For a supernaturalistic interpretation of Boyle and Newton, see: Hutchison, 'Supernaturalism and the mechanical philosophy', pp. 298-9, 318-9, 325; McGuire, 'Boyle's conception of nature', *Journal of the History of Ideas* 33 (1972) 523-42 *passim*. For a dissenting interpretation, see S. Jacobs, 'Laws of nature, corpuscles and concourse: Non-occasionalist tendencies in the natural philosophy of Robert Boyle', *Journal of Philosophical Research* 19 (1994) 369-89. For Newtonians who accepted a 'scholastic' interpretation of gravity, see Hutchison, 'Individualism and causal location', pp. 342, 350 n.48.

[46] For Descartes, see *Philosophical Works*, v.1, pp. 105, 158,301-2, v.2, pp. 77-8; for Glanvill, see: Glanvill, *Essays on Several Important Subjects in Philosophy and Religion*, [1676]. Facs. reprint edn (New York, 1970) pp. 20-1; H. Van Leeuwen, *The Problem of Certainty in English Thought 1630-1690* (The Hague, 1963) pp. 85-6.

[47] This is the central thesis of K. Hutchison, 'What happened to occult qualities in the Scientific Revolution?', *Isis* 73 (1982) 233-53.

KIRSTEN BIRKETT

EARLY ENGLISH REFORMERS AND MAGICAL HEALING

Keith Thomas wrote in 1971:

> We are, therefore, forced to the conclusion that men emancipated themselves from these magical beliefs without necessarily having devised any effective technology with which to replace them. In the seventeenth century they were able to take this step because magic was ceasing to be intellectually acceptable, and because their religion taught them to try self-help before invoking supernatural aid. But the ultimate origins of this faith in unaided human capacity remain mysterious.[1]

Why did magic decline? Where did people find the confidence to confront and examine the world, without fear of mysterious powers imbuing the objects they handled? Why did people begin to look to their own power without recourse to techniques which call upon higher powers? What is more, what was it that encouraged people to overcome their fear of the world in general and begin to look upon it positively, as a thing to be explored and utilised? Keith Thomas, in his book *Religion and the Decline of Magic*, surveyed a range of options for the decline of magic, but returned to this point: people in the sixteenth century found an acceptable alternative to magic, and Thomas did not know whence it came. Since Thomas, though other works have been written on magic, no one seems to have taken up this challenge.

This paper argues that perhaps part of the reason for Thomas' inability to find a solution is his incomplete analysis of the Protestant Reformers.[2] Thomas regarded the Reformation doctrine as cold and depressing, and argued that the Protestant doctrine of Providence, while it provided a world-picture of orderliness, could only lead to a passive fatalism. However this doctrine of Providence was not all the Reformers provided. The English Reformers presented a comprehensive and damning critique of magical practices, and moreover put together an alternative world view which made magic not only ineffective but redundant. This paper examines magical healing practices and

G. Freeland and A. Corones (eds.), 1543 and All That, 357 – 374

the way in which the Reformers' criticism of these undermined the beliefs that lay behind them. It will be seen that the religious instruction given by the Reformers created a whole new framework of thought, significantly different from the medieval framework, and one in which magic had no place.

1. BACKGROUND

The English Reformation can be traced back to the Lollards in the fourteenth century and further, but the period under question here is closely linked to the reign of Henry VIII. While the ideas of Luther's reaction against Papal power in Germany were being discussed in Cambridge in the 1520s, Henry was slowly beginning his campaign to be divorced from his wife Catherine of Aragon. This eventually led to his rejection of Papal control in England in 1533. Henry's concerns were primarily political, but a group of men who were interested in doctrinal reformation gained positions in or were propagandists for the new church. Prominent among these was Thomas Cranmer, Archbishop of Canterbury. After Henry's death in 1547 the English Reformation proceeded even more quickly under the reign of Henry's son Edward VI; the new Book of Common Prayer was written, and the new teaching was propagated by law throughout England.

This was happening against the background of European intellectual activity which comes under the heading 'Renaissance'. The Reformation had large debts to the humanist movement and its search for original, untampered texts—it was this desire that led to Erasmus' publication of the Greek New Testament in 1516. The printing press made possible new methods of study, and texts were available on a scale previously unthought of, while at the same time education was improving, creating a wider readership. Texts could now be compared and criticised. Old traditions were being challenged in any number of areas. In 1543, as Henry's reign was drawing to a close and the Reformers were battling to challenge the common view of God, Copernicus published his *De revolutionibus* challenging the commonly held view of the universe. At the same time, while Reformers were insisting on personal knowledge and research into Scripture, Vesalius published *De humani corporis fabrica* demonstrating the importance of research and observation in anatomy.

The intellectual climate was one of re-evaluation and reformation of long-held ideas. This process was to accelerate, particularly in the seventeenth century with what we call the Scientific Revolution. The Reformation was not an isolated religious movement, independent of the rest of intellectual history. This paper attempts to integrate the religious ideas of the Reformation into that intellectual history by seeing how commonly-held beliefs changed to accommodate a new way of looking at the world.

2. MAGICAL HEALING

What is magical healing? We have an immediate intuition about what it is, and may think of ancient books full of spells, robed old men going through arcane ceremonies, witch trials, or any number of other images associated with the word. The problem is, we may group practices together as 'magic' even though the practitioners may never have used the word, and indeed may have seen little similarity between such practices. However, for scholarly purposes we need some means of gathering together our data in a coherent manner, and the word 'magic' will do as well as any other as long as we are clear what we mean by it. I will therefore examine in a brief overview what kinds of practices were prevalent at the time involving healing, and then come to a working definition of the word 'magic'.[3]

By the middle ages, a whole gamut of traditions and superstitions surviving from pagan times were in use. As Christianity entered pagan countries, it had also lent some of its own terms and concepts to folk wisdom. All sorts of medieval people practised this kind of common magic—there is no reason to suppose it was restricted to special individuals.[4] In healing magic we can distinguish three major characteristics: the use of words, objects and actions. One of the most prominent features was the use of words in the form of set charms. Cures could be effected by words alone, or charms would be said over herbs, during the mixing of a recipe, or over the patient. Many charms survived from pagan times, and some were retained by Christian scholars with no change except the substitution of Christian names (God, Jesus, etc.) for pagan deities.[5] Later in the Middle Ages original charms were composed in Latin, drawing on Christian ideas.[6] There were charms for every conceivable ache or pain, and for different types of disease. There were charms to ensure conception, for safe pregnancy and safe delivery. Charms were used as preventative or protective measures.

Charms could be in verse, or in prose: some would tell a story from Christian legend and use the story as an analogy for the healing (as the Jordan stopped flowing when Christ was baptised, so may your blood stop flowing).[7] Diminishing charms would begin enumerating a number of things (warts, swellings) and subtract one each line until none were left.[8] The way in which the charms appeared to work varied. Some used prayers for the sick person, addressed to God or saints.[9] Psalms, masses or prayers could be said over ingredients to enhance their potency. Some charms were directly addressed to the devil, elf or fairy that was causing the disease, which entity would be told to flee: or the illness itself could be ordered to leave 'by the power of', or 'in the name of' some holy figure—God, Jesus, Mary or a saint.[10] Some charms used prayers or snippets of prayer in no coherent fashion, perhaps saying parts of a Latin prayer three times. Other charms used pure gibberish simply as 'magic words'.[11]

The second characteristic of healing magic was the significance given to particular objects. Things all of one colour were held to be particularly potent, for instance the milk from a cow of all one colour. Also particular colours would be associated with illness: a plant with red berries, flowers or roots could be held to be effective for bleeding.[12] Personal goods played an important role: these could be hair, blood, spittle, a footprint, or clothing.[13] These might be used as part of a recipe, or actions might be performed on them to produce an effect in the person by sympathetic magic. The gold of an alchemist, or earth from an ant hill, might cure leprosy.[14] Particular herbs were more significant than others: for instance, the mandrake which had a root supposedly in the shape of a person. Strong or fast animals were more desirable as ingredients than weak or gentle ones.[15] Unusual growths on plants or animals were also held to signify particular potency.[16]

Amulets were an important form of protection. These could be bits of animals, particular plants, or gems. For instance, amber was held to have medicinal or magical properties. The amulet would be worn on the person and would be protective against disease and attacks of enemies.[17] An amulet could even be a piece of paper with words written on it—bits of a gospel, a prayer, holy names.[18] Other objects that had magical potency were certain wells. Many wells held to have magical healing properties in pagan times were adopted by the Christian church and rededicated to a saint.[19] Religious objects especially were held to have potency. Holy water on its own could have healing powers, and it was an extremely common ingredient in healing recipes. Other sanctified objects were regarded in a similar light: a potion drunk from a church bell was made more effective,[20] and holy salt and the oil of unction were believed to have healing powers. The host was a very powerful object, and many precautions had to be taken by churches to prevent it from being stolen.[21] Even a person who saw the host would allegedly suffer no misfortune during the day.[22] The *agnus dei*, a wax image of a lamb supposedly blessed by the pope, was thought to have healing powers.[23] These beliefs were not necessarily endorsed by the church but did exist in popular perception.

The third major characteristic we can see in healing magic was its use of actions. There were many taboos involved in preparation of medicines: picking herbs in silence, not looking behind you, having bare feet, fasting beforehand. Personal purity was important. Sometimes plants could not be picked with iron: others could only be handled with iron. The left hand was significant, and some recipes demanded herbs be picked with the left hand. The sign of the cross was a very popular action and used in all sorts of healing rituals. The state of heavenly bodies could affect the preparation of healing mixtures: some herbs had to be picked with the moon on the wane, or before sunrise.[24]

There are a few underlying beliefs we can glean from the above practices. One is that illness was often held to be caused by an evil agent, called an elf or fairy in pagan folklore, and a demon in Christian language. Another is a belief in a power inherent in words. Some of the coherent prayers may indicate only a supplication to the deity for help, but in cases where the words are taken out of context and make no sense, or are simply gibberish, it seems clear that the words themselves were held to have potency. Similarly we can see a belief that power could reside in objects and actions, or at least could be directed by actions.

If we call these kinds of practices 'magical', how are we to define 'magic'? A suggestion that has been prevalent for some time is that magic is coercive, whereas religion is only supplicative.[25] That is, when the practitioner expects an automatic result from a ritual, that is magic: but when the practitioner is only supplicating the deity, with no certainty of results, that is religion. However this definition leaves us with slightly fuzzy edges. A religious person presumably expects supplicatory prayers to be answered—why else bother praying? A better definition is that magic assumes a technique which will somehow influence the arcane force. In magic, you *do* something, whether it be saying certain words, or inscribing a tablet, or carrying out a ceremony. It may not work, so it is not absolutely automatic: there may have been a mistake in the ritual, or the power addressed may not respond. However you certainly expect it to work, if the technique was carried out properly. Moreover, if the magic does work, you credit the technique with the success: you obviously did it properly. This seems to capture the similarity between all the activities we would regard as 'magical' from pagan propitiatory sacrifice to clerical necromancy.

3. THE MEDIEVAL ATTACK ON MAGICAL HEALING

The medieval Church condemned the practices outlined above and persecuted the practitioners for centuries. Using the Eucharist as a magical remedy for sickness, or inscribing Scripture on an amulet to ward off illness, was regarded as superstition and was vehemently criticised. Why, then, did such activities still proliferate though the Middle Ages? It was certainly not through lack of effort on the part of Church officials to stop them. The answer seems to lie in the way in which the medieval Church thought the divine related to the natural world. It was a relationship in which the supernatural was frequently obvious in the natural.[26]

Most medieval theologians worked on a very concise definition of magic—what was from God was religion, and good; any other supernatural activity was magic, and of the devil or his demons, whether the practitioner realised it or not.[27] However God worked through many different means, through the many diverse channels of the Church. Thus there were many legitimate ways to perform healing through religion.

One legitimate means of healing in the medieval Church was the miracle-working saint. 'Christianity', writes Ronald Finucane, 'was born into a world already familiar with wandering healers and soothsayers; it developed in an atmosphere heavy with magic and miracle.'[28] Miracles played an important role in the conversion of Europe: they were very effective in persuading potential converts. If pagans could not grasp theological principles, they could certainly see the power of Christianity in miracles—particularly if the Christian missionary won in a miracle competition against the local pagan priest.[29] The tombs of the saints were venerated from earliest times, springing from a desire on the part of the church community to honour its martyrs. The attribution of healing miracles to the saints was well established and began to be advertised by the fifth century.[30] Relics came to be elevated. Purification rituals were introduced which were necessary in order to approach the relics.[31] Battles were fought and huge amounts of money were at stake over the relics. By the sixteenth century healing shrines were very popular and pilgrimages were frequent.[32]

As well as this activity surrounding the saints, the Church also offered spiritual blessing on all sorts of secular activities. Not only could persons be blessed and exorcised but also houses, cattle, crops, ships, tools, armour, wells and kilns.[33] Holy water could be genuinely (that is, with Church sanction) used as a remedy for sickness, fertility for fields and animals, and protection against disaster.[34] Above all, the sacraments had particular significance. They were channels of God's grace to the people, and medieval theology gave them very strong potency.[35] Jesus was present in the mass—not just as an idea, but really and effectively. Many miracles were recorded in lives of the saints, in which men or women would see Christ, or his blood, or have various visions. These were seen as true manifestations of God, possible to the holy in connection with the presence of Christ on the altar. Baptism saved people from sins; it brought God's grace to a person. Ordination passed the spiritual authority that came through an unbroken tradition from the apostles. Theologians of the Medieval Church believed in real experiences of divine origin that happened around the physical objects that were particularly significant in their worship of God.

The official Church position was always against superstitious use of the relics and sacraments, and the Church tried to curb the abuses and excesses of relic-mongering.[36] Magic of all sorts was condemned, in penitentials and other official pronouncements.[37] However it proved very difficult for the Church to control local beliefs. The Church wanted the relics preserved, and encouraged the veneration of the saints. Also the Church was fully confident that miracles were done by saints, through the power of God, and applauded them. However to many of the pilgrims using the shrines the theological distinctions of the Church were irrelevant: the important thing was what they experienced at the shrine.[38] The problem was, where did legitimate religious use of relics cease and magic begin? Monks at a shrine could give a patient water to drink in which bones of a saint had been washed, for the healing power of the saint was held to be transferred to the water. Even when the water was poured on the earth, the dust there when the water dried was held to have the same healing power.[39] The problem existed with the distinction between religion and magic in general: Ælfric, an early medieval cleric, wrote 'No one shall enchant a herb with magic, but with God's word shall bless it, and so eat it'.[40]

Magic done by direct invocation of the devil, as heretical devil worship, was obviously evil in the Church's eyes, but even magic using herbs with a few muttered charms was evil—for either the devil taught the person the charms, or any efficacy they had was by the power of demons, even if unbeknownst to the magician.[41] However when there was so much legitimate divine activity, through the many heavenly beings with which the medieval universe was peopled, it was difficult to make the criticism as forceful as theologians would have liked. The matter was complicated further by Church emphasis on words and rituals. It came down to a matter of Church authority. Thus the very framework of thought on which the medieval Church rested made it impossible for an attack on magic to be successful. As long as the supernatural world was so prevalent in the natural, magic continued to be a popular recourse for people in pain.

4. THE PROTESTANT ATTACK ON MAGICAL HEALING

In the works of the English Reformers the word 'superstitious' is used again and again as a criticism of Church ritual. Which practices were superstitious? To a large extent, those which were used for the kinds of magical healing surveyed above. For instance, any belief that held there to be divine significance in *objects* was, in the eyes of the Reformers, superstitious. Thomas Cranmer, Archbishop of Canterbury, was quite passionate on the subject. He wrote, 'What thing can be more foolish, more superstitious, or ungodly, than that men, women and children, should wear a friar's coat to deliver them from agues or pestilence'.[42] He condemned the man who practised 'casting holy water upon his bed, or bearing about him holy bread, St John's Gospel, ringing of holy bells, or keeping of private holy days'.[43] Scattered throughout the documents that make up the official liturgy of the Edwardian reign are many orders to remove any objects that may be used superstitiously, to discontinue superstitious practices and to punish anyone who refuses to do so. Moreover, these commands are often accompanied by orders to explain precisely why the action is being taken and—if the minister himself has taken part in such practices—to recant publicly and repent of such actions.

This attitude is also reflected in the way ceremonies were described. The Reformers did not think ceremony itself was bad; what was bad was an attitude of blind trust in the power of the ceremony, rather than what it represented. William Tyndale, Protestant polemicist and Bible translator, wrote of the spiritual man, who 'ceaseth not to search the cause'[44] and 'in all ceremonies and sacraments, he searcheth the significations, and will not serve the visible things'.[45] Superstition was the opposite—believing there was power in the visible things. In Tyndale's opinion, anything used superstitiously by people was condemned. In this way he criticised practically the whole of medieval church practice—saints, images, holy days, all the sacraments, vestments, pilgrimages and so on. This 'superstitious' attitude was revealed in inordinate attention given to details, in a way that missed the entire significance of the ceremony:

> For if the priest should say mass, baptise, or hear confession, without a stole about his neck, [people] would think all were marred, and doubt whether he had power to consecrate, and think that the virtue of the mass were lost, and the child not well baptised, or not baptised at all, and that his absolution were not worth a mite.[46]

The saints and all the veneration surrounding them were dismissed wholesale. Cranmer complained of 'pilgrimages unto images ... kneeling, kissing, and censing of them' as superstitious. This was made official in several different injunctions. In Cranmer's *Visitation* articles for his diocese he ordered the parsons to be asked:

> ... whether they have not removed, taken away, and utterly extincted and destroyed in their churches, chapels, and houses, all images, all shrines ... all monuments of feigned miracles, pilgrimages, idolatry, and superstitions, so that there remain no memory of the same in walls, glass windows, or elsewhere.[47]

The parson is also ordered to have 'openly recanted and reproved the same'.[48] Edward's Injunctions of 1547 condemned images, relics and miracles, and instructed the minister that they should not entice people to Pilgrimages. Moreover the minister must preach a sermon at least every quarter in which, as well as exhorting the hearers to faith, mercy and charity, he should declare that 'works devised by mens phantasies' such as 'Wandring to Pilgrimages, offering of Money, Candles, or Tapers, or Relicks, or Images, or kissing and licking of the same; praying upon Beads, or such like superstition' are unacceptable.

The Books of Common Prayer reflected this opinion of the Reformers that Church objects were being used superstitiously and that this must be stopped. The communion services of 1549 and 1552 display deliberate efforts to remove superstitious practices. The way of handling the consecrated elements changed from medieval ceremony. There was no elevation or showing of the sacrament to the people (and this point was emphasised). This was perhaps a direct answer to beliefs that seeing the sacrament gave temporal benefit. Also in 1549 the notes of explanation insist that the sacrament be put directly in the communicant's mouth, instead of in their hands, to prevent people carrying the sacrament, for such people 'diversly abused it to superstition and wickedness'.[49]

Here we can see an important difference between the thought frameworks of the medieval Church and the Reformers. Both forbade 'superstitious' use of Church objects. What they meant by this, however, was completely different. The medieval Church believed there *was* potency in the objects of Church ritual—the holy bread, holy water and so on. They simply did not want illegitimate use of that power. The Reformers on the other hand thought that the belief that such things *could* have potency was itself superstitious. To them, any object or ceremony was only an outward sign or reminder of God. God acted and revealed his power in different ways—in the heart of the believer, for instance. Where did the Reformers get this radically different view of the world? To discover the framework that lay behind this belief that natural objects and human rituals had no power, we must look at the reasons for the condemnation.

The basis of the Reformers' rhetoric was a strong belief in the truth of the Bible. Thomas Cranmer complained about the people placing 'superstitious' trust in a friar's coat because such things are not the commandments of God, but of men. The failure to distinguish between these would in Cranmer's eyes, lead to 'error, superstition, idolatry, vain religion, preposterous judgement, great contention, with all ungodly living'.[50] Edward's Injunctions complained that these 'superstitious' things are wrong because they are not commanded in the Bible. This ran through the Reformers' works—Church ceremony and ritual were inherently human inventions, so without power.

This belief was reflected in the sacramental doctrine of the Reformers, which demonstrated a radical departure from medieval thought. Jesus was not present in the bread: it was natural bread, like any other bread. Cranmer argued that the passages in Scripture about eating and drinking Christ were figurative: they were references to the hunger and thirst of the soul for spiritual things. Such hunger and thirst are comforted by Christ. As Cranmer saw it, as we are 'a carnal people' Christ gave the sacrament that 'as surely as we see the bread and wine with our eyes ... so assuredly we ought to believe, that Christ is our spiritual life and sustenance of our souls, like as the said bread and wine is the food and sustenance of our bodies'.[51] Therefore the doctrine of transubstantiation was missing the point. Christ never said his body and blood would physically be in the bread and wine; it was merely a spiritual metaphor.

If this was true, the Eucharist could be of benefit only to Christians, and then only to remind them to trust Jesus completely for their salvation. It was not efficacious for salvation and most definitely not useful for anything else. If the bread and wine had none of God's spirit, of Jesus' body, or any spiritual power, any superstitions based on the presence of God in the actual bread collapsed. The superstition was made redundant, not just wrong.

At the centre of Tyndale's theology was the idea that God made promises. The proper response to promises is to trust them—Tyndale placed this in sharp opposition to trusting one's own actions, as one went about Church activities and ceremonies. The word 'trust' features very strongly in Tyndale's work, and can almost be seen as his touchstone for Christianity: do you trust God alone, or not? Anything that contributed to the 'or not' had to be done away with. This was behind Tyndale's vehement denunciation of Church ceremonies and theologians:

> For when they come to the point, that they should minister Christ's passion unto the salvation of our souls, there they poison altogether and gloss out the law ... and teach us to put our trust in our own works for the remission and satisfaction of our sins.[52]

This idea flowed on to Tyndale's view of sacraments. As he saw it, the sacraments were signs of the promises of God. The sacrament itself was not powerful—the promise it symbolised was. For instance, 'outward toil can neither heal the soul, nor make her feel, save as a sign ... neither is it a thing to put trust in'. The Lord's Supper is a sign given to make the conscience certain of forgiveness. Indeed, the sacraments preach, and serve the believer: it is not right to serve the sacrament or worship it. To separate the sacrament from its significance is extremely dangerous: 'But when he leadeth me by the darkness of sacraments without signification, I cannot but catch harm, and put by trust and confidence in that which is neither God nor his word'.[53] Such things as penance, matrimony and holy orders do not even have a promise attached: therefore they are not sacraments at all, and have no particular spiritual significance. The same could be said of other ceremonies. They should exhort you to trust in God, but Tyndale contended that priests distorted the ceremony so as to make you trust in it. The official Church doctrine may not have encouraged such trust. However Tyndale was not interested in this; rather he was interested in the way in which people actually used the ceremonies. He asked:

> How is it possible that the people can worship images, relics, ceremonies and sacraments, save superstitiously; so long as they know not the true meaning, neither will the prelates suffer any man to tell them.[54]

The Reformers also attacked contemperary miracles. Cranmer quoted many Bible verses predicting the advent of false prophets doing false miracles by the power of the devil. This criticism alone meant only that miracles could be of the power of the devil, not that all modern miracles were. However Cranmer was certain that any miracles done by the official Church must be of the Devil, as God could not possibly be working through what Cranmer saw as an heretical organisation. The interesting point is that Cranmer and Tyndale never denied the reality of modern miracles. The argument was not only over the occurrence of miracles, but their authority. In fact, in describing contemporary Church miracles, Cranmer moved between apparently real miraculous happenings to obvious frauds with no change in tone. In Cranmer's mind, it seems that whether the deed was done using sleight of hand or mysterious power was irrelevant, for it was all done by the devil. From our twentieth century perspective, this may not seem a disincentive for people who want to be healed at a shrine. However, to Cranmer, all things had an ultimately supernatural explanation so a miracle was nothing special in itself: to whom it testified was the important thing. His attack on miracles had nothing to do with their natural probability—in a world governed by God, in warfare with the devil, such a consideration was irrelevant. Rather, it was their significance that mattered. Whatever may have happened in saints' tombs God was not behind it.

The result was a general scepticism about all modern miracles. The possibility was left open that God might do a genuine miracle—but from the tone of the rhetoric, it is clear that Cranmer and Tyndale believed no such thing was going to happen. If pilgrimage only brought you close to the devil it was hardly an incentive to go. Certainly the Reformers denied that the outward signs of saints—relics and so on—had any power to heal.

5. THE PROTESTANT UNIVERSE

The Protestant thought-framework made it easy for Reformers to distinguish between acceptable religion and unacceptable magic. There was no gradation between heaven and earth: no saints who may intervene in daily life. There was humankind on earth, and God in heaven, and the only other power at work in the world (apart from angels, who only did the Lord's bidding) was the devil. It was a stark picture. So any miraculous occurrence was directly from God, or from the devil. This criticism of magical practices could be much more forceful than that available to the medieval Church. It did not need to leave room for legitimate religious practices which might be confused with illegitimate magical ones.

What is more, in Protestant theology the way in which God acted was much more circumscribed than in the medieval framework. He did not communicate with his people, or act in the world, through objects. He upheld and maintained every natural occurrence in the world: therefore no particular part of the natural world had special holiness. Thus there was no such thing as holy water, that is water more imbued with God's spirit than any other water. This was reflected in the views on the Eucharist. There were many different kinds of arguments used by Protestants to prove that the body of Jesus could not be present in the host. The fact is, such a doctrine simply would not fit into the Protestant framework. There was no place for God to be within a piece of bread—the Protestant God did not act that way.

Moreover, the way to approach their God was radically different. There was no technique—ceremonies, if they were retained at all, were only for the sake of keeping order amongst the people. Protestant prayer was direct conversation with God. There was expectation that prayer would work: but the expectation depended wholly on the character of God, and in no way on action (whether it be words, beads, or anything else). What the Reformers criticised as superstition was invariably religious technique, where the supplicant would rely upon the technique to sway God rather than relying upon God himself to act freely.

In place of the medieval framework, the Reformers presented a world view in which God was to be relied upon directly, without technique. It was a positive attitude towards the world. Trust in God was not obedience to an oppressive command, but a disposition to hope. God was repeatedly presented as the defence against evil. Edward VI's injunctions put complete trust in God as an official command. The people must be made to understand that God is to be trusted to help, and that despair is ignoring his plain offer of generosity. Prayers that occur throughout the whole liturgy emphasise the goodness of God. There are prayers that ask God to defend the believer against all manner of things: the assaults of enemies, fear of enemies, perils and dangers of the night, all adversity of body and soul, evil and mischief, lightning and tempest, plague, pestilence, battle, sudden death. There are prayers for morning, for evening, for getting up, for going to bed, for rain, fair weather, help from dearth, war, plague. The implication for the believer comes out clearly: 'defend us, thy humble servants, in all assaults of our enemies, that we, surely trusting in thy defence, may not fear the power of any adversaries'.[55]

Moreover, death and sickness were not the worst things, and should not persuade people to despair. Tyndale was eloquent in his exhortations to trust God in suffering. No suffering should ever persuade you to abandon God, he wrote: for God visits people with sickness, poverty and adversity out of fatherly love, as discipline. In the Prayer Book service for Visitation of the Sick God would be asked to give the sick comfort and confidence in him, and that the sick person may be restored to health or given grace to endure (this combination appears frequently concerning the sick in the Reformation liturgy).

This was not a passive fatalism. Quite the contrary: it was a faithful optimism that God would look after his children, that he wanted the best for them, and did not wish them to suffer. It was positive whatever happens—the sick person would be restored to health, or given grace to endure. As well as this, it was then quite all right to consult a doctor. If God worked in all the world, the work of a doctor was not opposed to God's will. God cared for his children: there was an emphasis on God's love, and his willingness to be generous in all areas of life. The sick person was assured that if he suffers, to remember that God only disciplines those he loves, and that suffering in the end will turn to profit, the great gift of eternal life. Instructions on how to pray showed an utter trust in God: 'Ask of him all things needful both for soul and body, privately for thine own self, and thy family, and generally for all the Christian congregation.'

6. CONCLUSION

What would the Protestant message say to the person who is relying upon magical healing? Essentially the Reformers demystified the world. They claimed that such things as the Eucharist had no essential power, and if used as a cure was no more effective than any other piece of bread. That condemned any magical healing charm you care to mention—the inscriptions used against disease or evil spirits that may cause disease, the charms recited to drive away illness, the wafers inscribed with Scripture that were swallowed as medicine. They did not work, said the Reformers—if God wanted you to be healed he was perfectly capable of providing a worldly remedy. The Protestant God himself did not appear in natural objects.

Obviously not everyone in England took on the Reformers' theology. However, they did not need to in order for this message to have an effect. Anyone could protest against the medieval Church without necessarily agreeing with the doctrine that the Reformers presented. The negative force of the Protestant attack—that is, the criticism of all they called superstition—was powerful in itself. They proclaimed all magical healing did not work, and so gave authority to the idea for anyone who suspected as much anyway. What is more, we cannot ignore the very practical action of removing shrines and discontinuing the mass.

However for the person who did accept the Reformed theology, there was a powerful religious impulse to trust God—an impulse quite powerful enough to banish magic. Such a person lived in a world without evil interference because God himself oversaw all things and had their good in mind. Objects and rituals did not and could not have special power: God did not work through them. The world was open to investigation and manipulation by ordinary people. There were no mysterious forces in the world that needed to be feared or controlled.

Could this be part of the reason for the decline of magic? Keith Thomas does not find any causal influence—and I would contend that this is perhaps because in his chapter on 'Providence' he misunderstands the Protestant message. He is quite right that the Reformation upheld a doctrine of general providence, which saw an ordered world. God's work in the world was moral providence—reward for virtue, punishment for evil, a view which became increasingly unpopular through the seventeenth century. However Thomas stops there, claiming that this only led to a pessimistic fatalism—'God controls when I die so why would I seek a doctor?' Thomas failed to see that the Reformers' message did not stop at Providence, which could after all be controlled by an impersonal force, but in fact involved Fatherhood. Sovereignty was more than moral providence. It was not 'an explanatory theory based on guilt'.[56] It was not 'a gloomy philosophy, teaching man how to suffer, and stressing the impenetrability of God's will'.[57] On the contrary, it taught man that regardless of any suffering, the almighty God loved him and would not remove his love from him. This was a basis for hope and forbearance. So it is not a paradox that 'those who did most to proclaim God's sovereignty were also those most active in helping themselves'[58]—this was rather a natural response to confidence in God. The Reformers' God was a loving father who looked after his children. Someone who believed that would have the confidence to put aside fear of suffering, of death or of evil spirits, and look boldly at the world that his God had made.

Matthias Centre for the Study of Modern Beliefs, Sydney

NOTES

[1] K. Thomas, *Religion and the Decline of Magic: Studies in Popular Beliefs in Sixteenth- and Seventeenth-Century England* (Harmondsworth, 1971) p. 794.

[2] Thomas describes aspects of the Protestant attack on magic, and concludes that this affected belief in the supernatural power in physical objects: 'Many men were now unwilling to believe that physical objects could change their nature by a ritual of exorcism and consecration' (p. 86). However Thomas focuses only on the negative aspect: the removal of means of supernatural aid for the individual: 'He could no longer rely upon the intercession of intermediaries, whether saints or clergy; neither could he trust in an imposing apparatus of ceremonial in the hope of prevailing upon God to grant his desires' (p. 87). The implicit question is 'Well then, on what could he rely?' Thomas highlights this at the end of his chapter 'The Impact of the Reformation' (p. 89) with the reminder that the problems of life remained—plague, disease, fear. If the Reformers had removed the magic of the church, what would take its place? I believe the Reformers answered with their doctrine of the sovereign fatherhood of God, which Thomas does not explore fully.

[3] The literature on magic is immense. Despite my difficulties with Thomas' view of the English Reformers (a small part of his work) his book *Religion and the Decline of Magic* is still arguably the best survey of magical practices in sixteenth- and seventeenth-century England. A few other works: Wilfrid Bonser, *The Medical Background of Anglo-Saxon England: A Study in History, Psychology, and Folklore* (London, 1963) deals with the English material until the eleventh century; Richard Kieckhefer, *Magic in the Middle Ages* (Cambridge, 1989), is a very helpful study of medieval practices; Richard Cavendish, *A History of Magic* (London, 1977) is a less scholarly general history of magic. I have not included in this paper the more learned magic of alchemy, astrology, Paracelsianism and the Hermetic tradition.

[4] Kieckhefer, *Magic in the Middle Ages*, p. 56.

[5] Bonser, *The Medical Background of Anglo-Saxon England*, p. 120.

[6] Kieckhefer, *Magic in the Middle Ages*, p. 72.

[7] Kieckhefer, *Magic in the Middle Ages*, p. 71; Bonser, *The Medical Background of Anglo-Saxon England*, pp. 241-243; Thomas, *Religion and the Decline of Magic*, p. 212.

[8] Bonser, *The Medical Background of Anglo-Saxon England*, p. 252

[9] Kieckhefer, *Magic in the Middle Ages*, p. 70; Thomas, *Religion and the Decline of Magic*, p. 211.

[10] Kieckhefer, *Magic in the Middle Ages*, p. 71; Dawson, George, *Healing: Pagan and Christian* (London, 1935) p. 162.

[11] Bonser, *The Medical Background of Anglo-Saxon England*, p. 245; Thomas, *Religion and the Decline of Magic*, p. 213.

[12] Bonser, *The Medical Background of Anglo-Saxon England*, p. 216.

[13] *Ibid.*, p. 221.

[14] Brody, Saul Nathanial, *The Disease of the Soul: Leprosy in Medieval Literature* (Ithaca, 1974) p. 72.

[15] Kieckhefer, *Magic in the Middle Ages,* p. 67.

[16] Bonser, *The Medical Background of Anglo-Saxon England*, p. 223.

[17] Kieckhefer, *Magic in the Middle Ages*, p. 75; Bonser, *The Medical Background of Anglo-Saxon England*, p. 231.

[18] Bonser, *The Medical Background of Anglo-Saxon England*, p. 236; Kieckhefer, *Magic in the Middle Ages*, p. 77; Thomas, *Religion and the Decline of Magic*, p. 212.

[19] Dawson, *Healing*, p. 171.

[20] Bonser, *The Medical Background of Anglo-Saxon England*, p. 225.

[21] Cavendish, *A History of Magic*, p. 51; Thomas, *Religion and the Decline of Magic*, pp. 38-9. The

1549 Book of Common Prayer warned against people who carried the sacrament away in their mouths and 'diversly abused it to superstition and wickedness' (p. 99).

[22] Kieckhefer, *Magic in the Middle Ages*, p. 79; Thomas, *Religion and the Decline of Magic*, p. 39.

[23] Kieckhefer, *Magic in the Middle Ages,* p. 78. See also Cavendish, *A History of Magic*, p. 50.

[24] Kieckhefer, *Magic in the Middle Ages*, p. 67; Bonser, *The Medical Background of Anglo-Saxon England*, pp. 223-24, 228-30.

[25] See Kieckhefer's discussion of definitions of magic, pp. 14-6. Thomas also rejects this hard distinction between magic and religion as being useful in theory but blurred in practice, p. 46.

[26] The word 'supernatural' is used in this paper in its general sense as a convenient label for spiritual and occult forces or entities, and does not reflect the kind of detailed analysis of medieval technical usage that Keith Hutchison has so ably done.

[27] For instance, this view is propounded by Augustine, *The City of God*, VIII.19.

[28] Finucane, Ronald C., *Miracles and Pilgrims: Popular Beliefs in Medieval England* (London, 1977) p. 18.

[29] Bonser, *The Medical Background of Anglo-Saxon England*, p. 127; Thomas, *Religion and the Decline of Magic*, p. 28, Finucane, *Miracles and Pilgrims*, p. 20.

[30] Bonser, *The Medical Background of Anglo-Saxon England*, pp. 172, 178f; Finucane, *Miracles and Pilgrims*, p. 17; Peter Brown, *The Cult of the Saints: Its Rise and Function in Latin Christianity* (Chicago, 1981) gives an excellent analysis of this topic.

[31] Finucane, *Miracles and Pilgrims*, pp. 27, 48.

[32] *Ibid.*, p. 202.

[33] Thomas, *Religion and the Decline of Magic*, p. 32.

[34] *Ibid.*

[35] Dawson, *Healing*, p. 163.

[36] Finucane, *Miracles and Pilgrims*, p. 19.

[37] Bonser, *The Medical Background of Anglo-Saxon England*, p. 129, pp. 148-50.

[38] Finucane, *Miracles and Pilgrims*, p. 38.

[39] Bonser describes these practices as they were used in conjunction with the relics of St Oswald and St Petroc (p. 187).

[40] Bonser, *The Medical Background of Anglo-Saxon England*, p. 119.

[41] Kieckhefer, *Magic in the Middle Ages*, p. 10.

[42] Cranmer, Thomas, *Works*, edited for the Parker Society in two volumes, Vol II, p. 147.

[43] Cranmer II, p. 158.

[44] Tyndale, William, *Works*, edited by Rev Henry Walter for the Parker Society in three volumes, Vol III, p. 7.

[45] *Ibid.*

[46] Tyndale III, p. 8.

[47] Cranmer II, p. 155.

[48] Cranmer II, p. 156

[49] *The Two Liturgies, A.D. 1549 and A.D. 1552, With Other Documents Set Forth by Authority in the Reign of King Edward VI*, edited by Joseph Ketley for the Parker Society, p. 99.

[50] Cranmer II, p. 148.

[51] Cranmer I, p. 20.

[52] Tyndale II, p. 12.

[53] Tyndale III, p. 149.

[54] Tyndale III, p. 62.

[55] *Two Liturgies*, pp. 35, 225.

[56] Thomas, *Religion and the Decline of Magic*, p. 130.

[57] *Ibid.*, p. 132.

[58] *Ibid.*, p. 131.

BARRY BRUNDELL

BELLARMINE TO FOSCARINI ON COPERNICANISM: A THEOLOGIAN'S RESPONSE

My purpose is to try to understand a little better the condemnation of Copernicanism in 1616 and of Galileo in 1633 by paying special attention to the theology upon which these judgements were based. I approach this task by studying the reply of Cardinal Robert Bellarmine (1542-1621) to the letter of Father Paolo Antonio Foscarini. Though Bellarmine replied briefly and in some haste, he provided a comprehensive exposition of the basic points of the Roman theology that conflicted with the Copernican system, consequently his letter to Foscarini furnishes valuable indications of the theological motivation behind the actions of the Roman authorities in the Galileo affair. Surprisingly little attention has been given to the theological issues of what was at least as much a theological event as a scientific one, while the historical and philosophical aspects have been thoroughly explored: Francois Russo, Ernan McMullin, Richard J. Westfall and Richard J. Blackwell are among the few scholars who have discussed the theological questions at any length.[1] I shall draw together a few strands of the discussion as it stands, make a few precisions of my own along the way, and move more expressly than hitherto beyond the particularities of the early seventeenth-century conflict in order to explore its significance for our understanding of the relationship between Catholic theology and the natural sciences.

The seventeenth century was not one of the better centuries for Catholic theology: the Counter-Reformation was moving into an advanced stage and the Church community as a whole had arrived at a degree of clarity about positions to be held and doctrines to be rejected that we find chilling. It is understandable that people should want total clarity in a time when it was important to know who was with you and who was not, who was inside your Christian community and a supporter, and who was outside and a danger to the integrity of your faith, but the effects on theology were unfortunate as theologians had a tendency to respond to that popular desire by drawing excessively clear lines.

G. Freeland and A. Corones (eds.), 1543 and All That, 375 – 393

The Counter-Reformation was a time, too, when it can be said that one person, Robert Bellarmine, *personified* the Catholic Church when he acted to safeguard doctrinal integrity. That may seem a large claim, but those were unusual times when impulses towards cohesion and uniformity of belief were exceptionally strong. Although he never was the Cardinal Prefect of the Holy Office nor of the Congregation of the Index of Forbidden Books as is sometimes asserted,[2] he was a key figure in the Counter-Reformation nonetheless. He was the outstanding Catholic theologian of his time, Professor of Controversial Theology (1576-94) and Rector (1592-94) of the Roman College before being made a Cardinal and called upon for special missions by successive popes. His theological status was mainly based on his *Disputations on Controversies concerning the Christian Faith against the Heretics of Our Time*,[3] a monumental work that was acclaimed throughout the Catholic world and used extensively as a basic text in controversies with Protestants. Richard Westfall has given good textual evidence to support the view that Bellarmine was a key figure also in the proceedings that led to the Catholic Church's condemnation of Copernicanism in 1616, and that his influence was still strong in the trial of Galileo in 1633.[4] The Copernican question was not Bellarmine's area of competence; when in 1611 he felt that he needed to know more about the issue he asked the views of Father Clavius and his colleagues at the Roman College.[5] Thus Bellarmine did not represent the best Catholic opinion available on the Copernican theory, nevertheless his personal authority was such that his response to the crisis was the dominant response. If we understand Bellarmine's actions and what motivated him we may feel confident that we have an accurate understanding of his Church's actions and motives.

Cardinal Bellarmine expressed his position on the Copernican question most explicitly in his letter to Father Paolo Antonio Foscarini (1580-1616).[6] But this letter, dated the 12th of April 1615, is doubly significant because his reply to Foscarini was at the same time an indirect warning to Galileo whose letter to Benedetto Castelli of the 21st of December 1613 had been delated to the Holy Office just two months earlier, in February 1615, by the Florentine Dominican Father Lorini. By early March Galileo's friend Monsignor Piero Dini had also given a copy of the letter to Bellarmine personally.[7]

It will become apparent as we examine the letter and try to put it into its context that the actual Copernican theory hardly came into Bellarmine's discussion at all, a fact which may legitimately be construed as an indication that he was not primarily concerned with astronomical and cosmological questions, not just because they were not his primary interest personally, but because they were not the official Church's primary interest either. Such in fact will be the conclusion I will offer from this brief study: that Bellarmine, the Counter-Reformation Church, in fact the Church at any time, has not been and is not primarily concerned with physical cosmologies or world-views.

By most people's standards, Foscarini's letter 'On the Opinion of the Pythagoreans and Copernicus concerning the Mobility of the Earth and the Stability of the Sun'[8] presented a good case for dropping the scriptural objections to Copernicanism. He wrote in a relaxed and urbane manner about psychological difficulties we experience when we are forced to adjust to new opinions; he clearly affirmed the absolute necessity of rejecting anything found contrary to divine authority and to the sacred words dictated by the Holy Spirit and its inspired interpretation by the Sacred Doctors, that is the Fathers of the Church; he recalled that many, including the Jesuit Father Clavius (+1612), had expressed dissatisfaction with the Ptolemaic astronomical system and had been searching for a better one; he stated that recent telescopic observations indicated that the Copernican system was the best alternative available, and noted that, alas, it was not in favour because of objections based on Scripture.[9] So, Foscarini wrote, he had decided to try to accommodate the problematic passages of scripture to the Copernican system, and he believed that he was the first to attempt to do this. He thereupon proceeded to deal with a whole range of biblical texts suggesting a number of commonsense principles for their reinterpretation such as the need to recognise that much of the Bible is written in figurative language[10] and that 'God teaches only the road to eternal life, not curiosities'.

Foscarini, pleased with his effort, sent a copy to Cardinal Bellarmine. He got back a reply that was courteous—Bellarmine said that he had read the letter with pleasure and found it full of skill and learning—but which was an uncompromising and thoroughly discouraging rejection not only of the main points of Foscarini's argument but also of the feasibility of his very project. What is more, to his indubitable dismay, Foscarini found his letter declared to be 'altogether prohibited and damned' a few months later in the same decree of the Congregation of the Index by which the *De revolutionibus* of Copernicus and the commentary on Job of Diego de Zuñiga S.J. were suspended until corrected.[11]

The fundamental problem with Foscarini's letter in Bellarmine's judgment was that there was not one mention in it of the Church's authority in interpreting the Bible, especially on matters of faith, and in determining what was taught in the Bible by divine authority. Foscarini, for Bellarmine, was a clear example of the type of person who most preoccupied Catholic Church authorities at that time: a freelance interpreter of the Bible.

In his reply to Foscarini Bellarmine prefaced his remarks by pointing out that he, Foscarini, had described many ways of explaining the scriptures without applying them in detail. Then he moved to what he considered to be the heart of the matter, making three essential points in which we note the Cardinal's emphasis on the authority of the Church. In summary form, Bellarmine's three points were: first, the instrumentalist interpretation of the Copernican theory is the only permissible interpretation; second, it is forbidden to interpret Scripture in a way that contradicts the common interpretation of the Fathers of the Church; third, the Copernican theory does not impose a reinterpretation of Scripture passages.

But the heaviest blow was Bellarmine's surprise assertion that these matters of the sun revolving around the earth and of the earth being motionless in the centre of the universe were matters of the faith; perhaps not *ex parte objecti* or by reason of the subject treated, Bellarmine conceded, but *ex parte dicentis* or by reason of Him who enounces it, since it was the Holy Spirit who had said it. Bellarmine based this claim on traditional exegetical principles, as we shall see, but even within that tradition his conclusion is strange and makes us question the logic of his argument or, at least, the single-mindedness with which he had pursued the logical path that led to such a conclusion. In effect he asserted that even trivial 'truths' were a crucial part of the Christian faith and to be believed under pain of exclusion from the Christian community. He would not find much warrant from theological tradition for his claim, but far from qualifying it or backing away from it he reaffirmed it with examples:

> He who should deny that Abraham had two sons and Jacob twelve would be just as much a heretic as a man who denied the Virgin Birth of Christ, because it is the Holy Spirit who speaks both truths by the mouths of the Prophets and Apostles.[12]

And thus Bellarmine concluded that divine authority, the Holy Spirit, had spoken in such terms in the text of the Bible that geocentrism and geostasis were to be believed as facts revealed by God.

Before turning to the premisses of Bellarmine's argument from which he derived this conclusion it will be helpful to review a little of the theological context. The Catholic Church in the late sixteenth and early seventeenth centuries was in an early stage of what was to be a long phase of reaction against the Reformers who had appealed to the authority of Scripture in their attacks on the Pope and the traditional Church. The watchword of the Reform movement was the principle 'Scripture alone', often applied in rejection of human authority in spiritual matters and above all of the claims to authority of the representatives of the Roman Church. In their efforts to counter these attacks from Luther and his supporters the Roman theologians did not want to deny the truth of the principle 'Scripture alone', but they needed to assert the authority of the pope, bishops, the Council of Trent, the Fathers of the Church, and Church tradition.[13] All this struggle between Reformers and Counter-Reformers forms the backdrop to Bellarmine's peremptory rejection of Foscarini's well-meaning efforts.

Then there was the Melchior Cano factor. One reason why theologians of the Counter-Reformation period had an exceptionally clear understanding of what was permitted and what was not in theology was that they accepted as their authoritative model the *De locis theologicis* (1564)[14] of Melchior Cano. This work is an enduring theological classic, a systematic treatise which sets out a critical methodology for theology in response to Luther's rejection of every theological norm except Scripture. Cano identified ten *loci* or grounds for theological argument, Scripture being the first of the ten; the second was apostolic tradition which Cano listed together with Scripture as integral to 'revelation' itself; then he listed five more *loci* in order of importance and authority: the Catholic Church, general councils, the Church of Rome, the Fathers of the Church, and scholastic theologians and canonists. All of these latter five *loci* were needed for the conservation, interpretation and transmission of revelation. These first seven *loci* were proper to theology; the three remaining of the ten were natural reason and the rational sciences, philosophy and the jurists, history, documents and oral traditions, and these were *loci* that provided confirmation of a Christian's faith. The probative value of these norms varied: the Catholic Church, general councils, and the Church of Rome (papal teaching) normally afforded principles of argument that were absolutely certain; the Fathers of the Church and the scholastic theologians and canonists normally provided probable arguments but on occasion, especially when they spoke with unanimity, they provided absolute certainty (see table).

MELCHIOR CANO, *DE LOCIS THEOLOGICIS* (1564)

1. Scripture 2. Apostolic Tradition	Revelation itself
3. The Catholic Church 4. General Councils 5. The Church of Rome (Papal Teaching) 6. The Church Fathers 7. Scholastic Theologians & Canonists	Conservation Interpretation Transmission
8. Natural Reason & the Rational Sciences 9. Philosophy & the Jurists 10. History, Documents, Oral Traditions	Confirmation

(3, 4 and 5 afforded principles of argument that were absolutely certain; 6 and 7 normally provided probable arguments, but could provide absolute certainty especially in cases of unanimity.)

As the Counter-Reformation progressed theologians were increasingly preoccupied with the disputes fomented by the Protestant revolt, with serious consequences for the quality of their theology. It was symptomatic of the narrowing effect of controversy that Bellarmine followed Cano's model, *e.g.* in his *Disputationes de controversiis*, but was less concerned with questions of method than with shoring up the disputed grounds of theological argument themselves, especially tradition and the papal *magisterium* or teaching office. And the full weight of Counter-Reformation theology lay behind Bellarmine's conclusion that the question of geocentrism and geostasis was 'of the faith'. He based his judgment on three presumptions which were not all equally transparent in his reply to Foscarini, but their relative significances can be appreciated by referring to his treatment 'On the interpretation of the word of

God and on the senses of scripture' in the *Disputationes de controversiis*.[15] The three more-or-less underlying presumptions were as follows:

1. The Catholic Church gives the correct interpretation of scripture in disputes on questions of faith.
2. The Fathers had been unanimous in interpreting the Bible text geocentrically and their unanimous consensus must not be contradicted.
3. The truth which the Holy Spirit has revealed in the Bible on cosmological questions is conveyed through the literal meaning of the text.

Now we need to take these presumptions one at a time.

First, that the Catholic Church gives the correct interpretation of scripture in disputes on matters of faith. Bellarmine stated rather baldly in the *Disputationes de controversiis* that 'the Church is the judge of the true sense of Scripture and of all controversies';[16] from the context it is clear that he was referring to interpretations of Scripture on matters of faith and other questions about matters of faith that were in dispute between the Reformers and the traditional Church. Bellarmine was echoing the Council of Trent which in its 'Decree on the editing and use of Sacred Scripture' (8th April, 1546) had not imposed in blanket fashion the Church's interpretation of Scripture but rather had forbidden interpretations of scriptural texts concerning matters of faith and morals which were contrary to the sense held and taught by the Church, whose role it is to judge the true sense and interpretation.[17] Bellarmine wrote by way of explanation in the *Disputationes de controversiis* that the scriptures are to be understood in the same spirit in which they were composed, so the question to be answered is, 'Where is that Spirit?' He answered his own question by saying that although this Spirit is often given to private individuals, it is certainly found in the Church *i.e.* the pope and a Council of bishops. He did not spell out, but certainly took for granted, the distinction between the positive action of the Spirit of God in inspiration and the providential action of the Spirit of God in guarding the Church from serious error: he did not wish to suggest that the Church is directly inspired by the Holy Spirit in its interpretation of Scripture.

Second, Bellarmine claimed that the Fathers had been unanimous in interpreting the Bible text geocentrically and that their unanimous consensus must not be contradicted. This was the weakest point in Bellarmine's response.

Bellarmine wrote to Foscarini that it is forbidden or, more exactly according to Bellarmine's text, that the *Council* forbids the exposition of Scripture in a way that contradicts the unanimous consensus of the Fathers. Bellarmine was again referring to the Council of Trent which had declared that no-one is permitted to interpret sacred scripture in matters of faith and morals contrary to the unanimous consensus of the Fathers,[18] and that included theologians such as Robert Bellarmine himself, theologically educated enthusiasts such as Foscarini, and natural philosophers such as Galileo who tried to defend himself against the scriptural objections to Copernicanism. Bellarmine was calling Foscarini to order by reminding him of what he was bound to accept by reason of the declaration of the Council of Trent, for as a Catholic he was committed to accept the decisions of General Councils, including the Council of Trent, and the Council of Trent had expressly forbidden the exposition of scriptures in a way that contradicts the unanimous consensus of the Fathers in matters of faith.

The argument from the unanimous consensus of the Fathers of the Church had a peculiar force in Counter-Reformation theology even though the Reformers also cited the Fathers.[19] It had been and still is customary in biblical exegesis to be guided by the expositions of scripture of the Fathers: the authority of the Fathers as expounders of Scripture was beyond question and part of the constant belief of the Church,[20] their wisdom being accepted as an effect of the permanent action of the Holy Spirit in the Church. Thus, St John Fisher, Bishop of Rochester could write in 1523:

> The Holy Spirit intervenes through the orthodox Fathers to root out heresy and enlighten the Church. Anyone who does not listen to the orthodox Fathers is spurning the Holy Spirit.[21]

But respect for the authority of the Fathers tended to increase and harden in the course of crises and controversies as, for example, people developed the habit of going more directly to the Scripture text when printed bibles became available, or in resistance to the Erasmian program of 'Preaching Christ from the sources'.[22] The trend towards rigidity was even more pronounced in the Council of Trent where the precision of 'the *unanimous* agreement of the Fathers' seems to make its first appearance in the report from the Committee on Abuses in Connection with Holy Scripture to the general assembly of the Council. The committee members declared:

> Neither the public nor the private interpretation of Holy Scripture can be left to individual good pleasure; on the contrary, this interpretation must conform to the Church's interpretation and the unanimous consent of the Fathers.[23]

The process of clarification and stiffening continued after the Council with the diffusion of the text of Melchior Cano who attributed to the Fathers a precise role in theological criteriology. As we have already noted, Cano explained that the teachings of the Fathers are not integral to revelation itself, but the Fathers themselves were intimately involved in the conservation, interpretation and transmission of revelation, and they normally furnished principles of theological argument that were probable, but which attained absolute certainty in the case of their unanimity on a question of faith, when their testimony participates in the infallibility of the Church.[24] Thus spoke Melchior Cano, and accordingly Bellarmine reminded Foscarini that it was not permitted to contradict the unanimous consensus of the Fathers in matters of faith and morals.

However, the major question about Bellarmine's appeal to the unanimous consensus of the Fathers was whether there had been any such consensus. Possibly all the Fathers had presumed the popular Ptolemaic-style picture of the universe; indeed, it would be surprising if they had done otherwise. Also, as Bellarmine stated, all the recognised exegetes of the first half of the seventeenth century, with the notable exception of the Spaniard Diego de Zuñiga who was duly condemned for his 'error' in 1616, had interpreted the relevant biblical texts in geostatic terms.[25] But Galileo raised the obvious objection in his letter to the Grand Duchess Christina which he was writing at the very time that Foscarini and Bellarmine were exchanging letters and which was in fact Galileo's answer to Bellarmine's letter to Foscarini, namely that in interpreting the Bible in geocentric and geostatic terms the Fathers were doing no more than rely upon a common unchallenged presumption, and that is not the same thing as a consensus.[26]

Bellarmine probably received a copy of Galileo's letter in due course, but he would have considered the objection to be irrelevant, a vain attempt to cut the Gordian knot, for he was convinced on other grounds that geocentrism and a motionless earth were matters of faith that had been taught by divine authority and that the 'consensus' of the Fathers was subsequent upon their recognition of that teaching: the Fathers had conserved and transmitted by their geocentric interpretation of the Scriptures the revealed truth that they had recognised. At this point Bellarmine took his stand on his third and clinching but least expressed presumption. He did not spell it out in his letter to Foscarini so we need to unveil it, as it were. In fact it was a traditional principle of biblical exegesis, namely that:

The truth which the Holy Spirit has revealed in the biblical texts that have a cosmological reference is conveyed through the literal meaning of the texts. This principle was adopted in the tradition of interpretation that runs from St Augustine through St Thomas Aquinas.[27] The meaning of 'literal sense' varied within this tradition: for Augustine the literal sense of the text included its theological significance; for Thomas Aquinas the literal sense was what the author (both the human and divine authors acting as one) intended to say. Bellarmine's understanding of the literal sense was simplistic by comparison: for him the literal sense meant the mere grammatical reading of the text, 'what the words immediately convey' (though often the words should be read figuratively).[28] But Augustine, Aquinas and Bellarmine were in agreement on one thing: that cosmological texts in the Bible were to be taken as referring to actual historical realities, events and persons—days of creation, trees, gardens, Adam and Eve, a serpent ...

This principle had been adopted down the centuries, in the first encounters between Christian theology and pagan philosophy in the early centuries of the Christian era, and in all later cases of conflict between what was stated in the biblical text and conclusions derived from other sources of knowledge. St Augustine's rule of thumb had been that the Bible is to be read literally unless there is adequate reason to cause us to adopt a metaphorical interpretation, and the only reason that could be deemed adequate in this sense was a fact demonstrated to be such by a true demonstration. Thus, Augustine wrote in the *De Genesi ad litteram*:

> But, someone may say, why does what is written in our Bible: 'He who stretches out the heavens like a tent' not contradict those who attribute spherical shape to the heavens? If what these latter said were to be false, then it would indeed be in contradiction. For that is true which rests on divine authority rather than that which proceeds from human weakness. But if on the other hand they can prove their claim with such evidence (*documenta*) that it is placed beyond any doubt, then it must be demonstrated that our speaking of a curtain does not in fact conflict with their true assertion. Otherwise, there

> would be a contradiction even with those other passages in Scripture where the heavens are said to be suspended like a dome.[29]

In other words, Augustine stated that when it is written in the Bible that the heavens are tent-shaped and philosophers claim they are spherical, the philosophers must prove their claim as indubitable. If they cannot produce certain proof, the Bible account is to be accepted as truth; if they can, then we must show that what is written in the Bible does not conflict with the truth they have established. For Augustine, the lightest word of God is a better witness to truth than the heaviest word of man, except where this latter is a demonstrated claim, *i.e.* science proper, and not mere opinion; certain and demonstrated scientific truth is the only kind that can be allowed to challenge the literal interpretation of Scripture.[30] (Certain and demonstrated truth, truth that was shown to be true of necessity, was the neo-Platonist and Aristotelian conception of and requirement for scientific truth: anything less than demonstrated truth was mere opinion.)

Accordingly, Bellarmine applied this exegetical principle in his reply to Foscarini; it was the last of his three points:

> I say that, when there is a true demonstration that the sun is in the centre of the world and the earth in the third heaven, and that the sun does not go round the earth but the earth round the sun, then we will need to proceed with great caution in explaining the Scripture passages which seem to be in contradiction and rather say that we do not understand them than say that what has been demonstrated is false.[31]

Bellarmine the theologian thus reminded Foscarini, Galileo and any who might have been in agreement with them of a basic principle of biblical interpretation which in his opinion they were tending to take a little too lightly and failing to apply consistently. Galileo also referred to Augustine's principle in his letter to Christina, but he tried to weaken its force by on the one hand quoting other statements from Augustine which depreciated knowledge of astronomical and other physical matters because the 'Holy Spirit did not desire that men should learn things that are useful to no-one for salvation' while on the other hand arguing for a re-evaluation of knowledge supported by 'experiments, long observation, and rigorous demonstration', such as whether or not the earth and heavens move, so that they would be reclassified as 'sure and demonstrated knowledge'.[32]

In 1615 Bellarmine was of the firm opinion that the Copernican theory had not been demonstrated as a certainly true picture of the world—in fact, he wrote to Foscarini that he very much doubted that there ever could be such a demonstration—hence there was not adequate reason to interpret metaphorically the geocentric passages in the Bible and, besides, the instrumentalist interpretation of astronomical hypotheses was still the accepted interpretation as it had been since Ptolemy. Consistent with accepted principles for the interpretation of the Bible, Bellarmine urged Foscarini: 'Now just reflect, prudent man that you are, whether or not the Church can accept that an interpretation be given to the Scriptures that is contrary to that of the holy Fathers and all the Greek and Latin interpreters'.[33]

Bellarmine's reply to Foscarini was thus based on theological principles or *loci*: to put them in their correct order according to the logic of his reply, these principles were: authority, rules of exegesis, the Fathers. It was the reply of a professional theologian, open to criticism as all theological assertions are. We can easily see the weaknesses in his argument. His argument from the universal consensus of the Fathers was unsustainable and needs no further comment. The principles for interpreting the Bible texts that have a cosmological reference were more significant for Bellarmine's stand. The only exegetical methods available at that time were effectively fundamentalist methods; this was still the precritical period of biblical interpretation, the modern critical period being usually dated from *ca.* 1650, and especially from the publication of Richard Simon's *Histoire Critique du Vieux Testament* (1678). The Bible was viewed at this time not as a literature with a history but as a collection of writings that came rather directly from heaven and which reported events in a factual style, independently of their cultural and historical milieux, and biblical interpretation, especially in the Reformation and Counter-Reformation, was dogmatic and theological; in such a context one should not expect a high standard of biblical interpretation.

Bellarmine and his contemporaries, following a hitherto unchallenged tradition of biblical interpretation, confused historically-conditioned forms of expression with revealed truth;[34] they were not in a position to distinguish the presuppositions, world views, images, categories and modes of expression found in the scriptures that had been more or less consciously borrowed and adapted from local cultures from the message itself that was conveyed by means of them. Following Augustine they assumed that Scripture contained a cosmology which had been taught by God, that that cosmology was an essential part of the truth of the Bible, and therefore that all the details of the cosmology of Scripture required assent until it should be proven otherwise by a certain demonstration, in the same way that everything else that was taught by Scripture demanded assent. According to these principles it was normal to settle astronomical questions with biblical evidence.

As would become increasingly obvious those principles were inadequate for the Church's encounter with the new sciences: they had been adequate enough up till the seventeenth century, but were now being shown to be inadequate. They were not adapted to meet challenges from a new type of *scientia* relying on well-founded hypotheses rather than strict logical demonstration, for they entailed that claims of natural knowledge have no claim to be taken seriously unless they can be demonstrated as certainly true; if they are less than certain, 'only' probable and even 'only' highly probable, the literal reading of the biblical text takes precedence over them.

A decree of the Congregation of the Index removed from the 1757 Index of Forbidden Books all writings that supported the heliocentric system.[35] It was considered that the heliocentric system had been 'proven' by that time, even though the criteria of proof acknowledged by the Catholic Church were still Aristotelian, but since it had been accepted as proven it was opportune to revise the Church's position according to the principle first enunciated by Augustine and repeated by Bellarmine in his letter to Foscarini:

> ... when there is a true demonstration that the sun is in the centre of the world and the earth in the third heaven, and that the sun does not go round the earth but the earth round the sun, then we will need to proceed with great caution in explaining the Scripture passages which seem to be in contradiction ...[36]

We now recognise the historical conditioning of Bellarmine's theology. It was the theology which not only explains Bellarmine's response to Foscarini, but also goes a long way towards explaining why the Catholic Church could ever have condemned Copernicanism in 1616 and put Galileo on trial and condemned him in 1633. It was a theology that had disastrous consequences, not just for Galileo but for the Catholic Church. In the Galileo affair we find:

a Copernican theory that was not entirely able to convince; a long tradition of interpretation of Scripture and of astronomical theory; rigidity in the Catholic Church's response to anyone who wished to promote a novel interpretation of Scripture; and a general mood that welcomed restrictive measures. And there was just a small amount of room to move: there was a traditional method for handling cases of apparent conflict between the teaching of the Scriptures and 'scientific' knowledge.

But Bellarmine's letter invites us to look beyond the impermanent features of his theology and identify his deeper theological concerns; in so doing we are enabled to discern basic features of theology as a discipline and clarify the origins and functions of theological concepts and assertions. Cosmology was one such impermanent feature: although Bellarmine very definitely insisted on the geocentric and geostatic system and consequently rejected a particular astronomical theory, *viz.* the Copernican system interpreted realistically, in the *final* analysis it was not the concern of Bellarmine or later of the ecclesiastical judges of the Holy Office that Galileo and other Copernicans were upholding a scientific theory or a cosmology that Christians did not agree with. Bellarmine's response to Foscarini helps to make that fact clear. Bellarmine was intent on correcting what he judged to be his correspondent's permissive and theologically deficient attitude to the teaching Church's role as the ultimate interpretative authority of Scripture and his presuppositions concerning the legitimate method for the interpretation of Scripture.

It was ultimately incidental that it was a humanist letter-writer such as Foscarini, or a Copernican natural philosopher such as Galileo who was under scrutiny because suspected of being a person who was teaching heresy. It could just as easily have been a follower of Martin Luther or of one of the other Reformers. What was the crux of the whole affair, the point of concern and the focus of consideration, for Bellarmine and for the judges in 1616, was the alleged fact that some persons wanted to interpret scripture on a matter that was believed to be 'of the faith' in a new way without authority and theological warrant. Again in 1633, although the question that the trial judges wanted to decide was whether or not Galileo had obeyed the injunction against teaching Copernicanism as anything more than an hypothesis in the traditional, instrumentalist, meaning of the word, their ultimate concern was still the unauthorised new interpretation of the Bible that would be necessary to accommodate a realist understanding of Copernicanism: they reacted to the fact that the Bible was being reinterpreted without authority on a point that was deemed to be revealed truth.

It was because he was heir to the exegetical principles of Augustine that Bellarmine considered the geocentric and geostatic cosmology to be imposed by the Bible as divine truth revealed in the Bible. God, according to his theology, had spoken through the literal text of the Bible, so he rejected the realist version of Copernicanism because he intended to protect the Christian faith from a theory that he judged was threatening it by jeopardising the status of the Scriptures as the infallible word of God; and since Foscarini's attempt to make the scriptural objections disappear threatened to favour the propagation of that system he rejected Foscarini's efforts. Bellarmine's concern about the realist version of the Copernican hypothesis was secondary and contingent.

This relationship to cosmology illustrated by Bellarmine's theology has been constant throughout the Judeo-Christian theological tradition, beginning with the very process from which the scripture texts themselves evolved: the Hebrew people borrowed freely from the cosmogonical myths of the ancient near-Eastern civilisations, specifically the Mesopotamian, Babylonian and Canaanite or Ugaritic myths and re-fashioned what they had appropriated in order to express their own distinctive theological interpretation of the world and their situation in it. The message was the thing, the medium was constructed from whatever lay to hand.

Subsequently, in encounters with the sciences, cosmological questions have been secondary in theology: theology does not claim to have an alternative scientific explanation or an alternative cosmology. On the contrary, theology is generally expressed in terms borrowed from standard cosmologies of the past that have been more or less successfully updated in keeping with popular acceptance of new information from the natural sciences. But theologians do have a theological truth which has ramifications that may conflict with statements made in the sciences, such as statements favouring certain reductionist, materialist or determinist theories and certain cosmological speculations that are presented as alternatives to the Christian doctrine of creation. Ideally it is only in such circumstances that theologians enter into the cosmological arena in a negative mode in order to clarify seeming conflicts with Christian truth. Theologians do not expect ultimate conflict with the natural sciences since they firmly hold to the conviction that truth is not self-contradictory; rather they hope to refine their own theological insights through interaction with other sources of truth.

Beyond all historical conditioning, Bellarmine's theology was regulated by fundamental theological principles of the Catholic tradition that are as relevant now as they were in the seventeenth century. I would express these principles as follows:

(a) that the Scriptures are the Word of God and contain the truth to be believed by all Christians, hence the Bible is the fundamental source for Christian faith and for the life of the Church;

(b) that to find the truth contained in the Bible we need the guidance of the Church and its tradition, hence the Scriptures are to be interpreted within the community of believers and within the mainstream of the Christian tradition of interpretation, and there is a need to be attentive to the contributions of patristic exegesis;

(c) that the Church is the ultimate interpretative authority of Scripture, and that the Church has the duty and the right to judge interpretations of Scripture.[37]

These, I suggest, were the basic principles which were of concern to Bellarmine in his response to Foscarini, and to the Holy Office acting in the name of the Catholic Church in its condemnation of Copernicanism in the following year, and its condemnation of Galileo in 1633; they are the basic principles of the discipline of Catholic theology. It was the particular historically-conditioned application of these principles by fallible theologians that led to disaster.

The formula implicitly followed by Bellarmine for finding the right way forward in a theological crisis was to respect tradition while at the same time leaving open the way to legitimate change. In Kuhnian terms we might say that Bellarmine was doing normal theology. For many even of his contemporaries Bellarmine was too cautious; fidelity to tradition in this case was the rigid imposition of the *status quo*. But Bellarmine acted from the conviction that tradition is normally in *status possidentis*, and that progress should be a development. One might agree with the formula, but unfortunately the seventeenth century was not the time for smooth transitions from the old to the new: the world around the protagonists in the whole Galileo/Copernican affair was moving too quickly.

School of Science and Technology Studies University of New South Wales

NOTES

[1] Francois Russo, 'Galileo and the theology of his time', in Paul Cardinal Poupard (ed.), *Galileo Galilei: Toward a Resolution of 350 Years of Debate, 1633-1983*, trans. Ian Campbell (Pittsburgh, 1987); Ernan McMullin, 'How should cosmology relate to theology?', in A.R. Peacocke (ed.), *The Sciences and Theology in the Twentieth Century* (London, 1981) 17-57; Richard S. Westfall, 'The trial of Galileo: Bellarmino, Galileo, and the clash of two worlds', *Journal for the History of Astronomy* 20 (1989) 1-23; Richard I. Blackwell, *Galileo, Bellarmine and the Bible* (Notre Dame, Ind., 1991).
[2] E.g., Westfall, 'The trial of Galileo'; some of Galileo's supporters thought Bellarmine held these posts but they were mistaken, cf. James Brodrick S.J., *Robert Bellarmine, Saint and Scholar* (London, 1961) p. 347.
[3] Robert Bellarmine, *Disputationes de controversiis christianae fidei adversus huius temporis haereticos*, 4 vols (Ingolstadt, 1601).
[4] Westfall, 'The trial of Galileo', pp. 12-4, 17-8.
[5] Antonio Favaro ed., *Le Opere di Galileo Galilei*, Edizione Nazionale, 20 vols, (Florence, 1890-1909) XI, pp. 87ff.
[6] Bellarmine to Foscarini, 12 April 1615; Galileo, *Opere*, XII, pp. 171-2. A translation can be found in James Brodrick, S.J., *The Life and Work of Blessed Robert Francis Cardinal Bellarmine, S.J. 1542-1621*, 2 vols (London, 1928) II, pp. 358-60; cf. *idem*, ch. 18 for other evidence of Bellarmine's personal position on the Copernican question. Cf. also Blackwell, *Galileo, Bellarmine, and the Bible*, ch. 4; Blackwell has correctly emphasised the significance of Foscarini's letter and Bellarmine's reply for a better understanding of the events of 1615 and 1633.
[7] Dini to Galileo, 7 March 1615; Galileo, *Opere*, XII, p. 151.
[8] 'Lettera sopra L'Opinione d'Pittagorici e del Copernico della Mobilità della Terra, e Stabilità del Sole, e del Nuovo Pittagorico Sistema del Mondo, al Reverendiss. P.M. Sebastiano Fantone, Generale dell'Ordine Carmelitano, nella Quale Si Accordano ed Appaciano i Luoghi della Sacra Scrittura, e le Proposizioni Teologiche, che Giammai Possano Addursi Contro di Tale Opinione', 6 January 1615. An English translation can be found in Blackwell, *Galileo, Bellarmine, and the Bible*, Appendix VI, pp. 217-51.
[9] The principal text was Joshua 10:12-13; other texts were Psalm 104:5; Psalm 19:4b-6a; Ecclesiastes 1:4,5 (cf. 2Kg 20:8-11); Job 9:6-7.
[10] Foscarini's principles were not new; they echoed the teachings of St Thomas Aquinas in the *Summa Theologiae*, I, q. 68, a. 3; I, q. 70, a. 1, ad 3; cf. T.F. Torrance, 'Scientific hermeneutics according to St. Thomas Aquinas', *Journal of Theological Studies* 13 (1962) 259-89.
[11] Galileo, *Opere*, XIX, p. 323.
[12] Bellarmine to Foscarini, 12 April 1615, in Galileo, *Opere*, XII, p. 172.
[13] Cf. Bernhard Lohse, *Martin Luther: An Introduction to His Life and Work*, (Philadelphia, 1986) pp. 153ff.
[14] Melchior Cano, *De Locis Theologicis*, Forzani et soc. ed. *Melchior Cano, Opera*, 3 vols, (Rome, 1564) III. The form of Gino's treatise was directly inspired by Rudolphe Agricola's De *Inventione dialectica* (Cologne, 1527) while the content was generally thomistic; cf. A. Gardeil, 'Lieux Theologiques' in *Dictionnaire de la Théologie Catholique*, Paris, IX, cols 712-47. Blackwell gives an account of Cano's doctrine in which the main purpose of the *loci* as the ten clearly identifiable foundations of Catholic theology tends to be lost to sight; it was Gino's further efforts to justify these foundations and his rules for argument on the basis of them that complicated his exposition; since Bellarmine was little troubled by these latter epistemological niceties, I believe we are better advised to leave them alone; cf. Blackwell, *Galileo, Bellarmine, and the Bible*, pp. 15-20.

[15] Bellarmine, *Disputationes de controversiis*, I, iii, 3.
[16] 'Iudicem veri sensus Scripturae et omnium controversiarum, esse Ecclesiam, id est, Pontificem cum Concilio' (*Disputationes de controversiis* I, iii, 3). In my opinion, Westfall and Blackwell do not make sufficient allowance for the context of this statement; cf. Westfall, 'The trial of Galileo', pp. 6-7; Blackwell, *Galileo, Bellarmine, and the Bible*, pp. 36-40.
[17] Praeterea ad coercenda petulantia ingenia, decernit, ut nemo suae prudentiae innixus, in rebus fidei et morum ad aedificationem doctrinae Christianae pertinentium, sacram scripturam ad suos sensus contorquens, contra eum sensum quem tenuit et tenet sancta mater ecclesia, cuius est iudicare de vero sensu et interpretatione scripturarum sanctarum, aut etiam contra unanimem consensum patrum, ipsam scripturam sacram interpretari audeat, etiamsi hujusmodi interpretationes nullo unquam tempore in lucem edendae forent. Qui contravenerint, per ordinarios declarentur, et poenis a jure statutis puniantur. (Johannes Domenicus Mansi *et al.*, *Sacrorum conciliorum nova et amplissima collectio*, 53 vols (Paris, 1902) XXXIII, col. 23.)

Which, in rather literal translation, reads:

To restrain the impudent [this Council] determines that, in matters of faith and morals that contribute to the establishment of Christian doctrine, no-one may dare to interpret sacred scripture while relying on his own discretion and twisting sacred scripture to his own sense and against that sense which holy mother Church, whose role it is to judge the true sense and interpretation of sacred scripture, holds and teaches, nor also against the unanimous consensus of the Fathers, even if said interpretations never are openly published. Let those who contravene be denounced to the Ordinaries and punished with those penalties determined by law.

[18] The 'Fathers of the Church' were the Christian preachers, writers and theologians of the post-canonical, or post-apostolic period; the Patristic age is the period that runs from near the end of the first century A.D. to around the middle of the eighth century.
[19] Cf. Yves Marie-Joseph Congar, *Tradition and Traditions: A Historical and a Theological Essay* (New York, 1967) pp. 183, 185; on the Reformers' use of the Fathers, cf. pp. 187-8.
[20] G. Geenan, 'The Place of tradition in the theology of St. Thomas', *The Thomist* 15 (1952) 110-35, Note 11.
[21] Proemium to 'Assertionis Lutheranae Confutatio' (1523) in *Ioannis Fisheri Opera Omnia*, Wurzburg, 1579, col. 279-96.
[22] 'Ex fontibus praedicare Christum'; cf. Congar, *Tradition and Traditions*, pp. 184, 196ff.
[23] Hubert Jedin, *A History of the Council of Trent*, trans. Dom Ernest Graf O.S.B., 2 vols (London, 1957) II, p. 71. No resistance was offered in the general congregation of the Council to any aspect of this statement. St Thomas Aquinas seems not to have heard of the expression 'unanimous consensus', as is pointed out by Geenan, 'The place of tradition'.
[24] Cano, *De locis theologicis*, XII, 5, 12.
[25] Rinaldo Fabris, *Galileo Galilei e gli orientamenti esegetici del suo tempo*, Pontifical Academy of Sciences, Scripta Varia 62 (Vatican City, 1986) pp. 23ff.
[26] Galileo, *Opere*, V, pp. 335-36.
[27] Torrance, 'Scientific hermeneutics according to St. Thomas Aquinas', pp. 282-5; also Brevard S. Childs, 'The *sensus literalis* of scripture: An ancient and modern problem', in Herbert Donner et al. (eds.), *Beiträge zur Altestestamentlichen Theologie: Festschrift fur Walther Zimmerli zum 70. Gebunstag* (Göttingen, 1977) 80-93.
[28] Bellarmine wrote that the Bible had two senses: literal/historical, and spiritual/mystical: 'Literalis est, quem verba immediata preferunt; spiritualis est, qui alio re fertur, quam ad id quod verba immediate significent' (*Disputationes de controversiis*, I, iii, 3). Cf. Blackwell, *Galileo, Bellarmine, and the Bible*, pp. 33-5, and Appendix III, pp. 187-93 for a fuller treatment of Bellarmine's views on the interpretation of Scripture.
[29] Augustine, *De genesi ad litteram*, 11,9. The translation, with minor changes, is from Ernan McMullin, 'How should cosmology relate to theology?', p. 19.

[30] Cf. McMullin, 'How should cosmology relate to theology?', p. 20.

[31] Bellarmine to Foscarini, *loc. cit.*, p. 172; cf. Augustine, *De genesi ad litteram*, I, 19, 21; II, 9.

[32] Galileo, *Opere*, V, pp. 325 ff.

[33] Bellarmine to Foscarini, *loc. cit.*, p. 172.

[34] Cf. Pontifical Biblical Commission, 'The interpretation of the Bible in the Church', I,F, *Origins* 23:29 (1994) 509-10.

[35] Galileo, *Opere*, XIX, p. 419.

[36] Bellarmine to Foscarini, *loc. cit.*; cf. M. D'Addio, *Considerazioni sui processi a Galileo*, Quaderni della Rivista di Storia della Chiesa in Italia no. 8, Roma, 1985.

[37] For an historical overview which shows clearly the perennial character of these principles, cf. Jean-Pierre Torrell, *La Theologie Catholique*, Que Sais-je? (Paris, 1994). The 'truth' revealed in the Scriptures and believed by Christans does have an intellectual component, but it is predominantly dialogical.

NOTES ON CONTRIBUTORS

KIRSTEN BIRKETT completed both her undergraduate and postgraduate study in history and philosophy of science at the University of New South Wales. Her thesis was entitled 'Early English Reformers and Magic: Reformation Ideas Concerning the Interaction Between the Natural and Supernatural Worlds'. Since graduating, she has been working as an editor for Matthias Media and teaching in Science and Religion at Moore Theological College, Sydney. In 1996 she launched the already highly successful religious journal *Kategoria* on behalf of the Matthias Centre for the Study of Modern Beliefs, of which she is the Director. '*Kategoria*' denotes, she tells us, the case for the prosecution, as opposed to the case for the defence signified by the familiar term 'apologetics'.

BARRY BRUNDELL studied theology in Rome and history and philosophy of science at the University of New South Wales. He is the author of *Pierre Gassendi: From Aritotelianism to a New Natural Philosophy* (Dordrecht, 1987). After lecturing in theology for many years at St Paul's National Seminary, Kensington, NSW, he served for seven years as Rector of the International Student House of his religious order in Rome and lecturer in Theology and Science at the Pontifical Gregorian University, his *alma mater.* He recently took up an appointment as Director of the Chevalier Centre in Australia. He is an Honorary Visiting Fellow in the School of Science and Technology Studies at the University of New South Wales. In his spare time he plays tennis and rides an exercise bike.

ANTHONY CORONES completed his undergraduate studies in history and philosophy of science and philosophy, and his postgraduate research in philosophy at the University of New South Wales, where he now lectures in history and philosophy of science. The title of his thesis was 'Naturalistic Epistemology: Overcoming the Dichotomy between the Normative and the Descriptive'. He is currently interested in postmodern philosophy of science, cognitive science, and philosophy of technology. He likes to relax of an evening over the kitchen stove, meditating on the philosophy of *haute cuisine* while preparing culinary delights.

JAMES FRANKLIN is a Senior Lecturer in mathematics at the University of New South Wales. After undergraduate and postgraduate work at Sydney University, he did research in algebra at Warwick University. He works on statistical aspects of neural nets, and on the philosophy of mathematics, where he defends a structuralist interpretation of mathematics and a logical view of probability. In the history of ideas, he has recently written the chapter on mathematics, logic and technology in the *Cambridge History of Eighteenth Century Philosophy*, and has completed a book on the history of probability before Pascal and a book on Australian philosophy.

GUY FREELAND read philosophy and psychology, and subsequently undertook research in psychology, at Bristol and studied history and philosophy of science at Cambridge. He has taught over a very wide area of history and philosophy of science for more than thirty years at the University of New South Wales, where he holds the position of Senior Lecturer. His earlier research interests were largely in philosophy of science, but the male climacteric shifted his attention in the direction of archaeology, landscape and iconography. These interests he has pursued in his teaching and in his most recent writing, a study of Canberra, entitled *Canberra Cosmos: The Pilgrim's Guidebook to Sacred Sites and Symbols of Australia's Capital* (Sydney, 1995), and papers on the embedding of cosmology within sacred iconography and architecture. In Australasia, he does his best to sublimate his passion for medieval churches by visiting Anzac memorials.

KEITH HUTCHISON, a Senior Lecturer at the University of Melbourne, began his long and relatively dull life (his words) in Strathfield, NSW. He read physics and mathematics at the ANU, followed by postgraduate studies in mathematics at Monash and research in history and philosophy of science at Oxford. He has a special interest in the history of European cosmology (teaching an introductory course in the history of astronomy), the relations between science, philosophy and political thought in Renaissance and Early Modern Europe, and the philosophy of probability. He has also published on nineteenth century thermodynamics. His article on 'Occult Qualities' won the Zeitlin-Verbrugge Prize for the best paper published in *Isis*, 1982-1985. On overseas jaunts, he likes collecting up examples of pre-Copernican heliocentric images.

JAMIE C. KASSLER studied musical composition at the University of Wisconsin and history and philosophy of music theory at Columbia University. In Australia she combined part-time non-academic work with part-time research fellowships in departments of music, English, and history and philosophy of science. In 1991 she was elected a Fellow of the Australian Academy of the Humanities. Her most recent book is *Inner Music: Hobbes, Hooke and North on Internal Character* (London, 1995). Although she now composes only in words, her first or 'natural language' was music. This 'natural language', which sometimes is referred to as musical ear, musicality or musical competence, is the subject of her current research.

MARTIN KEMP is that *rara avis*, a distinguished scholar who is both art historian and historian of science. He read the Natural Sciences Tripos at Cambridge, and history of art at Cambridge and the Courtauld. Recognised as an authority on Renaissance art, and in particular Leonardo da Vinci, he has explored in depth the historical interrelations between the visual arts and the natural sciences. In 1990 he published his masterly study, *The Science of Art: Optical Themes in Western Art from Brunelleschi to Seurat* (New Haven). Currently he is working on a companion volume on the human and natural sciences. Having occupied the chair in the History and Theory of Art at the University of St Andrews, he is currently Professor of the History of Art and British Academy Wolfson

Research Professor at the University of Oxford and a Fellow of Trinity College. In 1992 he visited Australia to deliver a keynote address at the 25th Anniversary Conference of the Australasian Association for the History, Philosophy and Social Studies of Science. He enjoys 'relaxing' on the hockey field.

JOHN SUTTON, who comes from central Scotland, read classics at New College, Oxford before moving for graduate work to the University of Sydney in search of sun. He has recently taken up a position as lecturer in philosophy at Macquarie University, after postdoctoral fellowships at UCLA (in the Center for Seventeenth and Eighteenth Century Studies) and at Sydney (as an ARC Research Fellow). His book, *Philosophy and Memory Traces: Descartes to Connectionism,* was published by Cambridge University Press in 1998. With Stephen Gaukroger, he is editing a collection of papers entitled *Descartes' Natural Philosophy, Vol. ii: Cognition and Physiology.* He's now working on the concepts of superposition, confusion, and mixture. Other passions include cricket, chess, and guitar bands.

NEIL THOMASON, who lectures in history and philosophy of science at the University of Melbourne, undertook his postgraduate study in philosophy at the University of California at Berkeley. He says that he seems 'to be interested in almost everything, except professional sports'. However, his immediate academic interests include the relevance of the history of science for the philosophy of science, and the foundations and sociology of statistics.

INDEX OF NAMES

A page number in *italics* refers to the first page of the 'Notes' section of a particular contribution. The relevant note number is indicated by (n. ##).

Abelard, 6, *245* (n. 155)
Abetti, G., 298
Ælfric, 364
Aesop, 238
Aetius, 216
Albategnius (Al-Battani), 39
Albert of Saxony, 293
Alberti, Leon Battista, 9, 31, 33, 37, 68-9, 71, 79-81, 96
Albertus Magnus (Albert the Great), 86, 93, 129, 338-9, *350* (n. 17)
al-Biruni, 292
Alexander of Aphrodisias, 134
Alexander VI, Pope, *245* (n. 132)
Alfonso of Castile, King, 325
Alhazen, 78, 81, 94
Ambrose, *245* (nn. 78, 106)
Anaxagoras, 24
Anderson, M.D., *245* (n. 66)
Andersson, G., 309, 312
Anselm, 6
Anthemius of Tralles, 79, *245* (n. 136)
Apianus, Petrus, 29, 31-2, 36-7, 45
Apollonius of Perga, 35, 79-80
Apuleius of Madaura, 57-8
Aquinas, Thomas, 87, 97, 99-100, 228, 241-2, *245* (nn. 167, 169, 173), 337, 339-340, 342-4, 346-8, 384, *391* (n. 23)
Archimedes, 1, 35, 71
Ariew, R., 293, 298
Aristarchus, 192-3, 234, *245* (n. 10)
Aristotle, 1, 74, 85-6, 93, 96, 102, 134, 157, 157-60, *178* (nn. 32, 46), 189, 207, 334-5
Aubrey, John, 151, 174, 178, *178* (n. 19)
Augustine, 86-8, *245* (nn. 43, 147), 345, 384-5, 387, 389
Aurelian, Emperor, 213
Averroës, 239
Avicenna, 87

Babb, L., *139* (n. 129)
Bacon, Francis, *139* (n. 15), 155, *245* (n. 160)
Bacon, Roger, See Roger Bacon
Bailly, J.S., 300
Barker, F., *139* (n. 121)
Bartolus, 75
Bastian, H.C., *178* (n. 75)
Beatus of Liébana, 202
Bell, E.T., *178* (n. 9)
Bellarmine, Cardinal Robert, 10, 15, *245* (n. 166), 317, 375-90, *391* (nn. 3, 14, 28)
Belloc, H., 192
Berengario da Carpi, 24-6, 27
Billingsley, Henry, 82
Birkett, K., 3, 15
Blackwell, R.J., 375, *391* (nn. 6, 14, 16)
Blumenberg, H., *245* (n. 8)
Bodin, Jean, 335-7, 343
Boethius, 60
Boyle, Robert, 156, 169, 171-4, 177, *178* (nn. 100, 105, 124, 125), 349
Brackenau, 'Dr. von', 22, 24-5
Brahe, Tycho, 31, 37-41, 45, 283-4, 314, 319
Bronowski, J., *245* (n. 166)
Brundell, B., 10, 15
Brunelleschi, Filippo, 72, 78-80
Bruno, Giordano, 127, 130, *139* (n. 54), 284
Burton, Robert, 136
Bylica, Martin, 37

Calvin, John, 338, 348
Camillo, Giulio, 91-2
Campanus, 96
Canano, Giovan Battista, 1
Cano, Melchior, 379-80, 383, *391* (n. 14)
Carey, J., 137
Carruthers, M., 122-6, 128-30, 132, *139* (nn. 24, 33)

Caspar, M., 298
Casserio, Giulio, 136
Cassian, see John Cassian
Castelli, Benedetto, 14, 297, 313-15, 376
Catherine of Aragon, 358
Cavendish, W. (Duke Of Newcastle), *178* (n. 58)
Cennini, Cennino, 80
Chalmers, A., 312
Chambers, G.F., 308
Chaucer, 94
Chesterton, G.K., 192
Christina, Grand Duchess, 383, 385
Chrysippus, 24, 157, *178* (n. 38)
Chrysostom, see John Chrysostom
Cicero, 90, 122, 170, 190
Cimabue, 9, *245* (n. 96)
Clark, S., 341
Clarke, S., 349
Clavelin, M., 309
Clavius, Christoph, 376-7
Cleanthes, *178* (n. 38)
Clement of Rome, 217
Clement of Alexandria, 197, 237
Cohen, I.B., 298
Coleman, J., 125
Colie, R., *178* (nn. 17, 99)
Columbus, Christopher, 55, 77, 324
Copernicus, Nicholaus, 1-2, 6, 8, 12, 14, 18, 20, 29-41, 53, 76, 100, 189-194, 201, 207, 208, 226, 230-7, 239-240, 242-4, *245* (nn. 20, 152, 161, 166, 174), 271, 274-85, 292, 294-7, 307-8, 313-4, 316, 318-23, *326* (n. 13), 358, 377
Corones, A., 7, 11, 13, 14
Cranmer, Thomas, 15, 358, 364-9
Crombie, A.C., 18, 305, 323
Curzon, R., *245* (n. 82)
Cusanus, see Nicholas of Cusa
Cyril of Alexandria, *245* (n. 78)

d'Alembert, Jean, 300
d'Abano, Pietro, 24

Damascene, see John Damascene
Dante, 63, 77, 237-239, 245, *245* (nn. 116, 162)
Darwin, C., 60
De Morgan, A., 306-8
Dee, John, 82
Delautre, H., *245* (n. 144)
Demus, O., *245* (nn. 142, 147)
Desargues, Gérard, 8
Descartes, René, 54, 87, 100-2, 128, 133, *139* (nn. 15, 86), 152, 160, 170, *178* (n. 95), 306, 334, 348-50
Digby, Kenelm, 125
Dini, Piero, 376
Dionysius the Pseudo-Areopagite, 228-9, *245* (n. 160)
Diophantus, 74
Dockery, D.S., *245* (n. 156)
Donne, John, 119, 136-8, 245
Doob, P., *245* (n. 95)
Dorn, Hans, 37
Drake, S., 298, 303, 304, 323, *326* (n. 41)
Dreyer, J., 303, 315, 323-5
Duby, G., 228-30
Duccio, 9, 79-80
Dürer, Albrecht, 22, 24, 40, 55, 67

Eco, U., *245* (n. 159)
Ecphantus, 190
Edgerton, S.Y., 67, 77
Edward VI, 358, 364, 366, 370
Einstein, A., 17, 19, 35, 44, 84, 316
Eisenstein, E., 3-4, 272-4, 285, *285* (nn. 5, 20)
Eliade, M., *245* (n. 147)
Elkana, Y., 321
Ennius, 195
Ephraem the Syrian, *245* (n. 106)
Erasmus, Desiderius, 358
Esson, M., 2
Estienne, Charles, 24, 25, 31, 33
Euclid, 1, 35, 53, 72-7, 78-80, 82, 96
Euler, Leonard, 60, *178* (n. 118)
Eustachio, Bartolommeo, 29-30

Fabricius, Ab Aquapendente, Hieronymus, *178* (n. 52)
Fahie, J., 302, *326* (n. 39)
Falloppio, Gabrielo, *178* (n. 45)
Faraday, Michael, *178* (n. 113)
Fermat, Pierre de, 74
Fernel, Jean, 1, 159, *178* (n. 29)
Feyerabend, P., *245* (n. 165)
Feynman, R.P., 98
Ficino, Marsilio, 43, 193
Finocchiaro, M., 312, 324
Finucane, R., 362
Fisher, John, 382
Fludd, Robert, 91
Fontenelle, Bernard de, 336
Foscarini, Paolo, 375-9, 382-90, *391* (nn. 3, 10)
Fra Angelico, 86
Francesco de Simone, *245* (n. 96)
Franklin, J., 11-12, 118
Franssen, M., 325
Frederick, Duke, 41-2
Freeland, G., 3, 10, 12-14, *285* (n. 53)
Frege, F., 84
Fries, Lorenz, 22, 23
Fuchs, Leonhart, 1, 17

Gaddo Gaddi, *245* (n. 109)
Gafurio, Franchino, 193
Galen, 17, 26, 31, 158-61
Galileo, 8-9, 11, 14, 44, 53, 63, 70, 72, 81, 96-9, 160, 172, *178* (nn. 102, 103), *245* (n. 150), 280, 291-2, 295-8, 306, 308, 313-25, *326* (n. 13), 375-6, 382-3, 385, 387-8, 390
Galton, F., 84
Gassendi, Pierre, 14, 170, 348
Gemperlin, Thobias, 38
Gennadius, Patriarch, *245* (n. 84)
Germanus, Patriarch, *245* (nn. 43, 75)
Gerson, Levi ben, 293
Ghiberti, Lorenzo, 78, 80-81
Giere, R., 312, *326* (n. 70)
Giese, Tiedemann, 276
Gingerich, O., 305, 317
Giotto, 9, 78-80, 86, *245* (n. 96)
Giovanni di Dondi, 70
Giuliano de' Medici, 295, 298
Glanvill, Joseph, 128, 173, *178* (n. 105), 350
Goodfield, J., 321
Grant, E., 82, 294
Greal, J., *245* (n. 144)
Greenblatt, S., 137
Gregory Palamas, 241, *245* (n. 169)
Grien, Hans Baldung, 22, 26
Grosseteste, 78, 96, 228, *245* (n. 134), 348
Guido da Vigevano, 66
Gutenberg, Johann, 66

Hallyn, F., 193-4
Harvey, William, 8, 12, 155, 160-61, 177, *178* (nn. 51, 58), 321-2
Helen of Troy, 31
Hempel, C., 323
Henry of Ghent, 94
Henry VIII, 358
Heraclides of Pontus, 190, *245* (n. 2)
Heraclitus, *178* (n. 37)
Hermes Trismegistus, 33, 193-4
Hesse, M., *178* (n. 113), *245* (n. 6)
Heytesbury, 94
Hicetus (incorrectly Nicetus), 190, *245* (n. 2)
Hilbert, D., 84
Hildegard of Bingen, 92
Hipparchus, 78
Hobbes, Thomas, 12, 134, 151ff., *178* (nn. 58, 74, 80, 82, 88-91, 95, 101, 103, 105, 112, 151-6, 162ff)
Hooke, Robert, 151, 156, 163, 174-7, *178* (nn. 71, 112-3, 117-8, 124)
Horace, 276
Hugh of St Victor, 76, 90
Huizinga, J., 60
Hume, David, *139* (n. 104)
Hundt, Magnus, 23
Hunt, H., *245* (n. 179)
Hutchison, K., 3, 14-15, *245* (n. 133), *373* (n. 26)
Huxley, T.H., *245* (n. 179)
Huygens, Christiaan, 67

Ibn al-Haytham, 293
Ignatius of Loyola, 92

Impyn, J. Christoffels, 64
Irving, W., 324
Isidore of Miletus, *245* (n. 136)
Isidore of Seville, 58-9

Jacob, J.R., *178* (n. 112)
Jacopo Torriti, *245* (n. 109)
Jacopo da Cremona, 1
James, J., *245* (n. 144)
Jastrow, R., 298
Joachim of Fiori (or Floris), 92
Johannes of Antwerp, 38
John Cassian, 237, *245* (n. 156)
John Chrysostom, *245* (n. 77)
John Damascene, *245* (n. 84)
John of Jaudan, 345
John of Salisbury, 123
John Pecham, 78
John Scotus Erigena, 228
Jonson, Ben, 138
Jordanus Nemorarius, 71
Justinian, Emperor, *245* (n. 136)

Kant, I., 72, 100
Kassler, J.C., 3, 12
Keill, J., 299
Kemp, M., 2-3, 9, 11, 15, 80, 118, 122
Kepler, Johannes, 17, 34-6, 41-4, 53, 79, 93, *245* (n. 168), 276, 284, 292, 296-7, 314, 319, 322, *326* (nn. 14, 26, 33, 66), 53, 79, 93
Koster, Laurens, 66
Krell, D., 124
Kuhn, T.S., 98, 303, 309, 316, 323, 325, *326* (n. 42)

Lakatos, I., 310-11, 323, *326* (nn. 64, 66)
Leach-Jones, A., 2
Lenat, D., 91
Leo I, Pope, *245* (n. 78)
Leo VI, Emperor, *245* (n. 136)
Leo X, Pope, 24
Leonardo da Vinci, 9-11, 17, 21, 23-4, 33, 37, 66, 159, 199
Lewis, C.S., 93
Locke, John, 174
Lodge, O., 302
Lorenzetti, See Pietro Lorenzetti
Lorini, Fr., 376
Louis XIV, King, *245* (n. 132)
Lull, Ramon, 58, 130
Luther, Martin, *245* (n. 166), 338, 341, *350* (n. 17), 358, 379, 388

Maas, J., *245* (n. 179)
Machiavelli, Niccolò, 5
Maestlin, Michael, see Mästlin, Michael
Mahoney, M.S., 67
Mâle, E., *245* (n. 42)
Mandeville, John, 77
Manutius, Aldus, 24
Mascall, E.L., *245* (nn. 167, 173)
Mästlin, Michael, 17, 42, 276
Matthew Paris, 68
Maurolico, Francesco, 1
Maver, J.R., *178* (n. 118)
Maximus the Confessor, 206-7, 228, 235, *245* (nn. 61, 63, 170)
McEvoy, J., *245* (n. 134)
McMullin, E., 375
Mercator, Gerardus, 69
Mersenne, Marin, 153, *178* (nn. 79, 88)
Metford, J.C.J., *245* (n. 124)
Metzler, J., 85
Michelangelo, 31
Millais, J., *245* (n. 179)
Miziołek, J., *245* (n. 110)
Moerbeke, 1
Montaigne, Michel de, *139* (n. 113)
Montanus (Giovanni Battista da Monte), 1
More, Henry, 131-2, 173, 176, *178* (n. 105), 349
Morell, T., 301
Münster, Sebastian, 43

Napier, John, 65
Nef, J.U., 64
Newton, Isaac, 9, 67, 171, 174, *178* (nn. 72, 95), 245, 349-50
Nicetus, see Hicetus
Nichol, J.P., 301
Nicholas of Cusa (Cusanus), 8, 228
Nicholas of Lyra, *245* (n. 156)
North, J., 305, 324

Oporinus, Johannes, 26
Oresme, 63, 70, 95-6, 97, 339, 346
Osiander, Andreas, 20, *245* (n. 161), 280-4, 313, *326* (nn. 18, 74), 347
Oughtred, William, 70

Pacioli, Luca, 60-1
Palamas, see Gregory Palamas
Pannekoek, A., 304
Papaioannou, K., *245* (n. 104)
Parkinson, James, 152, *178* (n. 5)
Parmenides, 78
Parrhasius, 31
Patin, Guy, *178* (n. 1)
Paul III, Pope, 14, 29, *245* (nn. 10, 161), 279
Pecham, See John Pecham
Pepin the Short, King, 228
Peruzzi, Baldassare, *245* (n. 96)
Petrarch, 6
Petreius, Johannes, *245* (n. 10), 282
Petty, William, 154, 174-5
Peurbach, Georg, 35
Phidias, *245* (n. 84)
Philo, 197
Philolaos, 190, 216
Piccolomini, A., 93
Pico della Mirandola, 193
Pietro Lorenzetti, 80
Plato, 25, 93, 100, 191, 198, 201, 217, *245* (nn. 2, 26), 276-80, 292, 334
Plethon, Gemistos, 201
Plutarch, 190, *245* (n. 2)
Pomponazzi, Pietro, 134, *139* (n. 107), 339
Popper, K., 309, 323
Porphyry, 60
Pozzo, Andrea, *245* (nn. 171, 177)
Proclus, 35, 93
Procopius, *245* (n. 136)
Pseudo-Aristotle, 217
Pseudo-Leo, *245* (n. 106)
Ptolemy, 31, 35, 38-9, 41, 45, 69, 74, 77, 217, 283, 319, 386

Quintilian, 122

Ramelli, Agostino, 73
Ramsey, I.T., *245* (n. 149)
Ramus, Petrus, 1
Raphael, 24, 226
Regiomontanus, Johannes, 35, 294
Reuterswärd, P., 224-6, *245* (nn. 93, 105, 112)
Rheticus, Georg Joachim, 31, 277, 281-4, *285* (n. 33)
Ricci, Matteo, 74, 91
Richard of Wallingford, 70
Roger Bacon, 69, 78
Rosen, E., 190-1, 308, *326* (n. 13)
Ross, Alexander, 340, 346
Russell, B., 84, 309
Russell, J., 324-5
Russell, J.L., *178* (n. 72)
Russo, F., 375
Ryff, Walther, 26
Rykwert, J., 195
Ryle, G., 83-4, 86

Sacks, O., *178* (nn. 66-7)
Sacrobosco, 70, 76, 348
Sambursky, S., 174
Scarry, E., 119
Schaffer, S., 171, *178* (n. 27)
Schmemann, A, *245* (n. 27)
Scholarios, George., 201
Schönberg, Nicolaus, 279
Segre, M., 304
Shakespeare, William, 88-9
Shapin, S., 171, *178* (n. 27)
Shapiro, A.E., *178* (n. 21)
Shea, W., 310
Shepard, R.N., 85
Sherrington, C., *178* (n. 75)
Sidney, Philip, 89
Sigismund I, King, *245* (n. 20)
Simón de Colonia, *245* (n. 146)
Simon, Richard, 386
Simonides of Ceos, 90
Singer, C., 2
Small, R., 301
Smith, Adam, 301, *326* (n. 33)
Small, R., 300
Snow, T.P., 297
Socrates, 277-8
Sophocles, 32, 191

Spieghel, Adriaan van den, 136
Spinoza, Benedict, 75
St Bernard, *139* (n. 41)
St Francis, 62
St Paul, *245* (n. 64)
Stensen, Niels, *178* (n. 45)
Stenwickel (or Steenwinckel), Johannes, 41
Stevin, Simon, 65, 71, 98
Stigler, S.M., 60
Suger, Abbot, 208, 228-30, *245* (nn. 136, 160)
Sutton, J., 12
Swerdlow, N.M., 313
Symeon of Thessalonica, *245* (n. 63)

Talley, T.J., *245* (n. 77)
Tartaglia, Niccolò, 1, 53
Teresa of Avila, 92
Theodoric of Freiberg, 78
Theodorus Lector, *245* (n. 84)
Theodosius, 77
Theophanes the Cretan, *245* (n. 95)
Theophylactus Simocatta, 31
Thomas, K., 348, 357, 372, *373* (nn. 2, 3)
Thomason, N., 3, 14
Thompson, M., 298
Torriti, see Jacopo Torriti
Turing, A.M., 84
Tynan, K., 93
Tyndale, William, 15, 365, 367-70

Valla, Lorenzo, *245* (n. 166)
Valverde, Juan de, 29
Van Helden, A., 298, 304
Vandermonde, A.-T., 61
Varro, 195
Vasileios, Archimandrite, *245* (n. 95)
Vesalius, Andreas, 1-3, 6, 8-9, 12-13, 15, 18, 21, 25-31, 39, 53, 66, 135, *139* (nn. 15, 112, 113), 153-4, 159-60, *178* (n. 45), 358
Villard de Honnecourt, 55-6
Vincino da Pistoia, *245* (n. 96)
Virgil, 26, 33
Vitruvius, 31, 199
von Humbolt, A., 307

Wächtlin (or Wechtlin), Hans, 22-5
Wagner, A., 59
Walker, D.P., *139* (nn. 15, 113)
Wallace, W., 312
Ward, 'The Hon. Mrs', 307-8, 326
Waugh, E., 192
Westfall, R.J., 298, 304, 375-6, *391* (n. 16)
Westman, R.S., 278-80, 313, *326* (n. 66)
Weyer, J., 347
Whewell, W., 309, *326* (n. 58)
White, A., 302, 315, 324
Wilberforce, S., Bishop, *245* (n. 179)
Wilkins, John, *178* (n. 13)
William of Ockham, *245* (n. 160)
Williams, J., *245* (n. 39)
Willis, John, 122, 125
Willis, Thomas, 160
Witelo, 35, 78, 88, 96
Wittgenstein, L., 132
Wycliffe, John, 6

Yates, F., 90, 122, 127, *139* (n. 74), 190, *245* (n. 6)

Zahar, E., 311, *326* (n. 66)
Zeno the Stoic, 24
Zuñiga, Diego de, 377, 383

AUSTRALIAN STUDIES
IN HISTORY AND PHILOSOPHY OF SCIENCE

1. R. McLaughlin (ed.): *What? Where? When? Why?* Essays on Induction, Space and Time, Explanation. Inspired by the Work of Wesley C. Salmon. 1982 ISBN 90-277-1337-5
2. D. Oldroyd and I. Langham (eds.): *The Wider Domain of Evolutionary Thought.* 1983 ISBN 90-277-1477-0
3. R.W. Home (ed.): *Science under Scrutinity.* The Place of History and Philosophy of Science. 1983 ISBN 90-277-1602-1
4. J.A. Schuster and R.R. Yeo (eds.): *The Politics and Rhetoric of Scientific Method.* Historical Studies. 1986 ISBN 90-277-2152-1
5. J. Forge (ed.): *Measurement, Realism and Objectivity.* Essays on Measurement in the Social and Physical Science. 1987 ISBN 90-277-2542-X
6. R. Nola (ed.): *Relativism and Realism in Science.* 1988 ISBN 90-277-2647-7
7. P. Slezak and W.R. Albury (eds.): *Computers, Brains and Minds.* Essays in Cognitive Science. 1989 ISBN 90-277-2759-7
8. H.E. Le Grand (ed.): *Experimental Inquiries.* Historical, Philosophical and Social Studies of Experimentation in Science. 1990 ISBN 0-7923-0790-9
9. R.W. Home and S.G. Kohlstedt (eds.): *International Science and National Scientific Identity.* Australia between Britain and America. 1991 ISBN 0-7923-0938-3
10. S. Gaukroger (ed.): *The Uses of Antiquity.* The Scientific Revolution and the Classical Tradition. 1991 ISBN 0-7923-1130-2
11. P. Griffiths (ed.): *Trees of Life.* Essays in Philosophy of Biology. 1992 ISBN 0-7923-1709-2
12. P.J. Riggs (ed.): *Natural Kinds, Laws of Nature and Scientific Methodology.* 1996 ISBN 0-7923-4225-9
13. G. Freeland and A. Corones (eds.): *1543 and All That.* Image and Word, Change and Continuity in the Proto-Scientific Revolution. 1999 ISBN 0-7923-5913-5
14. H. Sankey (ed.): *Causation and Laws of Nature.* 1999 ISBN 0-7923-5914-3

KLUWER ACADEMIC PUBLISHERS – DORDRECHT / BOSTON / LONDON

MIX
Papier aus verantwortungsvollen Quellen
Paper from responsible sources
FSC® C105338
FSC
www.fsc.org

If you have any concerns about our products,
you can contact us on
ProductSafety@springernature.com

In case Publisher is established outside the EU,
the EU authorized representative is:
Springer Nature Customer Service Center GmbH
Europaplatz 3, 69115 Heidelberg, Germany

Printed by Libri Plureos GmbH
in Hamburg, Germany